fire service hydraulics
SECOND EDITION

EDITED BY
JAMES F. CASEY
Editor *(retired)*, Fire Engineering

FIRE ENGINEERING "Leading the fire service since 1877"

Second Edition Copyright, 1970 The Reuben H. Donnelley Corporation
First Edition Copyright, 1941 Case-Shepperd-Mann Publishing Corporation

All rights reserved. No part of this work covered by the copyright
hereon may be reproduced or used in any form or by any means—
graphic, electronic, or mechanical, including photocopying,
recording, taping, or information storage and retrieval systems—
without prior written permission of the publisher.

Published by Fire Engineering Books & Videos,
a Division of PennWell Publishing Company
Park 80 West, Plaza 2
Saddle Brook, NJ 07662
United States of America

Printed in the United States of America
ISBN 0-912212-05-5
10 9 8 7 6 5 4 3

PREFACE

"Of all branches of fire fighting probably none has received more deserved attention than that of hydraulics; and it is to present the general knowledge current on this subject, as applied to fire fighting, that this treatise has been prepared."

These words formed part of the preface to "Practical Hydraulics for Firemen" that was written in 1917 by Fred Shepperd, B.Sc., M.E. Mr. Shepperd, who can rightfully be called the father of fire service hydraulics, also stated in his preface that "the sole object (of the text) is to serve as an aid in developing the judgment of firemen through constant application in solving imaginary problems. . ." He used almost the same words to introduce "Simplified Fire Department Hydraulics" in 1925 and again in 1941 with *Fire Service Hydraulics*. His words apply equally as well to this Second Edition of *Fire Service Hydraulics*.

The "general knowledge" current today is, understandably, different than it was in 1917 and even in 1941. And, of course, the general education of the men who will study this book today is on a much higher level than in those years. However, this second edition still contains all the basic information that Mr. Shepperd set forth in 1917 and the later revisions. What makes this edition different from its predecessors is the treatment in depth given to hydraulic theory and practice, and the elaboration on water systems and fire department pumps. Foam has come a long way since 1941—so much so that we have given a lengthy chapter to it alone.

This edition, as all the others have been, is geared to the ever-growing needs of the fire service. It was planned and written for the beginner and the advanced student, and for those who want to know "why" as well as "how-to."

James F. Casey

INTRODUCTION TO THE SECOND EDITION

According to Webster's Dictionary, hydraulics is that branch of science or engineering which treats of water, or other fluids, in motion and the works and machinery for conducting or raising it. *Fire Service Hydraulics* fits this definition except that it deals only with water and its various applications, including foam, and certain specialized machinery and equipment that includes hose and nozzles.

Fire Service Hydraulics also includes the action of the men who manipulate this water and machinery. It is therefore a unique branch of hydraulics that has been tailored to a unique service and to the men who provide it—fire fighting and fire fighters.

We have divided this Second Edition of *Fire Service Hydraulics* into four parts: *Theory, Water, Practice* and *Foam*.

Part 1—*Theory*, includes two chapters and was written by Robert L. Darwin. It could almost be called an introduction to fluid mechanics. In the first chapter, Mr. Darwin covers the physical laws relating to Hydrostatics (fluids at rest) and Hydrokinetics (fluids in motion). In the second he covers the relationships that control the performance of water when it is used in the typical fire fighting situations. All formulas that are used in pumping water, moving it in hose and discharging it from nozzles are developed and explained in considerable depth.

Mr. Darwin is a fire protection engineer with the Naval Material Command, Navy Department, Washington D.C. He holds a B. S. in fire protection engineering from the University of Maryland and is currently working on his master's degree in engineering administration.

Part 2—*Water*, is divided into two chapters: Water Distribution Systems and Fire Service Pumps.

INTRODUCTION TO THE SECOND EDITION

The first was prepared by Kenneth J. Carl, P.E., director of municipal surveys and assistant chief engineer of the Engineering, Safety and Research Department of The American Insurance Association. Mr. Carl is a well known figure in the American Water Works Association and speaks with an authority on water distribution systems that few possess. His thoughts on hydrants and mains and particularly on flow testing should be part of every fire fighters knowledge.

In the second chapter of *Water*, Hubert Walker, P.E., has covered the design and function of every pump used in the fire service. With a multitude of illustrations, and a clear and thorough text, Mr. Walker provides the pump operator with just about every bit of information he needs to know on the subject.

Mr. Walker is a former vice president of the American LaFrance Corporation and a monthly contributor to FIRE ENGINEERING through his "Apparatus Maintenance" column.

Part 3—*Practice*, takes the pump operator and company officer out into the field where the action is. In this part the laws and formulas governing hydraulics and the information supplied on water systems and pumps are put to use. From a practical viewpoint, Part 3 is probably the most important section of the Second Edition of *Fire Service Hydraulics*. It is divided into five chapters.

Richard P. Sylvia, associate editor of FIRE ENGINEERING, prepared the first two chapters of Part 3: Friction Loss Calculations, and Engine and Nozzle Pressure Calculations. In simple steps, with many short cuts and with appropriate problems, Sylvia spells out the operations of a fire company and its pump operator on the fireground.

In addition to his work on FIRE ENGINEERING, he is the former chief of the Noroton, Conn., Fire Department and more importantly for this edition he is chief instructor of the pump operation course of the Connecticut State Fire School.

The remaining three chapters of Part 3 cover Fire Streams, Standpipe Systems and Sprinkler Systems—all from the practical as opposed to the theoretical aspect. Material in these chapters was prepared by the editor. It has been adapted in part from the first edition of *Fire Service Hydraulics* and a number of articles that appeared in FIRE ENGINEERING over the years.

Part 4—*Foam*, was written by Dr. Richard L. Tuve, head of suppression research, U.S. Naval Research Laboratory, Washington D.C. In a strict sense, foam should not be included in a book that carries the title of Hydraulics. However, foam is such a part of fire suppression and so closely linked to water and fire apparatus that we felt that it should be included in *Fire Service Hydraulics*.

INTRODUCTION TO THE SECOND EDITION

In this section Dr. Tuve has just about covered everything the fire fighter should know about foam. He writes specifically for this audience—and in straightforward nontechnical language. He begins with the general characteristics and uses of foam and goes on through its methods of generation, foam proportioning devices and foam systems.

Dr. Tuve has been in fire suppression research for most of his working life and is most notable for his development of Purple-K and Light Water. He is at present chairman of the Foam Committee, National Fire Protection Association.

It is our hope that the fire fighter will study this book as a whole. However, in preparing it, we kept in mind the fact that not all who read it will have the time or the desire to go into the theory as deeply as Mr. Darwin has, nor will all desire the knowledge of pumps that Mr. Walker gives.

It is for this reason that we divided *Fire Service Hydraulics* into four parts—each of which can be studied independently. For instance, Part 3—*Practice*, could have been titled Fireground Hydraulics. It contains all that the pump operator needs to know when operating at a fire and, with Walker's chapter, all that a pump operation instructor needs to know to conduct a short course.

Grateful acknowledgment is made to the American Insurance Association and to the National Fire Protection Association for permission to use various tables from their publications. And we extend particular thanks to Dorothy P. Ferguson, managing editor of FIRE ENGINEERING, who labored hard and long in the preparation of all copy for this edition.

J. F. C.

FIRE ENGINEERING
New York, N.Y.
August 1970

CONTENTS

Chapter	Page
Preface	III
Introduction	V

PART 1—THEORY

1. Water—at Rest and in Motion 2
   Hydrostatics or Liquids at Rest
   Liquids in Motion, Hydrokinetics
2. Velocity and Discharge 43
   Flow Velocity and Nozzle Discharge
   Nozzle Reaction
   Friction Loss in Water Conductors

PART 2—WATER

3. Water Distribution Systems 74
   Water Supply Systems
   Fire Flow Tests
4. Fire Service Pumps 138
   Centrifugal, Piston and Rotary Pumps
   Booster and High Pressure Pumps
   Portable and Back-Pack Pumps
   Tanker Pumps
   Municipal and Industrial Fire Pumps (Stationary)

PART 3—PRACTICE

5. Friction Loss Calculations 226
   How Friction Loss Works

CONTENTS

   Friction Loss in Small Hose
   Flow Advantages of Big Hose
6 Engine and Nozzle Pressures 239
   Parallel Lines
   Aerial Stream Calculations
   Relay Pumping
7 Fire Streams 293
   Nozzle Performance
   Heavy Streams
8 Standpipe Systems 308
   Water Supplies
   Operating at Fires
9 Automatic Sprinkler Systems 324
   Classifications of Sprinkler Systems
   Water Supplies for Sprinkler Systems

PART 4—FOAM

10 Fire Fighting Foams and Foam Systems 344
   Characteristics of Foam
   Foam Generation Methods
   Proportioning Devices
   Foam Systems

Appendix 393
   Fire Stream Tables, Friction Loss Table, Conversion Tables (Cubic Feet to Gallons, Gallons to Cubic Feet), Contents of Cylinders, Circumferences and Areas of Circles, Square Roots

Index 423

fire service hydraulics
SECOND EDITION

PART ONE: **theory**

CHAPTER ONE

Water—at rest and in motion

A full and complete knowledge of the characteristics of water and fluid systems which move and utilize water is of prime importance in the science of fire engineering. Without this economical and efficient fluid, fire protection of all kinds would be severely handicapped. However, with 75 percent of the earth's surface consisting of liquid water, salt and fresh, we need only to learn how to use it most efficiently for our needs as a fire fighting agent. Just remember the cardinal rule that fire protection requires the delivery of the needed material or agent at the right time, in the right volume, in the right form, in the desired place to accomplish a needed result.

A study of the science of hydraulics provides the professional fire officer with these necessary facts. This science deals with the properties of fluids at rest in tanks or reservoirs and the characteristics demonstrated by the flow of fluids in motion when they traverse pipes or conduits and issue from nozzles.

An important feature of the science of hydraulics and dynamic flow is connected with its capability for quantization, that is, the use of numbers and the assignment of values to important physical relationships. Perhaps nowhere else in the fire protection profession are numbers and mathematical expressions used so extensively. This must not deter the eager student in his pursuit of the necessary knowledge for achieving excellence.

WATER—AT REST AND IN MOTION

Definition of hydraulics

Hydraulics is a branch of applied mechanics that deals with the physical characteristics exhibited by fluids at rest and in motion. That branch of hydraulics concerned with the study of fluids at rest is called *Hydrostatics*, while the study of fluids in motion is called *Hydrokinetics*. For instance, a study of static pressure in a water supply system would be an exercise in Hydrostatics; a study of flowing pressure at a fire hydrant or at various types of nozzles would be an exercise in Hydrokinetics. The term "fluid" as used in the definition of many studies of physical properties such as hydraulics may mean either liquid or gas. However, we will concern ourselves primarily with liquid water in the following pages.

The history of hydraulics

The recognition of hydraulics as a separate study of physical principles dates back to ancient times. There is evidence of attention to hydraulic principles as early as 4000 B.C. The Bible contains numerous references to problems in hydraulics such as the flood control measures following the great flood in Genesis and the construction of irrigation ditches in ancient Egypt. In approximately 310 B.C. the Romans began construction of large masonry aqueducts to supply water to the city of Rome. In spite of a lack of modern construction equipment and techniques, they were able to build over 350 miles of aqueducts. By 130 A.D. this water system was capable of supplying 50 million gallons of water per day to the city of Rome. Large portions of the original Roman water supply system are still in existence.

The first detailed scientific publication on hydraulics is credited to Leonardo Da Vinci in Italy, who, in the latter part of the 15th century, produced a treatise entitled "On the Motion and Measurement of Water." The science of hydraulics was further advanced in the 17th century by the writing of Torricelli and Pascal. During the 18th century, fundamental theories of hydraulics were developed by scientists such as Pitot, Bernoulli and Venturi. We will study some of their contributions later in this chapter.

The first hydraulic experiments in the United States directly related to fire protection were conducted by John R. Freeman in 1888. Freeman determined some of the basic characteristics of water flow through fire hose and nozzles. He also developed a standard nozzle known as the "Underwriters Playpipe," and is responsible for the basic nozzle formula, $Q = 29.71cd^2 \sqrt{P}$, which will be discussed in detail later on.

Another significant contribution to the fire protection field was the development of basic flow formulas showing the relationship of various factors governing the flow of water in pipes by G. S. Williams and Allen Hazen. The Hazen and Williams formula is widely used by fire protection engineers in the design and evaluation of water supply systems.

Early in the 1900s, the National Board of Fire Underwriters (now the American Insurance Association) established a fire insurance grading schedule for municipalities. This grading schedule focused on the importance of a city's water supply system.

Water consumption

The following interesting statistics issued by the American Water Works Association emphasize the importance of water for domestic consumption and for fire protection:

1. U.S. cities now produce and distribute approximately 25 billion gallons of water per day.

2. For every 1,000 persons in the United States, there is about 3 miles of water main.

3. The per capita consumption of water is 150 gallons per day. This figure includes total domestic, industrial, commercial and public use.

4. The average per capita use of water for fire protection is 10–15 gallons per day.

The average annual use for fire protection purposes is small when compared to other uses of water. However, very large fires can cause tremendously high demands for short periods. This demand for large fires is determined by the "fire demand" formula of the American Insurance Association which serves as the basic water supply criteria of their municipal grading schedule.

The storage of water in tanks and reservoirs, the pumping and distribution of water through water supply mains, the flow of water from a fire hydrant, the pumping of water by a fire truck, the flow of water through sprinkler piping, the flow of water through a fire hose, and the discharge of water from sprinkler heads and nozzles are all based on important well-known hydraulic principles. Since water is the primary fire fighting agent employed by the fire service, an understanding of water characteristics and hydraulics is essential to anyone engaged in the field of fire protection.

WATER—AT REST AND IN MOTION

Characteristics of water

Water exists abundantly as a solid, liquid, or gas. Below 32°F it exists in the form of ice. From 32°F to 212°F it exists as a liquid. Above its boiling point of 212°F, it exists in the form of a gas known as water vapor. Since it is primarily used for fire fighting in the liquid form, we are most concerned with the laws governing water as a liquid.

For practical purposes, water may be considered incompressible. Even at the great pressure of 65,000 pounds per square inch, the volume of water is reduced by only 10 percent. Water has a maximum density at 39.9°F (4°C). Its weight varies from 62.4 pounds per cubic foot at 40°F to 60.1 pounds per cubic foot at 200°F. Unless otherwise specified, we will consider the density of water to be 62.4 pounds per cubic foot for fresh water, and 64 pounds per cubic foot for sea water. It should be noted that when speaking of density we mean weight per unit volume, or more commonly, pounds per cubic foot.

The ability of water to extinguish fires is a result of its capacity to absorb great quantities of heat before changing to steam. One gallon of water before changing to steam can absorb all of the heat given off by the burning of 1 pound of ordinary (Class A) combustibles. A summary of the significant properties of water and some commonly used conversion factors is shown below:

$$1 \text{ gallon of water weighs } 8.3 \text{ lb } (8.3453)$$
$$\text{Density of water: } 62.4 \text{ lb/ft}^3 \text{ (fresh)}$$
$$64 \text{ lb/ft}^3 \text{ (sea)}$$
$$1 \text{ gal} = 231 \text{ in}^3 = .1337 \text{ ft}^3$$
$$449 \text{ gpm} = 1 \text{ ft}^3/\text{sec}$$
$$1 \text{ ft}^3 \text{ contains } 1728 \text{ in}^3$$
$$1 \text{ ft}^3 \text{ contains } 7.48 \text{ gal}$$

HYDROSTATICS OR LIQUIDS AT REST

As we have mentioned previously, that branch of hydraulics dealing with liquids at rest is called hydrostatics. As the name implies, hydrostatics deals only with the properties of water while in a "static" condition (no water flowing). A hose line charged with water is in a static condition when the nozzle is fully closed and no water is flowing. Once the nozzle is open and water begins discharging from the nozzle the hose line ceases to be in a static condition.

FIRE SERVICE HYDRAULICS

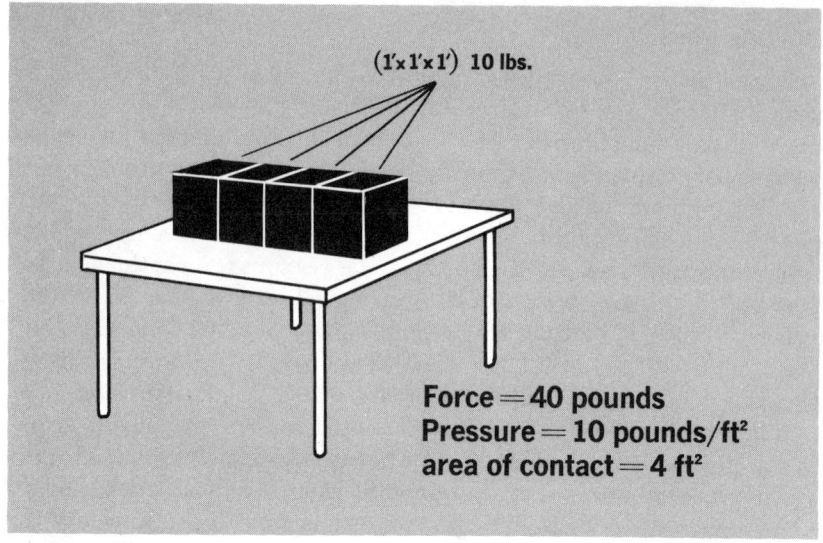

Figure 1

Pressure vs. force

Very often reference is made to water pressure in a hose, pressure in a tire, pressure at a nozzle, and so on. The word pressure has some meaning to nearly everyone. However, the term is often used incorrectly and pressure is frequently confused with force. Many texts on fire department hydralics fail to stress the difference between these important terms.

Force is a measure of weight and is commonly expressed in pounds. *Pressure* is force per unit area and is commonly expressed in pounds per square foot (psf) or pounds per square inch (psi). The relationship between force and pressure is shown in the formula $F = P \times A$, where F is force in pounds, P is pressure in pounds per square foot or pounds per square inch, and A is area in square feet or square inches.

To illustrate the difference between pressure and force, assume we have four blocks setting side by side on a table (Figure 1).

Assume further that each block weighs 10 pounds and is $1 \times 1 \times 1$ foot. The total force acting on the table would be determined as follows:

$$F = 4 \text{ blocks} \times 10 \text{ lb/block} = 40 \text{ lb}$$

In other words, force may be regarded as the total weight pressing

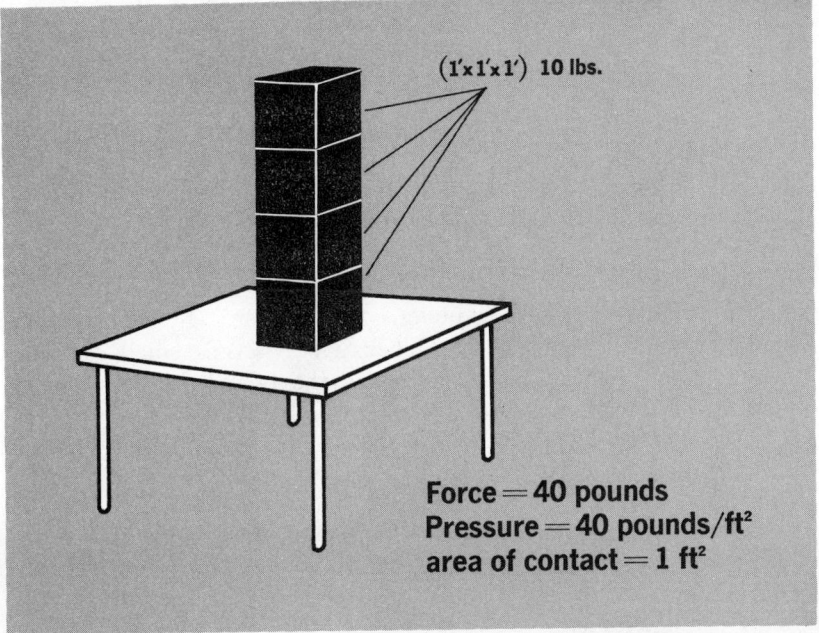

Figure 2

down on the table. The area of contact between the blocks and the table would be determined as follows:

$$A = 4 \text{ ft} \times 1 \text{ ft} = 4 \text{ ft}^2$$

To find the pressure, we merely rearrange the formula, $F = PA$, to solve for P. Hence,

$$P = \frac{F}{A} = \frac{40 \text{ lb}}{4 \text{ ft}^2} = 10 \text{ lb/ft}^2 = 10 \text{ psf}$$

and we would say that 10 pounds of force is acting on every square foot of table area on which the blocks rest. We could easily determine the pressure in pounds per square inch rather than pounds per square foot by using the conversion factor between square feet and square inches as shown below:

$$P = \frac{F}{A} = \frac{40 \text{ lb}}{4 \text{ ft}^2} = 10 \text{ lb/ft}^2 \times 1 \text{ ft}^2/144 \text{ in}^2 = .069 \text{ lb/in}^2 = .069 \text{ psi}$$

Now, suppose we stack the same blocks on top of each other (Figure 2).

FIRE SERVICE HYDRAULICS

The total force would remain the same:

$$F = 4 \text{ blocks} \times 10 \text{ lb/block} = 40 \text{ lb}$$

However, the area of contact between the table and the blocks has been reduced:

$$A = 1 \text{ ft} \times 1 \text{ ft} = 1 \text{ ft}^2$$

This will cause the pressure to change as shown below:

$$P = F/A = 40 \text{ lb}/1 \text{ ft}^2 = 40 \text{ lb/ft}^2 = 40 \text{ psf}$$

and we could now say that 40 pounds of force is acting on every square foot of table area on which the blocks rest. Even though the total force in both illustrations is the same, the pressure differs because pressure is always a measure of force per unit area.

Problem No. 1

Problem: A water tank contains 1000 gallons of fresh water. If the bottom of the tank is rectangular having dimensions of 4 × 5 feet, determine the total force and the pressure acting on the bottom of the tank.

Answer:

$$F = 1000 \text{ gal} \times \frac{8.3 \text{ lb}}{\text{gal}} = 8300 \text{ lb}$$

$$A = 4 \text{ ft} \times 5 \text{ ft} = 20 \text{ ft}^2$$

$$P = \frac{F}{A} = \frac{8300 \text{ lb}}{20 \text{ ft}^2} = 415 \text{ lb/ft}^2 = 415 \text{ psf}$$

or

$$P = \frac{415 \text{ lb}}{\text{ft}^2} \times \frac{1 \text{ ft}^2}{144 \text{ in}^2} = 2.88 \text{ lb/in}^2 = 2.88 \text{ psi}$$

It should now be obvious that commonly heard phrases such as "100 pounds of hose pressure" or "25 pounds tire pressure" are incorrect. Pressure must always be stated as pounds per square inch or pounds per square foot, provided we are using pounds as our force measurement and square inches or square feet as our measure of area.

As a mental exercise to test comprehension of the pressure and force concepts presented above, the student might decide for himself why snowshoes enable a person to walk over snow, or likewise why a woman wearing high-heeled shoes will sink into soft ground while a man wearing wide-heeled "loafers" will not!

WATER—AT REST AND IN MOTION

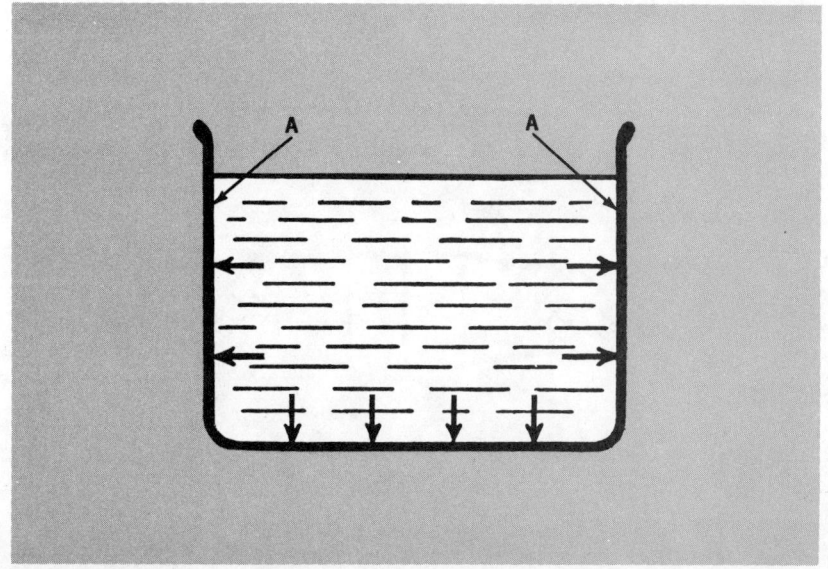

Figure 3

Six principles of fluid pressure

There are six basic principles that apply to fluid pressure. They are:
1. Liquid pressure is exerted in a perpendicular direction to any surface on which it acts.
2. At any given point beneath the surface of a liquid, the pressure is the same in all directions—downward, upward and sidewise.
3. Pressure applied to a confined liquid from without is transmitted in all directions without diminution (reduction in intensity).
4. The pressure of a liquid in an open vessel is proportional to the depth of the liquid.
5. The pressure of a liquid in an open vessel is proportional to the density of the liquid.
6. Liquid pressure on the bottom of a vessel is unaffected by the size and shape of the vessel.

An explanation of the first principle is indicated in Figure 3. Suppose a vessel having flat sides contains water. Then the pressure on the various sides due to the weight of the water is in the direction shown by the small arrows, perpendicular to the wall of the containing vessel. If it were at any other direction, as indicated by arrows AA, then the water would start moving downward along the sides, and rising in the center of the container. This is not the case, for obser-

FIRE SERVICE HYDRAULICS

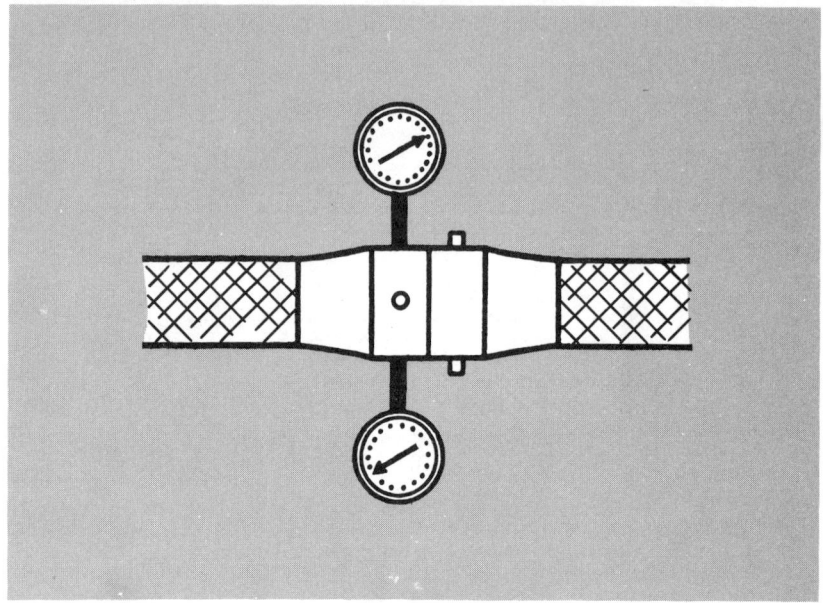

Figure 4

vations show that a fluid placed in a vessel tends to assume a position of rest, all currents ceasing.

The second principle is illustrated in Figure 4. This illustration represents a coupling on a line of hose into which are inserted two pressure gages. When the water in the hose is at rest, such as when the shut-off nozzle is closed, both gages will indicate the same pressure, for the pressure at a point in a fluid at rest is of the same intensity in all directions.

The third principle is illustrated by Figure 5. This illustration represents a hollow sphere to which a water pump is attached. Set into the sphere around its circumference are a series of gages.

When the apparatus is filled with water, and the pressure is applied by means of the pump, all gages show the same reading. This illustrates the principle that pressure applied to a confined fluid from without is transmitted in all directions without diminution.

Principle 4 is illustrated by Figure 6. Here are shown three vertical containers each 1 square inch in cross-section area. The depth of the water in A is 1 foot; in B, 2 feet; in C, 3 feet. The pressure at the bottom of A is .434 psi; at the bottom of B, .868 psi; and at the bottom of C, 1.302 psi.

WATER—AT REST AND IN MOTION

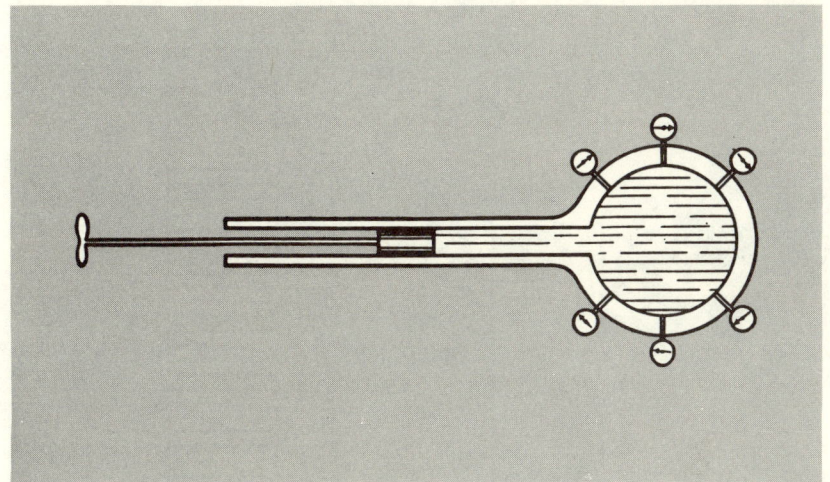

Figure 5

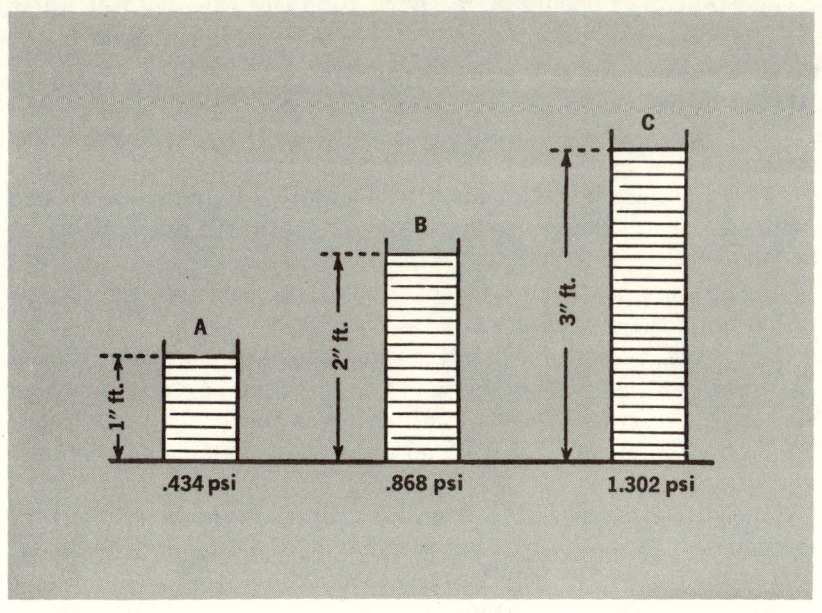

Figure 6

FIRE SERVICE HYDRAULICS

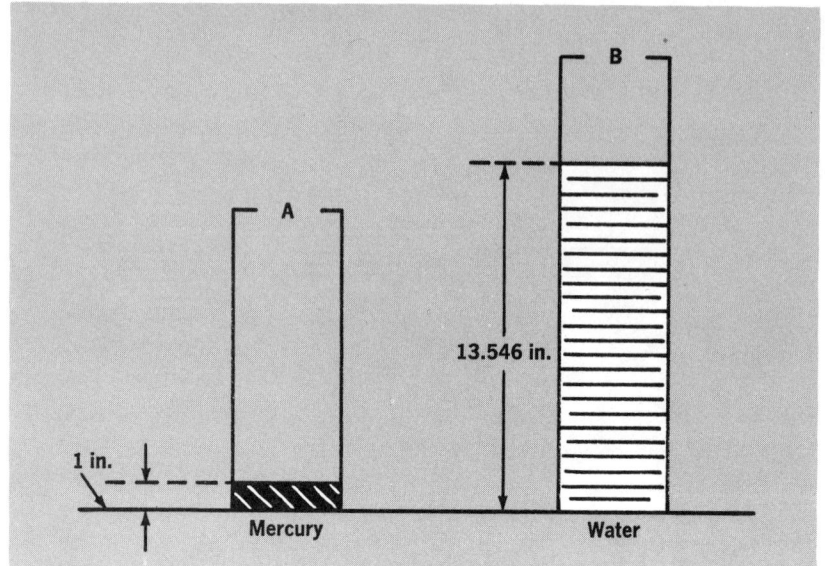

Figure 7

Note that the pressure in B, which has twice the depth of water of A is just twice the pressure in A. Likewise the pressure in C, which has three times the depth of water in A is three times the pressure in A.

Thus the downward pressure of a liquid in an open vessel is proportional to its depth.

The fifth principle is illustrated by Figure 7. Here are shown two containers, one holding mercury of 1-inch depth, the other, water at 13.546 inches depth.

The pressure at the bottom of each is the same, as mercury is 13.546 times the weight of water.

If container A were filled to the same height as container B, then the pressure at the base of container A would be 13.546 as much as that in B, as mercury weighs 13.546 times as much as water. Hence the downward pressure of a liquid in an open vessel is proportional to the liquid density.

Principle 6 is illustrated by Figure 8. Here are shown a number of containers of various shapes but each having the same cross-sectional area at the bottom, namely 1 square inch.

The pressure at the bottom of each is exactly the same, for the downward pressure of a liquid on the bottom of a vessel is independent

WATER—AT REST AND IN MOTION

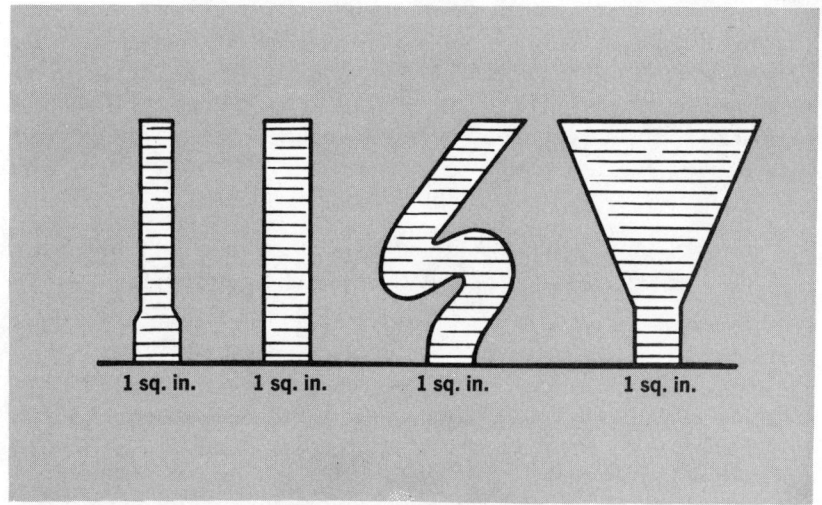

Figure 8

of the shape of the vessel itself, as long as the depth of liquid (or head) is the same in each.

Pressure as a function of height and density

Principles 4 and 5 from the preceding section can be combined to establish a relationship between pressure, height, and density. This relationship can be expressed in equation form as follows:

$$P = hD \text{ or } lb/in^2 = in \times lb/in^3$$

where P is pressure in pounds per square foot or pounds per square inch, h is height (or depth of liquid) in feet or inches, and D is density in pounds per cubic foot or pounds per cubic inch. By using appropriate conversion factors, we can simplify these equations if we are dealing with water, so that:

$$P = .434h \text{ (where } P \text{ is in psi and } h \text{ is in ft)}$$

and

$$h = 2.31P \text{ (where } P \text{ is in psi and } h \text{ is in ft)}$$

The height of water, h, is often referred to as *head*. Head is the vertical height to which a given pressure will elevate a column of liquid, or, stated in other words, it is the equivalent height of water necessary to create the given pressure at its base. For every pressure there is a corresponding head. In an open tank, depth of liquid and head are

FIRE SERVICE HYDRAULICS

synonymous. For a piped system under pressure, head is the height to which the liquid could be raised by the pressure. Hence, any water pressure may be given in psi (or psf) or in head, since pressure and head (or height) are related by the formulas shown above. Remember that head is always expressed in feet or inches.

Problem No. 2

Problem: A standpipe contains water to a height of 100 feet. What is the pressure (in psi) at the base of the standpipe?
Answer:

$$P = .434h = .434(100) = 43.4 \text{ psi}$$

Problem No. 3

Problem: The static pressure in a fire hose is 150 pounds per square inch. To what head, h, is this pressure equivalent to?
Answer:

$$h = 2.31P = 2.31(150) = 346 \text{ ft}$$

Atmospheric pressure and hydraulics

No study of fluid pressure would be complete without a discussion of atmospheric pressure. An understanding of atmospheric pressure is essential to any pump operator because of the use of atmospheric pressure in drafting operations (this will be discussed in detail in Chapter 2).

To understand atmospheric pressure, we must realize that the earth is enveloped in an environment of air. Everything we do on the surface of the earth must be relative to this environment. This environment of air has depth and density and therefore, as with water, it creates a constant pressure on everything around us, including, of course, ourselves. The atmosphere exerts pressure on us the same as water does when we swim to the bottom of a swimming pool. Atmospheric pressure is about 14.7 psi at sea level.

Most pressure gages read pressure relative to existing atmospheric pressure. When a gage reads 5 psi it is actually reading 5 psi greater than atmospheric or $14.7 + 5 = 19.7$ psi greater than "absolute zero." Absolute zero is total lack of pressure or a perfect vacuum. Gages that read pressures relative to atmospheric pressure actually indicate what is called "gage pressure." Therefore, atmospheric pressure is actually 0 psi gage or 14.7 psi absolute. When a gage reads -5 psi gage, it is actually reading 5 psi less than atmospheric or 14.7 minus 5 which

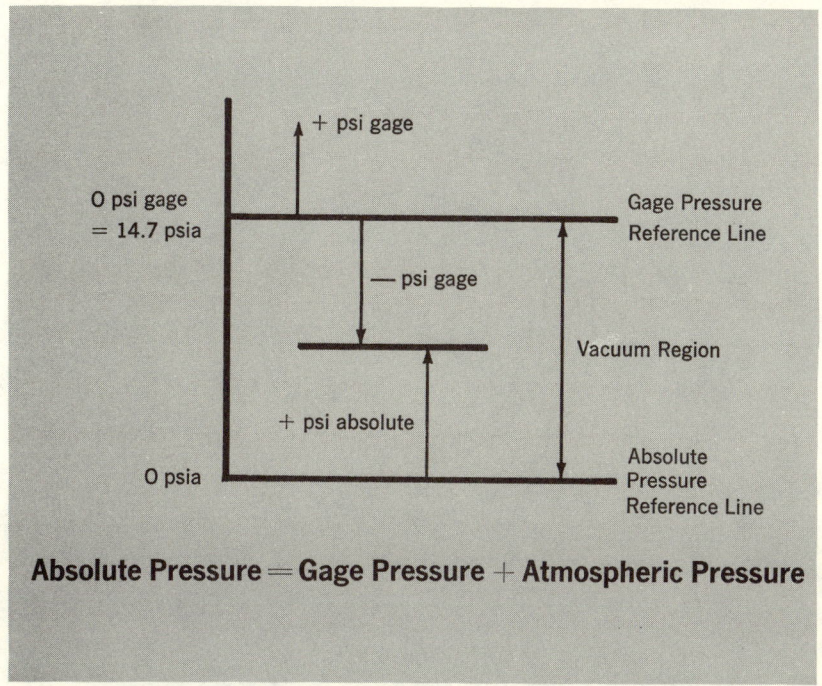

Figure 9

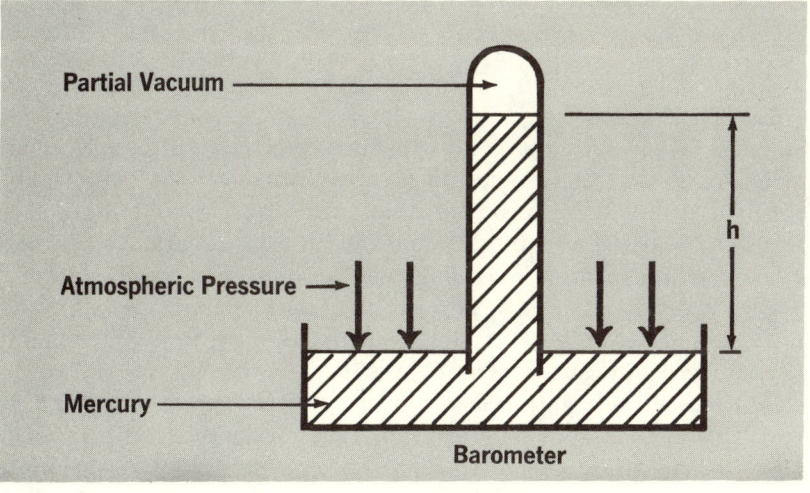

Figure 10

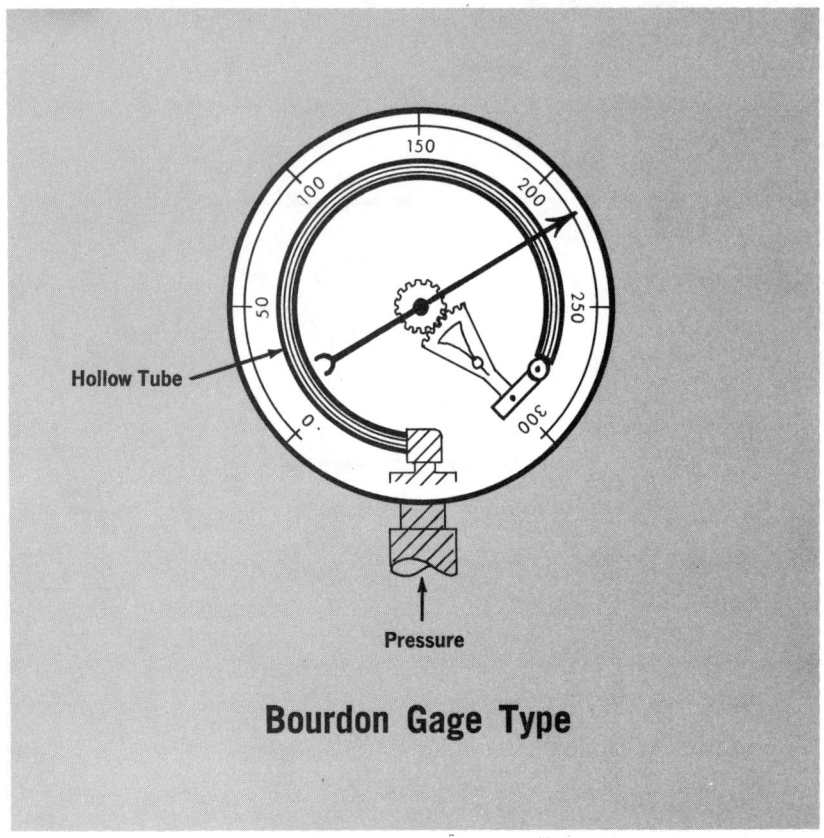

Figure 11

equals 9.7 psi absolute. Any pressures less than atmospheric are called vacuums. As is common practice, whenever the abbreviation psi is used in this book it shall mean gage pressure. Absolute pressures will be designated by the abbreviation psia. A graph summarizing the relationship between gage and absolute pressure is shown in Figure 9.

Figure 10 shows a barometer, which is the instrument commonly used for measuring atmospheric pressure.

Atmospheric pressure is determined by measuring the height, h, to which atmospheric pressure forces mercury up the tube. By applying the formula relating height and density to pressure, $P = hD$, the atmospheric pressure is determined. At sea level, one atmosphere (14.7 psi) is equivalent to 29.92 inches of mercury. The equivalent

WATER—AT REST AND IN MOTION

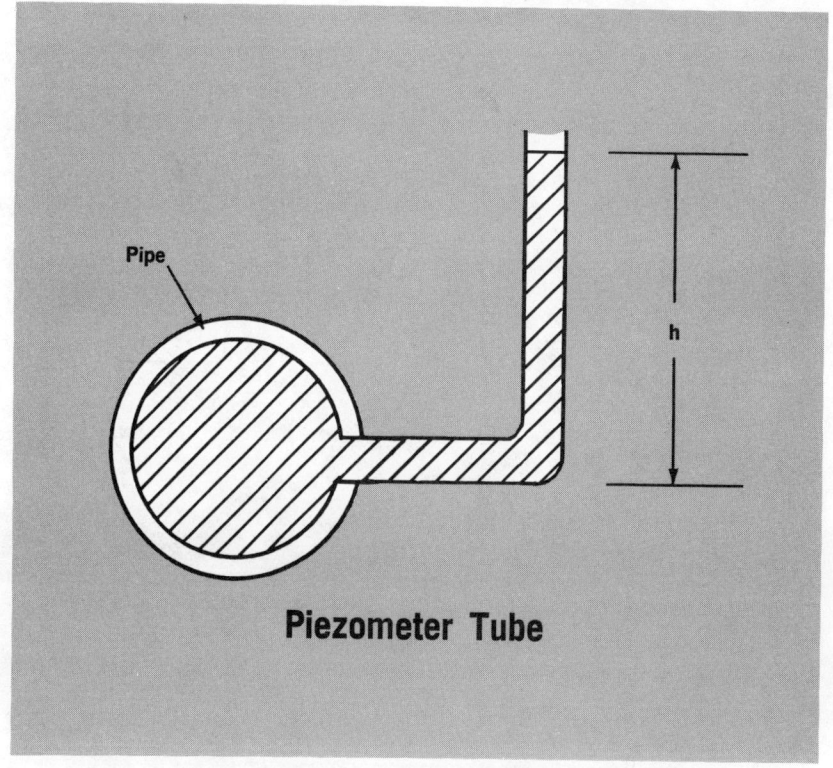

Figure 12

height of water can be determined by $h = 2.31P$, where P is 14.7 psi:

$$h = 2.31P = 2.31 \times 14.7 = 33.92 \text{ ft}$$

Hence, one atmosphere is equivalent to 29.92 inches of mercury or 33.92 feet of water (at sea level). The reason for using mercury rather than water in the barometer should be obvious to the student.

Devices for measuring static water pressure

A common device for measuring static pressure is the Bourdon gage shown in Figure 11.

As discussed above, the Bourdon gage measures gage pressure. As pressure is introduced into the hollow metal tube, the tube tends to straighten. As it does, the pointer is forced to move so as to indicate the pressure on the face of the gage.

FIRE SERVICE HYDRAULICS

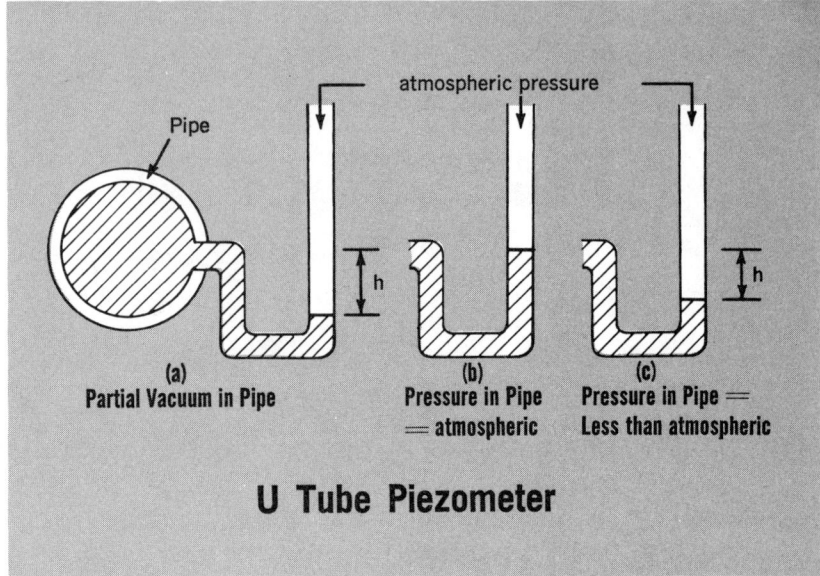

Figure 13

Another device useful for measuring static pressure is the piezometer tube (Figure 12).

This device is often used for determining pressure in a tank or pipe. It is generally used for relatively low pressures. The straight piezometer tube shown in Figure 12 can only read positive gage pressures.

A modification of the piezometer tube, known as the U tube piezometer (sometimes referred to as a simple manometer), can read negative pressure (Figure 13).

Figure 13(a) shows a U tube piezometer indicating a pressure less than atmospheric, which means that a partial vacuum exists in the pipe. Figure 13(b) shows how a U tube piezometer would look if the pressures on each side were equal (if the pressure in the pipe was exactly equal to atmospheric pressure). When the liquid level is lower on the right, as in Figure 13(c), it means that atmospheric pressure is greater than the pressure within the pipe.

A U tube piezometer can also be used to illustrate the effect of density on head as previously discussed. Since mercury is 13.6 times heavier than water, 1 inch of mercury will support 13.6 inches of water (Figure 14).

A mercury manometer also utilizes this principle. In order to allow

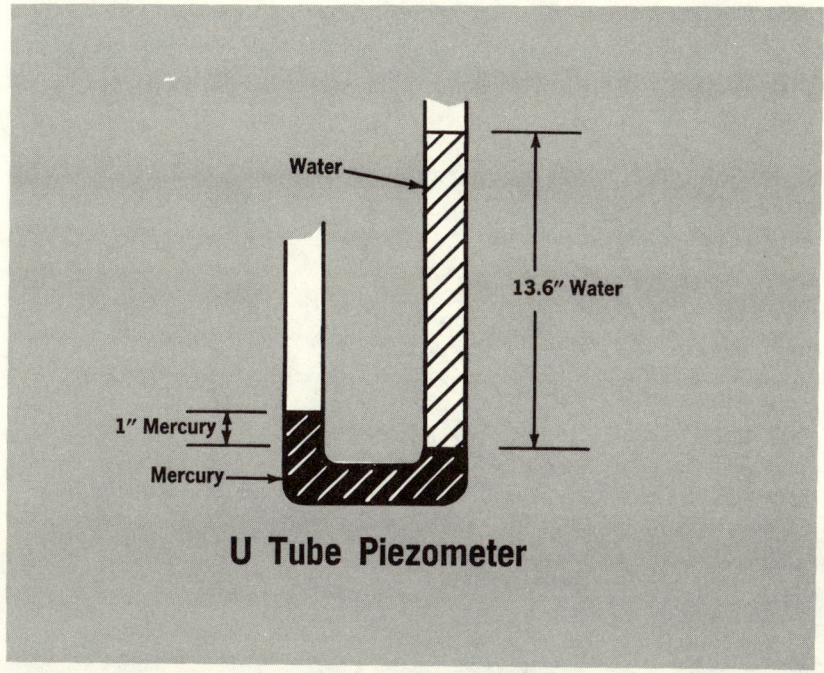

Figure 14

for shorter legs on each side of the U, which permits higher pressure readings, mercury rather than water is used in the tube. A mercury manometer is shown in Figure 15.

A differential manometer (Figure 16) can be used to measure the difference in pressure between two pressure sources.

The illustration shows that the pressure in the pipe on the right is greater than the pressure in the pipe on the left. The difference in pressure is calculated the same as before, by using $P = hD$.

LIQUIDS IN MOTION, HYDROKINETICS

Thus far, we have been concerned primarily with the characteristics of liquids at rest. For the balance of this chapter, we will concern ourselves primarily with hydrokinetics which is the study of liquids in motion.

The basic theorem governing fluids in motion is called Bernoulli's Equation. This equation was initially developed in the 18th century by Daniel Bernoulli, a Swiss physicist. It is impossible to fully master

FIRE SERVICE HYDRAULICS

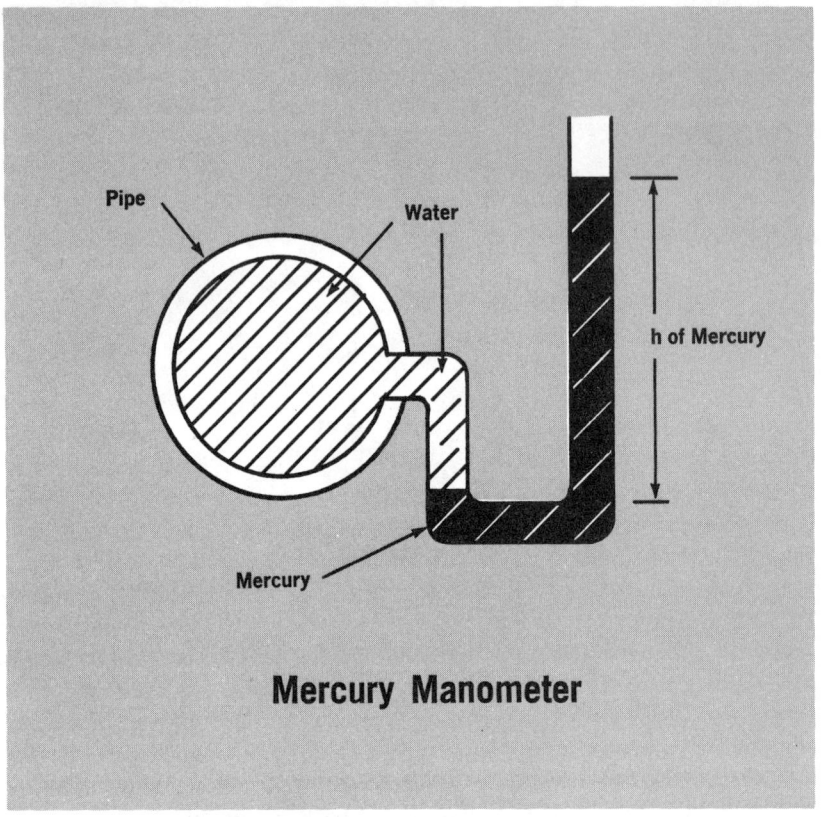

Figure 15

hydraulics without having an understanding of Bernoulli's Equation. While one could certainly learn to operate a fire pump or to conduct routine waterflow tests without knowing all about this equation, one could never be regarded as being knowledgeable in hydraulics if he couldn't use Bernoulli's Equation for solving hydraulic problems.

Energy

In order to derive and use Bernoulli's Equation, we must first discuss energy and its relationship to hydraulics. By energy, we mean the ability to perform work, or more specifically, the ability to apply sufficient force so as to cause motion. Hence, if a weight lifter has the ability to lift a heavy bar-bell, we may say that he has the energy to lift the bar-bell. If a gasoline-powered engine is able to move a

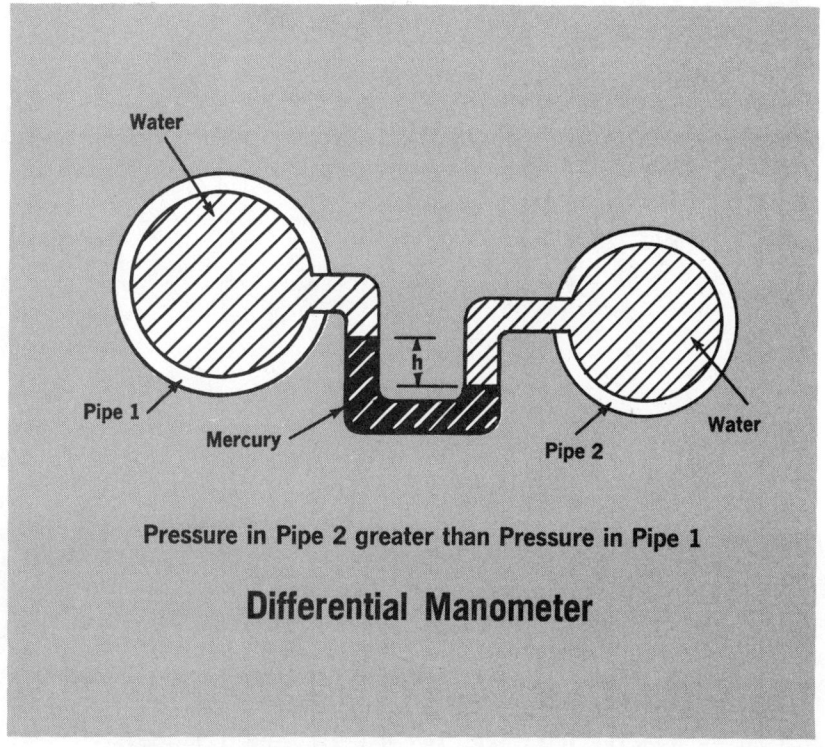

Figure 16

vehicle, we may say that the engine has sufficient energy output to propel the vehicle.

In hydraulics, we are interested in two basic types of energy: *potential energy* and *kinetic energy*.

Potential energy is energy that is stored, or energy that has the potential to do work. A flashlight battery, a coiled spring, a stretched rubber band, and a stick of dynamite, all possess potential energy. Likewise, water can have potential energy due to elevation or due to static pressure. This may be illustrated as follows:

The ball on the left in Figure 17 possesses potential energy relative to the ground due to its elevation, h, because it can do work as soon as it is released to fall. In the same way, the water in the storage tank possesses potential energy relative to the ground as a result of its elevation, h. As a further example of potential energy resulting from elevation, consider a large volume of water at the top of Niagara Falls.

FIRE SERVICE HYDRAULICS

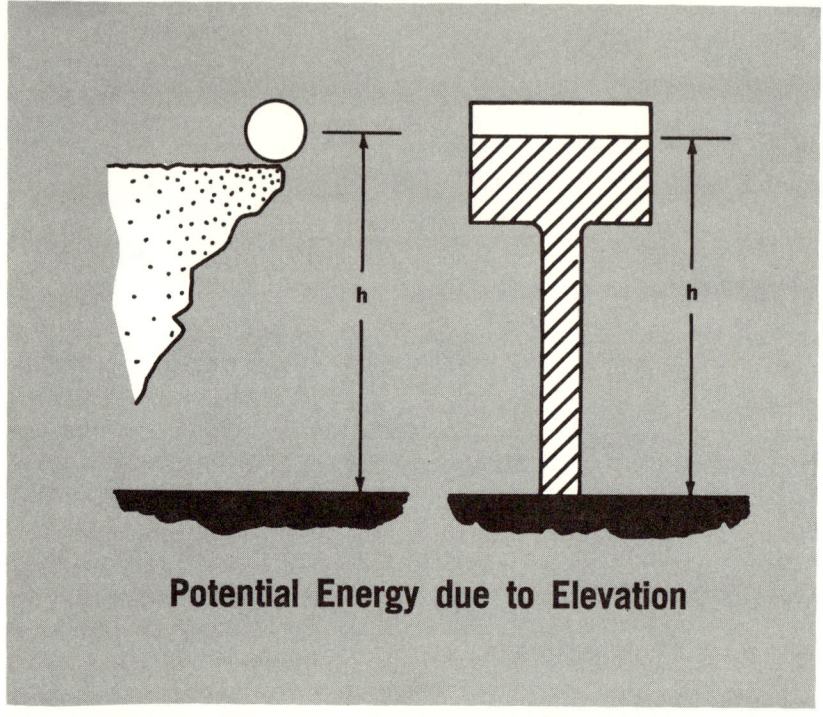

Figure 17

This water at the top of the falls possesses considerable potential energy relative to the bottom of the falls. This is emphasized by the ability of this water (in falling from a high elevation to a lower one) to turn turbogenerators. This, of course, is the principle behind hydroelectric power plants.

Another form of potential energy is shown in Figure 18.

The compressed spring shown on the left has potential energy because it can do work as it springs from A to B. Only at B, where it is no longer under compression, will the spring not have potential energy. Likewise, water under pressure in a pipe or hose line has potential energy as shown on the manometer or the piezometer, because it exerts a pressure greater than atmospheric and hence can do work until the pressure drops to zero gage pressure. Anyone who has witnessed the bursting of a fire hose should be aware of the potential energy possessed by a liquid under pressure.

It should be apparent from the above discussion that potential energy as used in the study of hydraulics is really the potential ability

WATER—AT REST AND IN MOTION

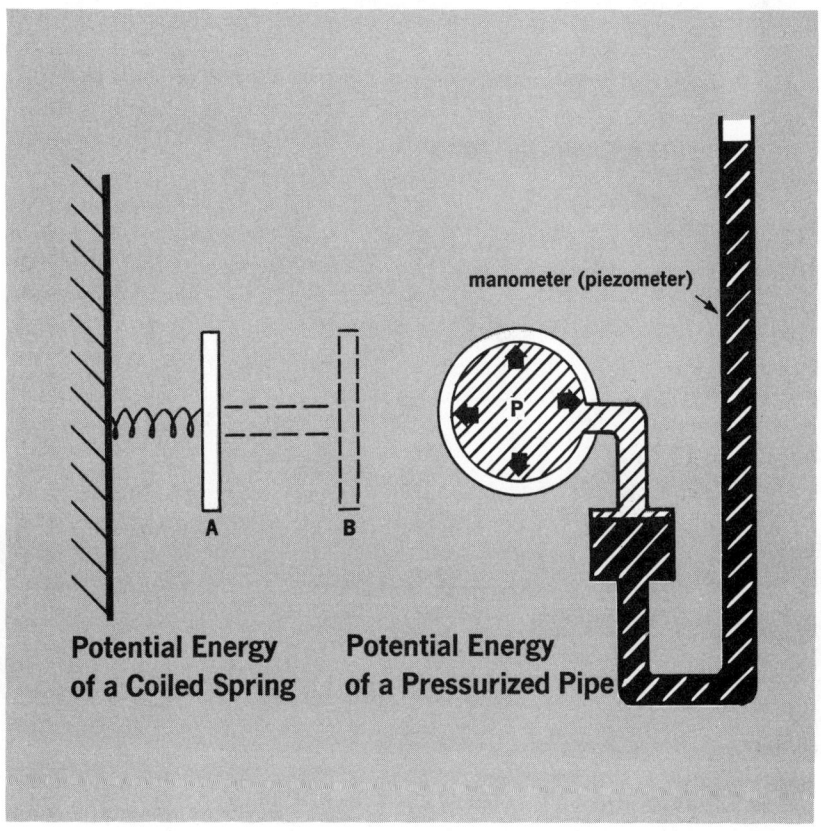

Figure 18

to cause water to flow. As soon as the water actually starts to flow, then the potential energy that caused the flow ceases to exist. In other words, we will consider that the potential energy of water is the same thing as static pressure which we discussed earlier in this chapter. We may summarize this definition of potential energy as follows:

Total Potential Energy = Potential Energy + Potential Energy
                          Due to Pressure   Due to Elevation

Total potential energy is equal to total static pressure or total static head.

We will call "potential energy due to pressure" pressure head since it is the head derived from using the equation $P = hD$, where h is the

FIRE SERVICE HYDRAULICS

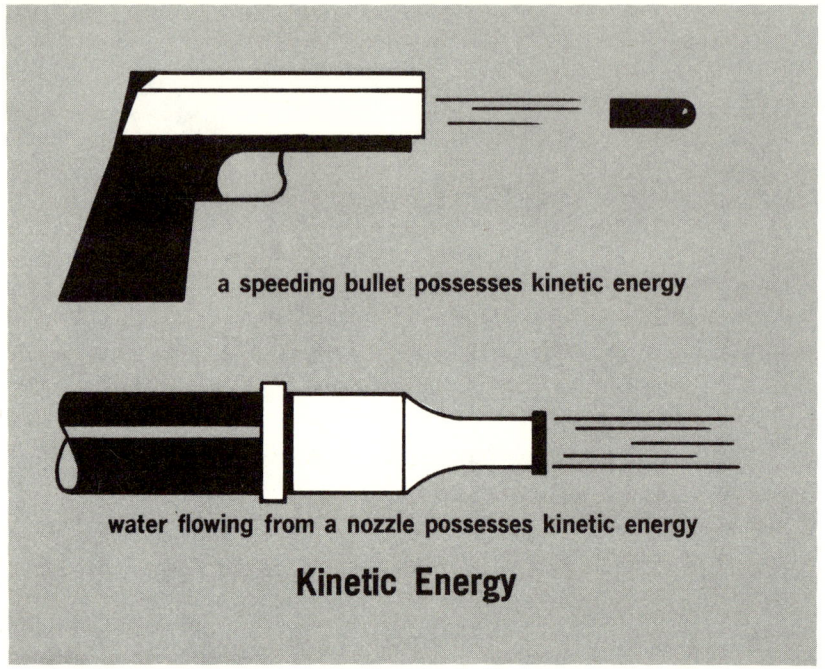

Figure 19

pressure head. Pressure head is therefore a measure of how much the water pressure at a particular location exceeds atmospheric pressure.

We will use the term elevation head to mean potential energy due to elevation. Elevation head exists only if we have a difference in elevation between two points being considered. Elevation head is simply the vertical distance between two reference points.

By using these terms, the equation for potential energy may be rewritten as follows:

Total Potential Energy = Potential Energy + Potential Energy
$\downarrow$ Due to Pressure   Due to Elevation
                                                            $\downarrow$ $\downarrow$
Total Static Head   = Pressure Head   + Elevation Head

In some hydraulic textbooks, total static head (or total potential energy) is called "piezometric head" since it is the type of head measured with a piezometer.

Now that we have an understanding of potential energy, let's discuss the second type of energy, kinetic energy. Kinetic energy, as the name implies, is the energy possessed by a body in motion. A speeding bullet, water actually falling from the top of Niagara Falls, an automobile in motion, and water flowing from a nozzle, all possess kinetic energy because they can perform work while in motion (Figure 19).

For ages man has taken advantage of the work-performing ability of water in motion. The water wheel mill and the hydroelectric generation of power are two common examples of man's use of the kinetic energy of water. Since we used the terms pressure head and elevation head to denote potential energy, we will use the term velocity head to denote the kinetic energy of water. Obviously, velocity head cannot exist if water is in a static condition.

At this point, we can present a general equation for the total energy possessed by water (or by other liquid) from a hydraulic standpoint:

Total Energy = Potential Energy + Kinetic Energy

which may be rewritten:

(Total Energy = Pressure Head + Elevation Head + Velocity Head)

where Pressure Head = Potential energy resulting from water being under pressure
Elevation Head = Potential energy resulting from a difference in elevation when two reference points are being compared
Velocity Head = Kinetic energy resulting from water flow

Note: Remember that pressure head, elevation head and velocity head are expressed in units of height, generally in feet. As stressed before, any head can be converted to pressure by using $P = hD$.

Conservation of energy

The next step toward deriving Bernoulli's Equation involves a fundamental law of physics known as "The Law of Conservation of Energy." This law states that "Energy can neither be created nor destroyed since the total amount of energy available is constant." This means that the creation of one form of energy results from the destruction of another form of energy. In other words, man does not have the ability to actually create energy—we may only convert energy from one form to another. Hence a weight lifter doesn't create the energy to lift a bar-bell, he merely transforms the energy he receives from food into lifting energy. An automobile engine doesn't create the energy to move the automobile, it merely transforms the

FIRE SERVICE HYDRAULICS

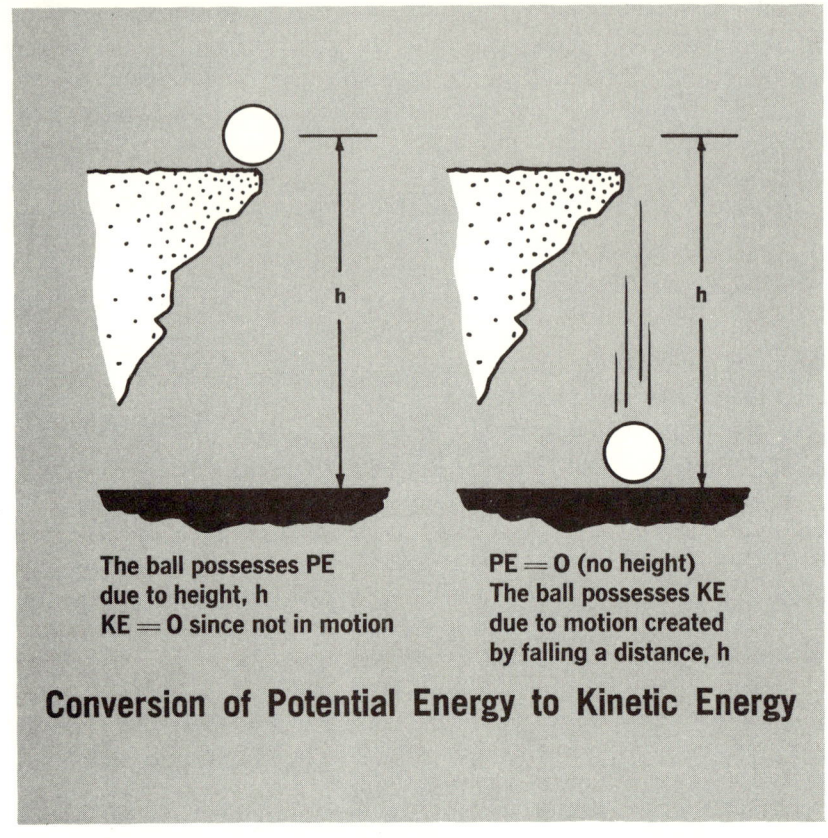

Figure 20

potential fuel energy of gasoline into the kinetic energy of the automobile. In the same way, a steam locomotive converts the potential fuel energy of coal into the energy of steam and then converts steam energy into the kinetic energy of the locomotive.

It is apparent therefore, that an increase in one form of energy always involves a decrease in another form of energy. As an example, consider the ball shown in Figure 20. If the ball were to fall, all of its potential energy would be converted to kinetic energy by the time it struck the ground. Likewise, if a valve was opened at the bottom of the tank (Figure 21), the potential energy of the stored water resulting from head, h, would be converted to kinetic energy (velocity head) as the water flowed from the valve.

WATER—AT REST AND IN MOTION

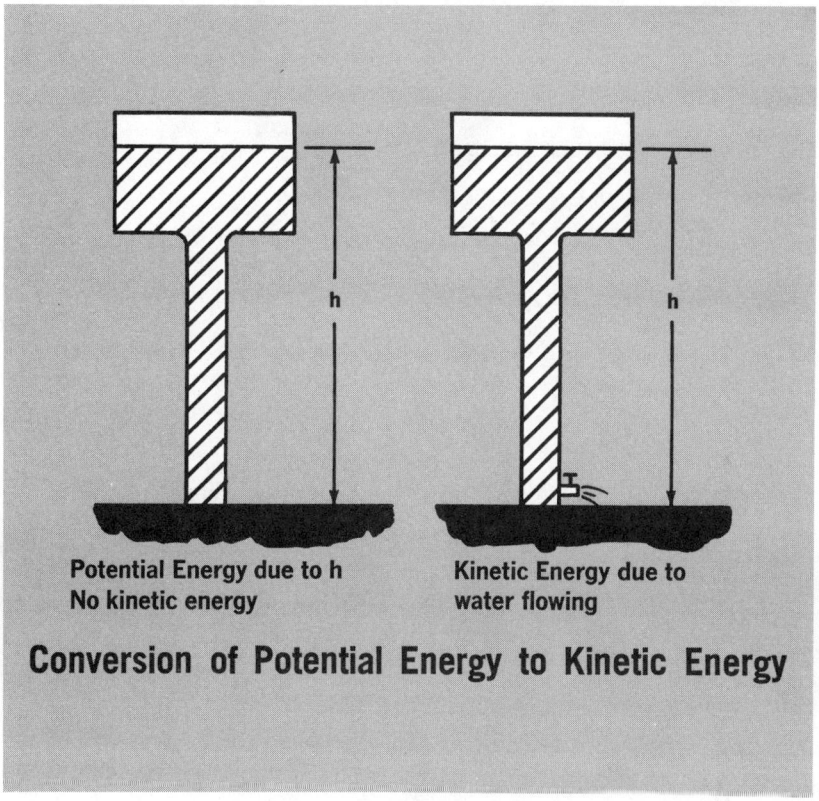

Potential Energy due to h
No kinetic energy

Kinetic Energy due to
water flowing

Conversion of Potential Energy to Kinetic Energy

Figure 21

Bernoulli's Equation

Bernoulli applied the Law of Conservation of Energy to hydraulics. His theorem states that the sum of potential energy and kinetic energy must remain constant; or more specifically, that the sum of pressure head, elevation head, and velocity head must remain constant. This permits us to compare any two reference points in a common system by the following relationship:

(Pressure Head$_①$ + Elevation Head$_①$ + Velocity Head$_①$ =
  Pressure Head$_②$ + Elevation Head$_②$ + Velocity Head$_②$)

which means the sum of all types of heads at any reference point$_①$ must equal the sum of the heads at any other reference point$_②$. This is the general form of Bernoulli's Equation.

As an exercise in applying Bernoulli's Equation, consider the open

FIRE SERVICE HYDRAULICS

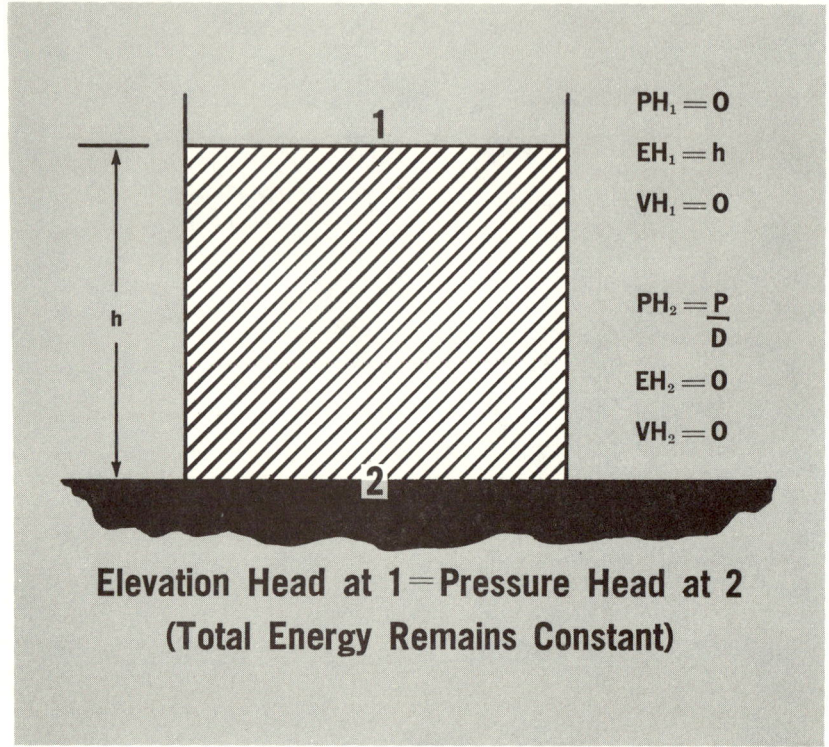

Figure 22

storage tank shown in Figure 22. We will compare two reference points—point₁ at the top surface of the liquid and point₂ at the bottom of the tank. We will assume that no water is flowing.

At reference point₁
  Pressure Head₁ = 0 since the top surface of the liquid is open to the atmosphere
  Elevation Head₁ = h, which is the height of point₁ above point₂
  Velocity Head₁ = 0 since no water is flowing and hence no kinetic energy

At reference point₂
  Pressure Head₂ = $h = \dfrac{P}{D}$ from our static pressure formula
  Elevation Head₂ = 0 since we are considering the elevation above point₂
  Velocity Head₂ = 0 since no water is flowing.

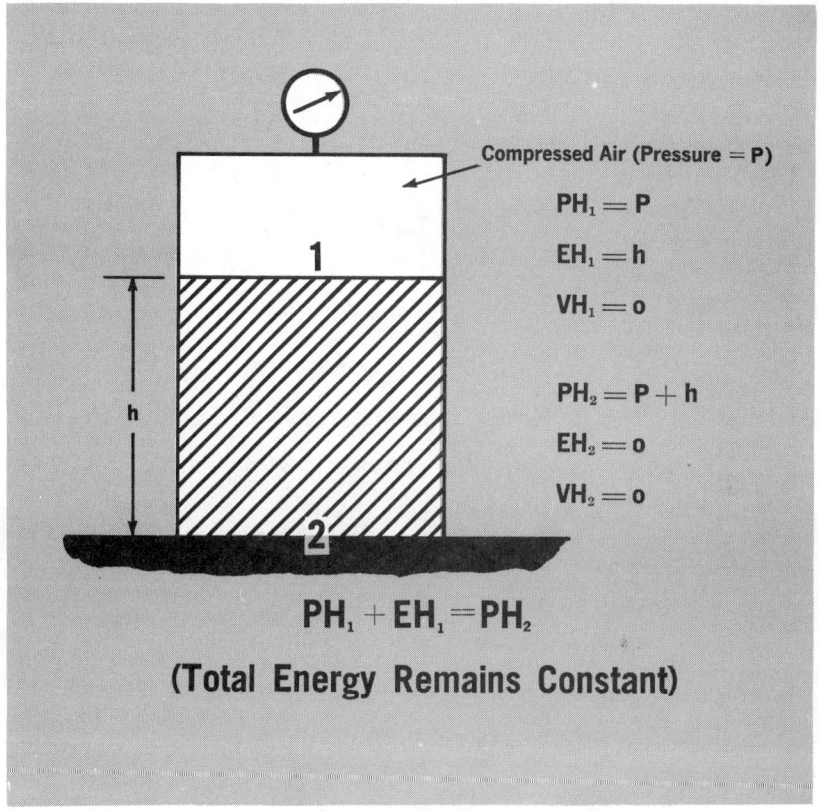

Figure 23

Bernoulli's Equation tells us that total head must remain constant, or that:

(Pressure Head$_{①}$ + Elevation Head$_{①}$ + Velocity Head$_{①}$ =
 Pressure Head$_{②}$ + Elevation Head$_{②}$ + Velocity Head$_{②}$)

which becomes

Elevation Head$_{①}$ = Pressure Head$_{②}$

or:

$h = h$ (since other heads = 0)

This shows that total head *does* in fact remain constant.

Now consider the closed storage tank (Figure 23) where the water tank is pressurized by a head of compressed air, which we shall call P.

This might represent the storage of water in a standard pressurized-

FIRE SERVICE HYDRAULICS

water fire extinguisher. We will again compare a reference point at the top of the liquid surface with a reference point at the bottom.

At reference point₁
  Pressure Head₁ = P since the top of the liquid surface is exposed to the compressed air head
  Elevation Head₁ = H which is the elevation of₁ with respect to₂
  Velocity Head₁ = 0 since no water is flowing
At reference point₂
  Pressure Head₂ = ?
  Elevation Head₂ = 0 since point₂ is our reference point for elevation
  Velocity Head₂ = 0 since no water is flowing

To determine Pressure Head₂, we simply set up Bernoulli's Equation:

$$PH_① + EH_① + VH_① = PH_② + EH_② + VH_②$$

which now can be reduced to

$$PH_① + EH_① = PH_② \text{ (since other heads } = 0)$$

Then, by substitution, $P + h = PH_②$ which shows that the pressure at point₂ results from the height of water above₂ and the compressed air head at the top surface of the water.

Problem No. 4

Problem: The static discharge pressure at a pumper is 150 psi. A hose line extending from the pumper connects to a nozzle 50 feet below the pumper (Figure 24). Assuming the nozzle is closed (no water flowing) what is the pressure head at the nozzle?
Answer:

$$PH_① = 2.31 \times 150 = 347 \text{ ft.}$$
$$EH_① = 50 \text{ ft.}$$
$$VH_① = 0$$
$$PH_② = ?$$
$$EH_② = 0$$
$$VH_② = 0$$
$$PH_① + EH_① + VH_① = PH_② + EH_② + VH_②$$
$$PH_① + EH_① = PH_② \text{ since other heads} = 0$$
$$347 + 50 = PH_②$$
$$PH_② = 397 \text{ ft}$$

(To find pressure at₂, if desired, use $P = hD$, or $P = .434h$. Hence, pressure at₂ is: $.434 \times 397 = 172$ psi)

WATER—AT REST AND IN MOTION

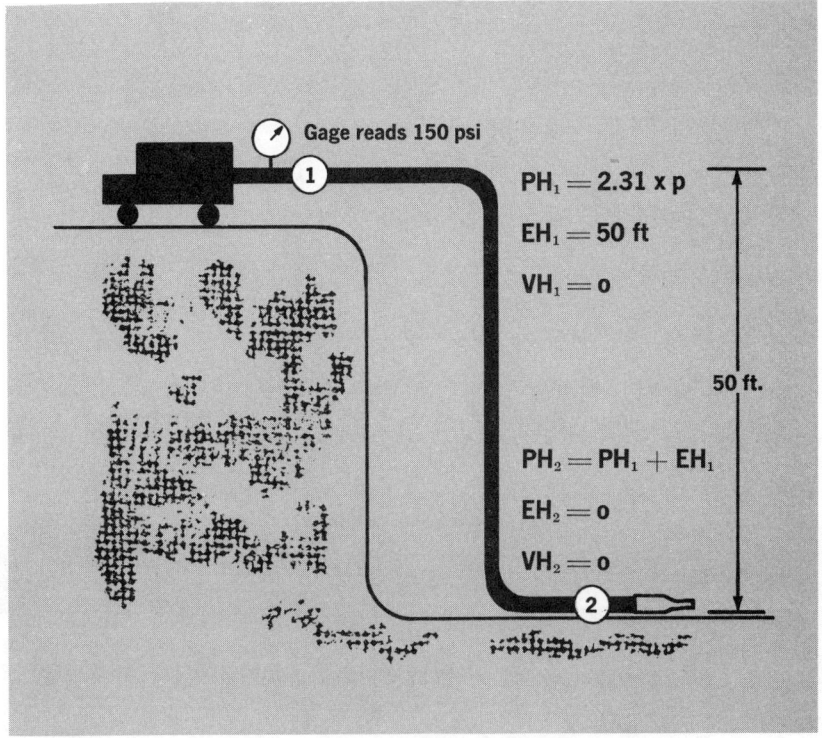

Figure 24

Relationships of discharge velocities from orifices and nozzles

The applications of Bernoulli's Equation discussed previously, involved static conditions. A more common application of Bernoulli's Equation involves hydraulic problems where water is flowing. In fact, we can use Bernoulli's Equation to develop a common formula for calculating the velocity of flow from orifices or nozzles.

Consider the system shown in Figure 25, where we have water discharging from a hole at the base of a large storage tank. To derive a formula for the velocity of discharge, we will compare reference point ① at the top surface of the liquid with reference point ② in the stream just beyond the hole.

$PH_①= 0$ (since the surface is open to the atmosphere)
$EH_① = h$ (difference in elevation)
$VH_① = 0$ (since we may assume that the tank is so large that a flow at the base doesn't cause velocity at the top)

FIRE SERVICE HYDRAULICS

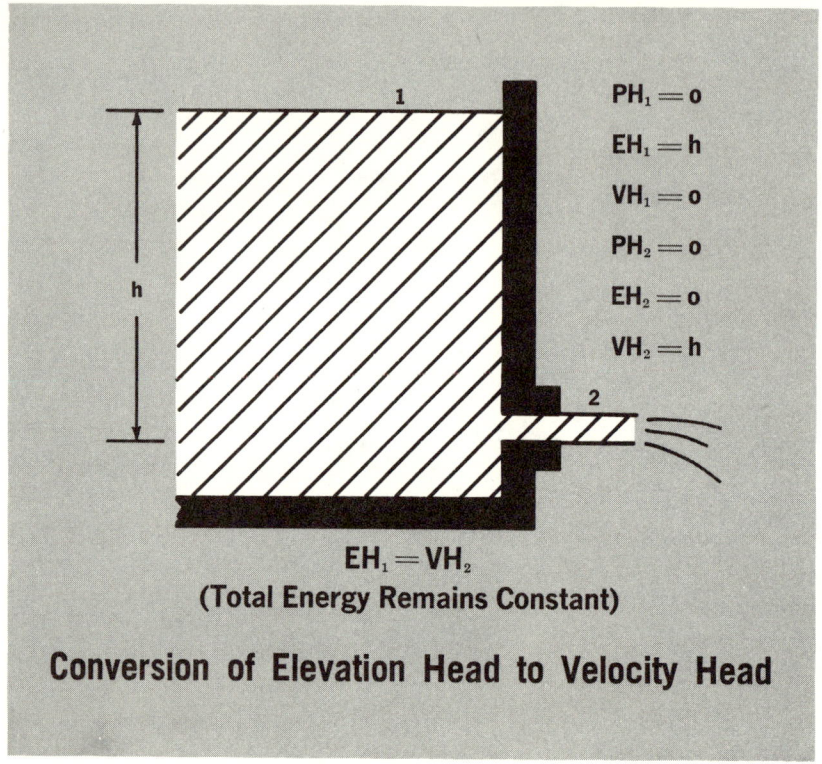

Figure 25

$PH_{②} = 0$ (since the stream is open to the atmosphere)
$EH_{②} = 0$ (since ② is our reference point for elevation)
$VH_{②} = ?$

Applying Bernoulli's Equation, we get

$$PH_{①} + EH_{①} + VH_{①} = PH_{②} + EH_{②} + VH_{②}$$

which can be simplified to:

$$EH_{①} = VH_{②} \text{ since other heads} = 0$$

or

$$h = VH_{②}$$

This shows that the energy due to velocity (velocity head) at reference point ② results from the water falling a distance, h. Therefore, we can relate discharge velocity to velocity head, h, by using a basic

WATER—AT REST AND IN MOTION

law of physics which says that the velocity reached by any object, subject to the force of gravity, g, falling a height, h, is governed by the formula

$$V = \sqrt{2gh}$$

where V is velocity (usually in ft/sec)
      g is acceleration due to gravity (32 ft/sec^2)
and h is velocity head (usually in ft)

This formula gives us a method of determining discharge velocity whenever we know the velocity head (or pressure) acting at an orifice or nozzle. If the velocity head is in feet, velocity of discharge is determined by using:

$$(V = 8\sqrt{h})$$

where h is in feet, V is ft/sec since

$$V = \sqrt{2gh} = \sqrt{2(32)h} = \sqrt{64h} = 8\sqrt{h}$$

If the velocity head is known in terms of pressure, velocity of discharge is determined by using:

$$(V = 12.1\sqrt{P})$$

where P is in psi, V is ft/sec since

$$V = \sqrt{2gh} = \sqrt{2(32)(2.31P)} = 12.1\sqrt{P}$$

When using either of the above formulas, it must be remembered that the velocity head at the nozzle (expressed either in head, h, or pressure, psi) is the pressure at the nozzle or orifice when water is flowing.

Problem No. 5

Problem: The depth of water in a tank is 50 feet. What is the velocity of discharge through a hole in the bottom of the tank?
Answer:

$$V = 8\sqrt{h} = 8\sqrt{50} = 56.6 \text{ ft/sec}$$

Problem No. 6

Problem: The flowing pressure at a nozzle is 60 psi. What is the velocity of discharge?
Answer:

$$V = 12.1\sqrt{P} = 12.1\sqrt{60} = 94 \text{ ft/sec}$$

FIRE SERVICE HYDRAULICS

Another form of Bernoulli's Equation

By solving the above equation, $V = \sqrt{2gh}$, for h, we can derive an equation for velocity head:

$$V = \sqrt{2gh}$$
$$V^2 = 2gh$$
$$\left(h = \frac{V^2}{2g}\right)$$

We can use this to express Bernoulli's Equation in a form commonly found in hydraulics text books. We do this by starting with our previously developed expression for Bernoulli's Equation,

$$PH_① + EH_① + VH_① = PH_② + EH_② + VH_②$$

and then substituting for each term as follows:

$$PH = \frac{P}{D} \quad \text{(from } P = hD\text{)}$$

$$EH = Z \quad \text{(where } Z \text{ is vertical distance or elevation)}$$

$$VH = \frac{V^2}{2g} \quad \text{(as developed above)}$$

Hence, the now familiar equation:

$$PH_① + EH_① + VH_① = PH_② + EH_② + VH_②$$

may be substituted and rewritten as:

$$\frac{P_①}{D} + Z_① + \frac{V_①^2}{2g} = \frac{P_②}{D} + Z_② + \frac{V_②^2}{2g}$$

This is the form of Bernoulli's Equation used in the "Fire Protection Handbook" of the National Fire Protection Association.

Equations for determining discharge from a nozzle or orifice

The discharge of water through a pipe or hose line is equal to the area of the outlet multiplied by the velocity of flow. This is shown by the following formula:

$$Q = AV$$

where Q = quantity per minute
A = area of outlet
V = velocity of flow

WATER—AT REST AND IN MOTION

If the quantity is to be in cubic inches per minute (in³/min), the area must be in square inches (in²) and the velocity must be in inches per minute (in/min):

$$Q = AV$$
$$\text{in}^3/\text{min} = \text{in}^2 \times \text{in/min}$$

If the quantity is to be in cubic feet per minute (ft³/min), the area must be in square feet (ft²) and the velocity in feet per minute (ft/min):

$$Q = AV$$
$$\text{ft}^3/\text{min} = \text{ft}^2 \times \text{ft/min}$$

If the quantity is to be in gallons per minute (gpm), then it is best to have the area in square inches (in²) and the velocity in inches per minute (in/min) and divide the product, AV, by 231, which is the number of cubic inches in a gallon.

$$Q = AV$$
$$\text{gpm} = \text{in}^2 \times \text{in/min} \times 1 \text{ gal}/231 \text{ in}^3$$

As an illustration of the formula, $Q = AV$, assume we have a square pipe (Figure 26), and that through this pipe is being forced a stream of blocks—four abreast and each 1 cubic inch in size. The cross sectional area of the pipe must then be 2 inches × 2 inches or 4 square inches. Now suppose that the velocity of flow is 4 inches per minute; then the number of blocks discharged per minute will be 4 × 4 or 16 blocks per minute, or 16 cubic inches of blocks per minute. But, as mentioned above, the area of the pipe is 4 square inches and the velocity of flow is 4 inches per minute. Thus:

$$Q = AV$$
$$Q = 4 \text{ in}^2 \times 4 \text{ in/min}$$
$$Q = 16 \text{ in}^3/\text{min}$$

This illustrates the formula. In the next chapter this formula will be used for determining nozzle and hose discharge rates.

The "continuity" of flow

By applying another fundamental law of physics known as the "Law of the Conservation of Matter," we can show that $Q = AV$ must be constant throughout a flowing system.

The Law of Conservation of Matter states that "Matter can neither be created nor destroyed." This means that the quantity of water

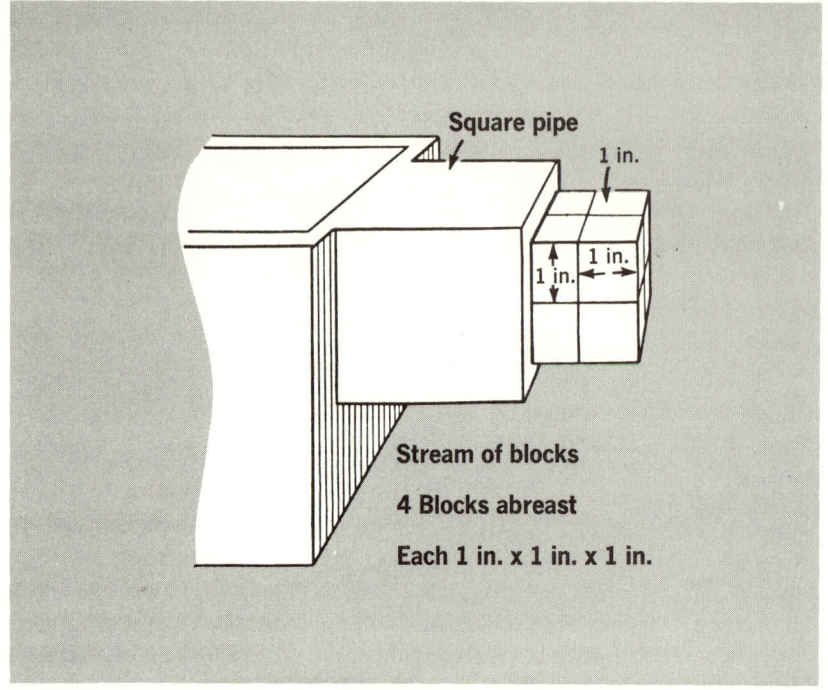

Figure 26

entering one end of a pipe during a given time interval must be equal to the quantity leaving the other end of the pipe in the same time interval. Hence, if the quantity flowing must be constant from section to section in a hydraulic system, then the product AV must remain constant since $Q = AV$. It should be apparent then, that since the product AV must remain constant, where A is large, V must be small, and likewise where A is small, V must be large.

This principle is illustrated in Figure 27, where we have a section of 2½-inch hose coupled to a section of 1½-inch hose. By the Law of the Conservation of Matter if 100 gpm is entering the 2½-inch hose, then 100 gpm must be leaving the 1½-inch hose. This means that the product A_1V_1 for the 2½-inch hose must equal the product A_2V_2 for the 1½-inch hose. Since the area, A_1, of the 2½-inch hose is greater than the area, A_2, of the 1½-inch hose, it should be obvious that the velocity of flow, V_2, through the 1½-inch hose must be somewhat greater than the velocity of flow, V_1, through the 2½-inch hose. In other words, since the area of the 1½-inch hose is smaller, the water

WATER—AT REST AND IN MOTION

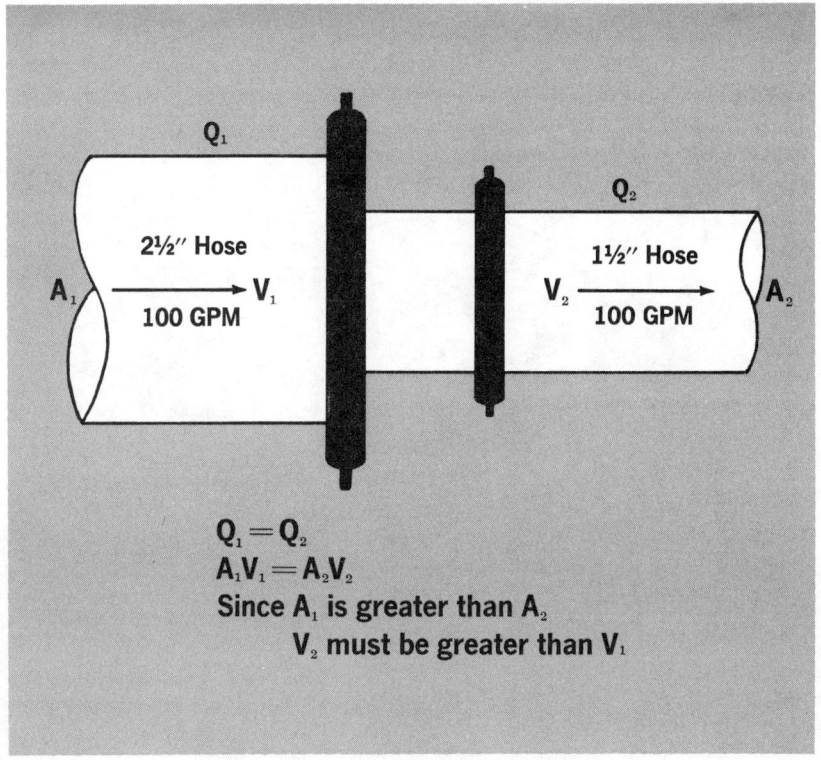

Figure 27

must flow faster through 1½-inch hose in order to flow the same quantity in the same amount of time.

Problem No. 7

Problem: The velocity of flow through a 1¼-inch nozzle (Area = 1.23 square inches) is 100 feet per second. The nozzle is supplied by a 2½-inch hose (Area = 4.92 square inches). What is the velocity of flow through the 2½-inch hose?

Answer:

A_1 = area of 1¼-inch nozzle = 1.23 in²
V_1 = velocity through 1¼-inch nozzle = 100 ft/sec
A_2 = area of 2½-inch hose = 4.92 in²
V_2 = velocity through 2½-inch hose = ?
$Q = AV$ must be constant, hence: $A_1V_1 = A_2V_2$

FIRE SERVICE HYDRAULICS

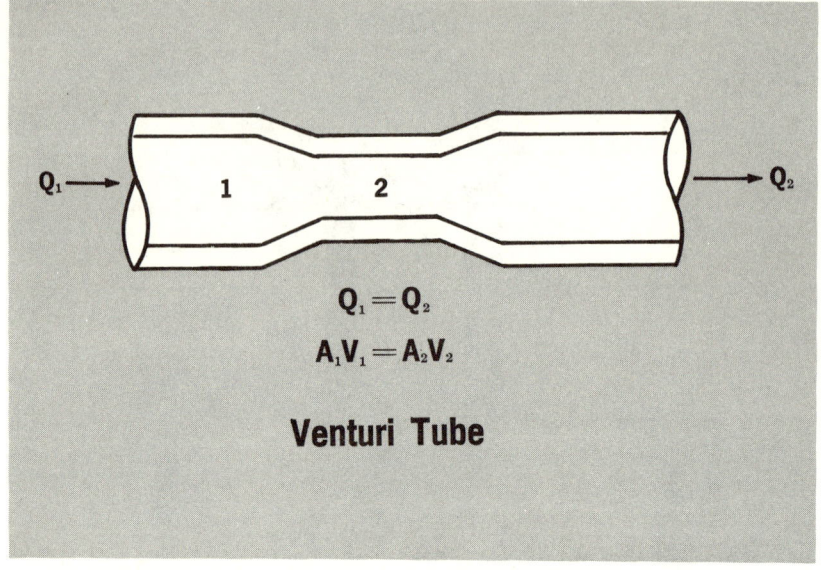

Figure 28

or

$$V_2 = \frac{A_1}{A_2} V_1$$

$$V_2 = \frac{1.23 \text{ in}^2}{4.92 \text{ in}^2} (100 \text{ ft/sec})$$

$$V_2 = 25 \text{ ft/sec}$$

From the example above, it should be apparent that a common rule of hydraulic flow is: "The quantity of flow varies inversely with area."

Devices for measuring flow

One method of measuring flow in pipe lines or hose lines is by the use of a *venturi tube*. The venturi tube (Figure 28) is based on both Bernoulli's Equation and the $Q = AV$ flow formula. Consider reference points ① and ② as shown, where reference point ② is located in the center of the constricted area of the tube. As discussed in the preceding section, since the area at point ② is less than the area at point ①, the velocity head must be greater at point ②. If we connect a mercury manometer (Figure 29), we create a venturi meter. By applying Bernoulli's Equation, we can use the venturi meter for

WATER—AT REST AND IN MOTION

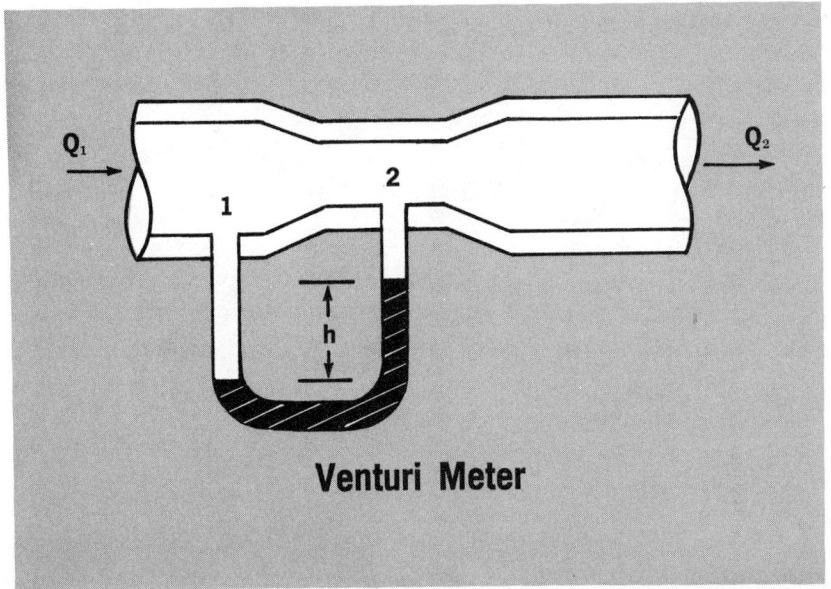

Figure 29

determining Q:

$$PH_{\text{①}} + EH_{\text{①}} + VH_{\text{①}} = PH_{\text{②}} + EH_{\text{②}} + VH_{\text{②}}$$

Since we may assume that the tube is level, there is no difference in elevation between points ① and ② so Bernoulli's Equation reduces to:

$$PH_{\text{①}} + VH_{\text{①}} = PH_{\text{②}} + VH_{\text{②}}$$

Now, since we determined above that the velocity head at ② is greater than the velocity head at ①, it follows that the pressure head must be less at ②. This explains why the manometer shows a lower pressure at point ②, as evidenced by the higher level of mercury beneath point ②. Venturi meters are usually calibrated in advance so that for any h we can readily determine V and Q.

In addition to measuring flow rates, a venturi tube may also be used as a device for inducting gas or liquid into the flowing stream because of the suction created at the throat. A venturi tube is often used as a "pick-up" tube for injecting a fire foam concentrate into a hose line.

Another device for measuring flow is the *pitot tube*. The principle of the pitot tube is illustrated in Figure 30. When the small opening in the pitot tube is inserted in the center of the stream, the tube

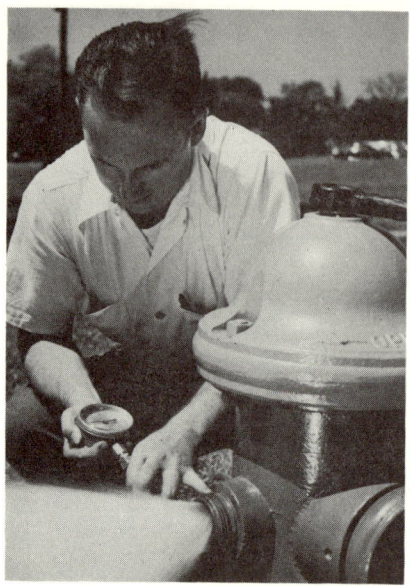

Figure 30

measures the total head at that point. The head is usually indicated on a piezometer or Bourdon gage attached to the top of the tube. Since the stream is open to the atmosphere, there can be no pressure head so the indicated reading will be velocity head only. After reading the head, h, from the piezometer or Bourdon gage, the velocity of flow can be calculated by using $V = \sqrt{2gh}$ and the quantity of flow can be determined from

$$Q = AV$$

As explained in Chapter 2, tables are available that give values of Q for various pitot readings and various nozzle (or orifice) sizes. Figure 30 shows a typical pitot gage being used to determine the flow rate from a straight stream nozzle.

SUMMARY OF CHAPTER 1

(1) Important Terms:
    Hydraulics   Manometer
    Hydrostatics   Bernoulli's Equation
    Hydrokinetics   Energy
    Density   Potential Energy
    Pressure   Kinetic Energy

WATER—AT REST AND IN MOTION

Force
Head
Static Pressure
Residual Pressure
Atmospheric Pressure
Vacuum
Gage Pressure
Absolute Pressure
Barometer
Bourdon Gage
Piezometer

Pressure Head
Elevation Head
Velocity Head
Total Head
Conservation of Energy
Reference Points
Velocity
Discharge Rate
Conservation of Matter
Venturi Tube
Pitot Tube

(2) Important Equations

(a) Force = Pressure × Area
$$F = PA$$

(b) Pressure = Height × Density
$$P = hD$$

(c) Absolute Pressure = Gage Pressure + Atmospheric Pressure

(d) Total Potential Energy = Potential Energy due to Pressure + Potential Energy due to Elevation
↓ ↓ ↓

(e) Total Static Head = Pressure Head + Elevation Head

(f) Total Energy = Potential Energy + Kinetic Energy

(g) Total Energy = Pressure Head + Elevation Head + Velocity Head

(h) Bernoulli's Equation

$$PH_① + EH_① + VH_① = PH_② + EH_② + VH_②$$

$$\frac{P_①}{D} + Z_① + \frac{V_①{}^2}{2g} = \frac{P_②}{D} + Z_② + \frac{V_②{}^2}{2g}$$

(i) $V = \sqrt{2gh}$ V = velocity
g = acceleration due to gravity (32 ft/sec)
h = velocity head

(j) $Q = AV$ Q = quantity
A = area
V = velocity

(k) $Q_1 = Q_2$
$A_1 V_1 = A_2 V_2$

A SELECTED REFERENCE LIST

Introduction to Fluid Mechanics, R. W. Henke, Addison-Wesley Publishing Co., Reading, Mass., 1966.

Fire Protection Handbook, Thirteenth Edition, National Fire Protection Association, Boston, 1969.

Fluid Mechanics for Engineers, M. L. Albertson, J. R. Barton and D. B. Simons, Prentice Hall, Inc., Englewood Cliffs, N.J., 1960.

Water Supply and Sewerage, Ernest M. Steel, McGraw-Hill Book Co., New York, 1960.

Flow of Fluids, Technical Paper No. 410, The Crane Co., Chicago, 1957.

Physics for the New Age, R. H. Carleton, H. H. Williams and M. H. Buell, J. B. Lippincott Co., New York, 1954.

CHAPTER TWO

Velocity and discharge

This chapter deals with some important fluid dynamic considerations of water which are important to fire engineering. Chapter 1 gives the basic physical relationships which apply to water in a static or non-flowing condition and the important laws governing water systems when they are in a moving or flowing condition. We must now consider the relationships which control the performance of water when it is used in typical fire fighting conditions of pumping, transporting in hoses, and discharging from nozzles.

To utilize water properly we must provide the necessary conditions for its correct flow. This chapter relates hydraulic conditions with the results that may be expected from them in terms of fluid system characteristics and performance.

In general, water is available for fire fighting operations from central water system mains or grids which terminate at hydrants. The large-diameter pipes which supply these hydrants allow water at the pressure of the main system to be supplied to the fire pump in sufficient volume of flow through connections of a length or more of large-diameter hose.

Frequently, however, the fire officer finds that his only source of water is an open body of water such as a pond or ground reservoir. In order to employ this water, which is essentially at rest, he must supply power to it in the form of pressure or "head." The water must

FIRE SERVICE HYDRAULICS

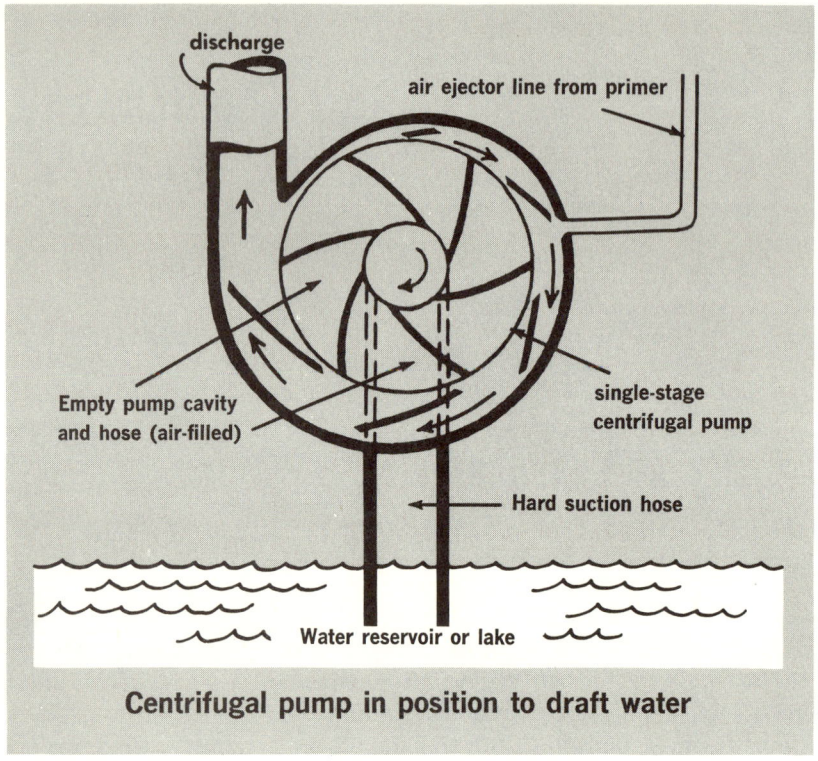

Figure 1

first be "lifted" from its lower level in the pond to the pump where it can be acted on and moved by the pressurizing action of the pump.

The initial "lifting" of the water to a point where it can be pumped is called "drafting" and is accomplished by employing atmospheric pressure in a "suction" manner.

Consider the centrifugal pump of Figure 1 with its noncollapsible "hard suction" hose connected to the inlet of the pump and extending down below the level of the water that we wish to force into the pump by "lifting."

If we now employ a device for evacuating or removing the air in the cavity of the hose and the closed pump, known as a primer or ejector (Centrifugal pumps are provided with such priming devices as shown in Figure 2), we will create a *negative* pressure difference in the

VELOCITY AND DISCHARGE

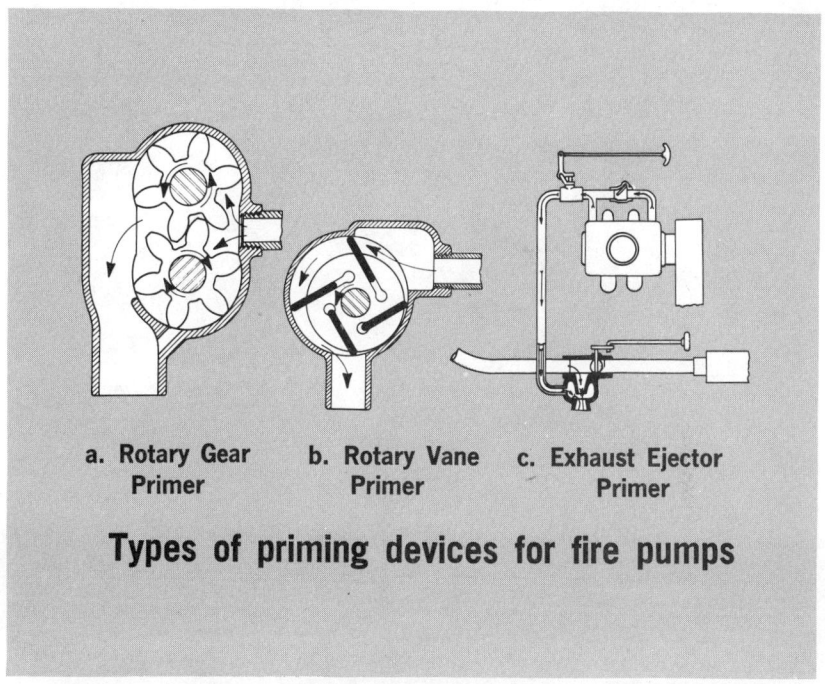

a. Rotary Gear Primer b. Rotary Vane Primer c. Exhaust Ejector Primer

Types of priming devices for fire pumps

Figure 2

system (a vacuum). The *positive* atmospheric pressure on the body of water in the lake or pond will force water up into the evacuated volume of hose and into the rotating pump as in Figure 3.

The pump can now pressurize the water which has been "lifted" by the "drafting" action and water may now be moved continuously from the level of the pond or reservoir into the fire hose lines.

The vacuum, or negative pressure, on the suction side of a pump is often measured in inches of mercury, usually written as in. Hg. or Hg (Hg is the chemical symbol for mercury). A vacuum of 1 inch of mercury is equal to a negative pressure of 0.49 pounds per square inch, or 1 inch Hg = 0.49 psi = 1.13 feet of water.

The difference in elevation between the water level and the center of the pump is known as "lift." The maximum lift that is possible under ideal conditions may be calculated by using one of the basic equations from Chapter 1, Pressure = Height × Density ($P = hD$).

Since atmospheric pressure = 14.7 psi, our maximum lift (or height)

FIRE SERVICE HYDRAULICS

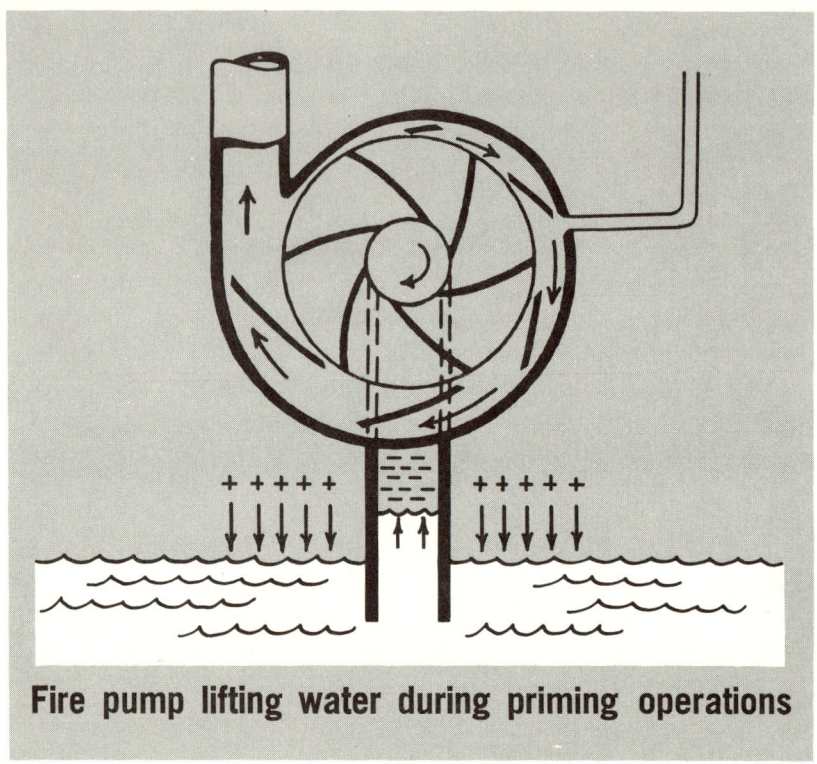

Fire pump lifting water during priming operations

Figure 3

is determined as follows:

$$P = hD$$
$$h = \frac{P}{D}, \text{ or}$$
$$h = 2.31P \text{ (where } P \text{ is in psi and } h \text{ is in feet)}$$
$$h = 2.31\,(14.7)$$
$$h = 33.9 \text{ ft} = \text{maximum lift}$$

This maximum lift of 33.9 feet is dependent upon the creation of a perfect vacuum. Since no pump is able to create a perfect vacuum, our actual "lifting ability" is somewhat less than 33.9 feet. In fact, a fire pump in fairly good condition can lift water only about 25 feet by the suction process. This decrease in the theoretical maximum lift results from several other factors aside from the inability to create a perfect vacuum. The factors limiting lift distances to a pump are: altitude, weather, water temperature and hydraulic losses.

Effects of altitude

As the elevation above sea level of the pumping site increases, the atmospheric pressure decreases, which in turn causes a reduction in lifting ability. The loss of lift at various elevations is as follows:

Elevation Above Sea Level (feet)	Loss of Lift (feet of water)
1,000	1.22
2,000	2.38
3,000	3.50
4,000	4.75
5,000	5.80
6,000	6.80
7,000	7.70

The power generated by internal combustion engines also decreases with increased altitude, which further impedes the lifting ability of pumps driven by internal combustion engines.

A decrease in barometric pressure can also be caused by poor weather conditions, but the reduction is hardly significant. The difference in barometric pressure due to the presence of a rainstorm means a reduction in lift of only about 1 foot.

Water temperature

Hot water reduces the drafting ability of a pump. This is due to the tendency of hot water, when placed under a partial vacuum, to give off water vapor, thus destroying or greatly reducing the ability of the pump to create the vacuum necessary for raising water. If water is fairly hot, it will be found that the pump will be unable to lift at all. At 32 degrees, the vapor pressure (feet loss) is .204. At 100 degrees, it is 2.19. At 212°F it should be noted that the vapor pressure is equal to atmospheric pressure, and a suction lift is impossible. For practical purposes when water temperature exceeds 160°F, drafting should not be attempted.

Hydraulic losses

Hydraulic losses in the suction pipe (or hose) and minor pressure losses through strainers and similar components must also be con-

FIRE SERVICE HYDRAULICS

sidered. Fire departments use large-size suction hose to ensure large water flow with reduced friction loss. For a given pump discharge, the maximum lift varies with the size of the suction hose. For instance, a pump can lift 500 gpm through a vertical lift of 23 feet if a

Table Showing Minimum Discharge Which Should be Expected of a Pumper in Good Condition Operating at Draft at Various Lifts.

CONDITIONS: Operating at Net Pump Pressure of 150 psi; Altitude of 1000 feet; Water Temperature of 60°F; Barometric Pressure of 28.94" Hg (poor weather conditions).

Rated Capacity, Pump		500 gpm		750 gpm		1000 gpm		1250 gpm	1500 gpm	
Suction Hose Size		4"	4½"	4½"	5"	5"	6"	6"	Dual 5"	Dual 6"
Lift in Feet	4	590	660	870	945	1160	1345	1435	1735	1990	2250
	6	560	630	830	905	1110	1290	1375	1660	1990	2150
	8	530	595	790	860	1055	1230	1310	1575	1810	2040
	10 (20' Suction Hose, Two Sections)	500	560	750	820	1000	1170	1250	1500	1720	1935
	12	465	520	700	770	935	1105	1175	1410	1615	1820
	14	430	480	650	720	870	1045	1100	1325	1520	1710
	16	390	430	585	655	790	960	1020	1225	1405	1585
	18	325	370	495	560	670	835	900	1085	1240	1420
	20 (30' Suction Hose, Three Sections)	270	310	425	480	590	725	790	955	1110	1270
	22	195	225	340	375	485	590	660	800	950	1085
	24	65	70	205	235	340	400	495	590	730	835

NOTES: 1—Net pump pressure is 150 psi. Operation at a lower pressure will result in an increased discharge; operation at a higher pressure, a decreased discharge.

2—Data based on a pumper with ability to discharge rated capacity when drafting at not more than a 10-foot lift. Many pumpers will exceed this performance and therefore will discharge greater quantities than shown at all lifts.

Courtesy of American Insurance Association.

5-inch-diameter suction hose is employed, whereas it can lift the same amount only 12½ feet if 3½-inch suction is used. The "discharge-lift" table on the preceding page shows the relationship between water flow quantity, suction hose size and lift:

FLOW VELOCITY AND NOZZLE DISCHARGE

Equipped with the hydraulic theory and basic equations presented in Chapter 1, we are now ready to examine the formulas used to calculate nozzle discharge and the flow characteristics of nozzles and orifices.

Basic flow formulas

In Chapter 1, we used Bernoulli's Equation to show that discharge velocity and velocity head are related by the equation:

$$V = \sqrt{2gh}$$

We also determined that this equation could be written in the form

$$V = 12.1\sqrt{P}$$

where V is velocity in feet per second, and P is velocity head (pressure) in pounds per square inch. (Remember that when using either of these equations, the velocity head at the nozzle or orifice expressed either as head, h, or pressure, psi, is the pressure at the nozzle when water is flowing.)

Additionally, we found that the discharge of water through a pipe or hose line is equal to the area of the outlet multiplied by the velocity of flow. This is expressed by the formula

$$Q = AV$$

where Q is quantity per unit of time, A is the area of the outlet, and V is the velocity of flow. By substituting terms in the above equations it follows that:

$$Q = AV, \text{ or}$$
$$Q = A\sqrt{2gh}, \text{ or}$$
$$Q = A \times 12.1\sqrt{P}$$

Since Q is commonly expressed in gallons per minute in the fire protection field and since the diameter of a nozzle or orifice is usually expressed in inches, we convert the above equation to:

FIRE SERVICE HYDRAULICS

$$Q = A \times 12.1 \sqrt{P}$$

$$Q = \frac{\pi d^2}{4} \text{ in}^2 \times 12.1 \sqrt{P} \text{ ft/sec}$$

$$Q = \frac{\pi d^2}{4} \text{ in}^2 \times \frac{1 \text{ ft}^2}{144 \text{ in}^2} \times 12.1 \sqrt{P} \text{ ft/sec} \times$$

$$60 \text{ sec}/1 \text{ min} \times 7.48 \text{ gal/ft}^3$$

$$Q = \frac{\pi d^2 \times 12.1 \sqrt{P} \times 60 \times 7.48}{4 \times 144}, \text{ or}$$

$$Q = 29.71 d^2 \sqrt{P}$$

where Q is the discharge rate from the nozzle or orifice in gpm; d is the diameter of the nozzle or orifice in inches, and P is the nozzle or orifice pressure when water is flowing in psi.

$Q = 29.71 d^2 \sqrt{P}$ is the basic nozzle flow formula developed by John Freeman in 1888, and is known appropriately as the "Freeman Formula." By using this formula, the discharge from any orifice can be calculated if we know the diameter of the orifice and the discharge pressure (velocity head).

To simplify calculations, the figure "30" is often substituted for 29.7. This creates a negligible error. The American Insurance Association publishes the most extensive flow tables in use today. Their tables are derived from a slight variation of the Freeman Formula, $Q = 29.83 d^2 \sqrt{P}$. As is readily apparent, the AIA formula varies with Freeman's formula by less than 1 percent.

Discharge coefficients

It should be stressed that both Freeman's formula ($Q = 29.71 d^2 \sqrt{P}$) and the AIA formula ($Q = 29.83 d^2 \sqrt{P}$) are strictly theoretical as shown, since no hydraulic losses are considered. In actuality, certain head losses, mainly due to friction, do occur as water discharges from any orifice. For the best designed fire hose nozzles, the difference between theoretical and actual discharge is almost negligible. For orifices not specifically designed for efficient discharge, the reduction may be as high as 50 percent.

An orifice with a sharp entrance edge as shown in Figure 4 is known as a "standard orifice" and is commonly used for measuring flow.

VELOCITY AND DISCHARGE

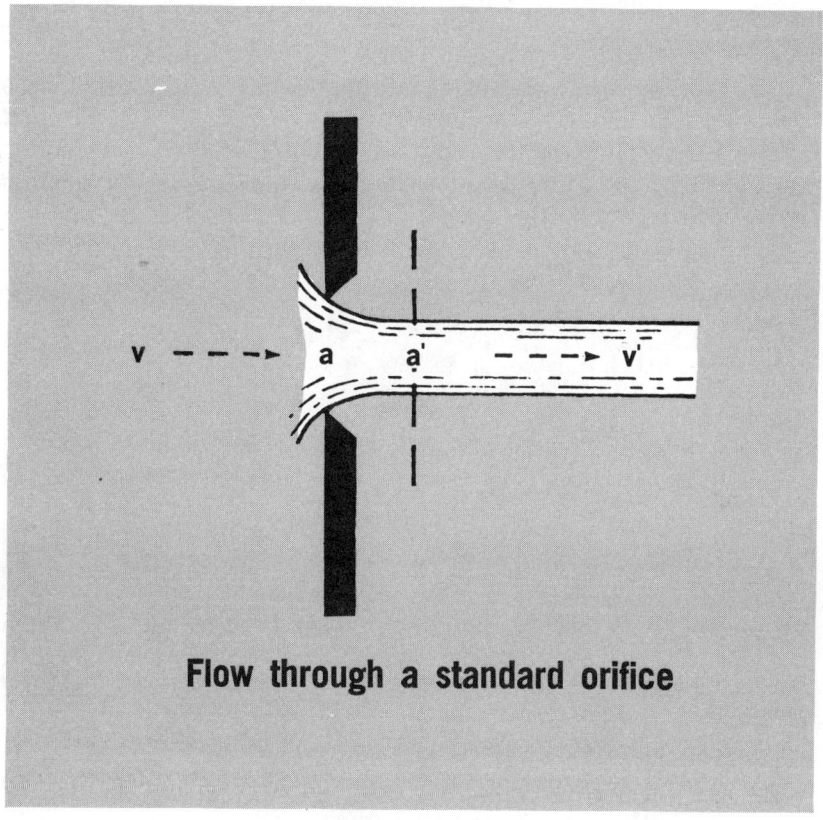

Figure 4

As the water leaves the orifice, it contracts to form a jet whose cross-sectional area is less than that of the orifice. At a distance from the orifice of about one half the diameter of the jet the contraction is complete (plane a').

Actual measurement has shown that the velocity of the stream produced by the orifice, V', is always less than the theoretical value, $V' = \sqrt{2gh}$, primarily because of fluid friction. This loss of velocity is expressed by a *coefficient of velocity* which is designated by the term C_v. For the standard orifice, the loss of velocity is only about 2 percent, so $C_v = .98$.

In using the orifice as a measuring device for applying $Q = AV$ (or the Freeman or AIA equations), the area is calculated at a in Figure 4, the smallest diameter of the orifice. The velocity is measured at

FIRE SERVICE HYDRAULICS

a' where velocity has its highest value. To correct for this contraction, we use another coefficient known as the *coefficient of contraction* which expresses the ratio of a'/a. The coefficient of contraction is written as C_c. For the standard orifice, $C_c = .62$.

The actual discharge from any orifice is the product of C_v and C_c multiplied by the theoretical discharge. As an example of this, the actual discharge from the standard orifice would be

$$Q = C_v C_c A \sqrt{2gh}, \text{ or}$$
$$Q = (.98)(.62) A \sqrt{2gh} = .61 A \sqrt{2gh}$$

In general then, for any orifice:

$$Q_{(actual)} = C_v C_c A \sqrt{2gh}$$
$$= C_v C_c 29.71 d^2 \sqrt{P} \text{ or (Freeman)}$$
$$= C_v C_c 29.83 d^2 \sqrt{P} \text{ (AIA)}$$

The product of $C_v \times C_c$ is usually expressed as a single term, C_d, which is known as the overall *coefficient of discharge*, ($C_d = C_v \times C_c$). Hence

$$Q_{(actual)} = C_d A \sqrt{2gh} \text{ or}$$
$$Q_{(actual)} = 29.71 C_d d^2 \sqrt{P} \text{ (Freeman) or}$$
$$Q_{(actual)} = 29.83 C_d d^2 \sqrt{P} \text{ (AIA)}$$

Accordingly, when calculating actual discharge by use of the above formulas the coefficient of discharge for the nozzle or orifice in question should be known. Typical coefficients of discharge for common orifices used in the fire protection field are shown below:

Typical Coefficients of Discharge (C_d)

Standard orifice	.61
Representative sprinkler (½ inch ring orifice)	.78
Hydrant butt	.70–.90
Smooth cone nozzles	.96–.997

Note that the coefficient for smooth flow nozzles is nearly equal to 1. This explains why some basic fire hydraulics textbooks ignore the

VELOCITY AND DISCHARGE

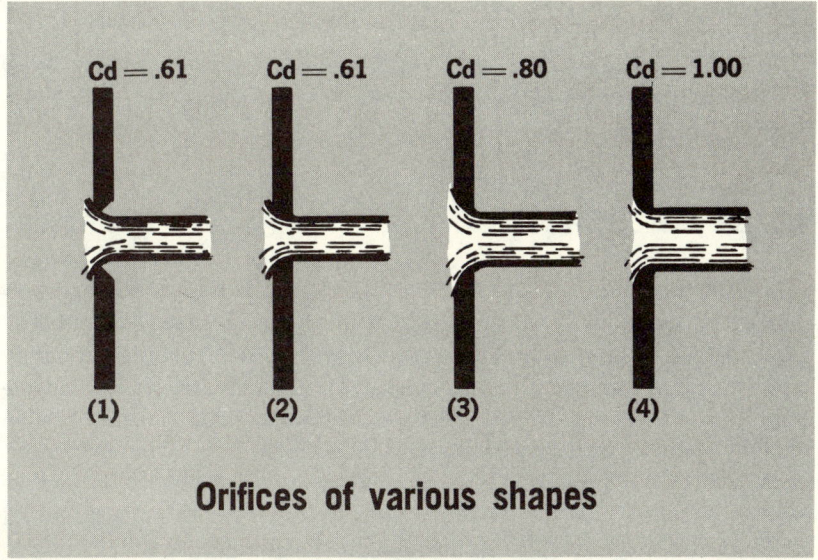

Figure 5

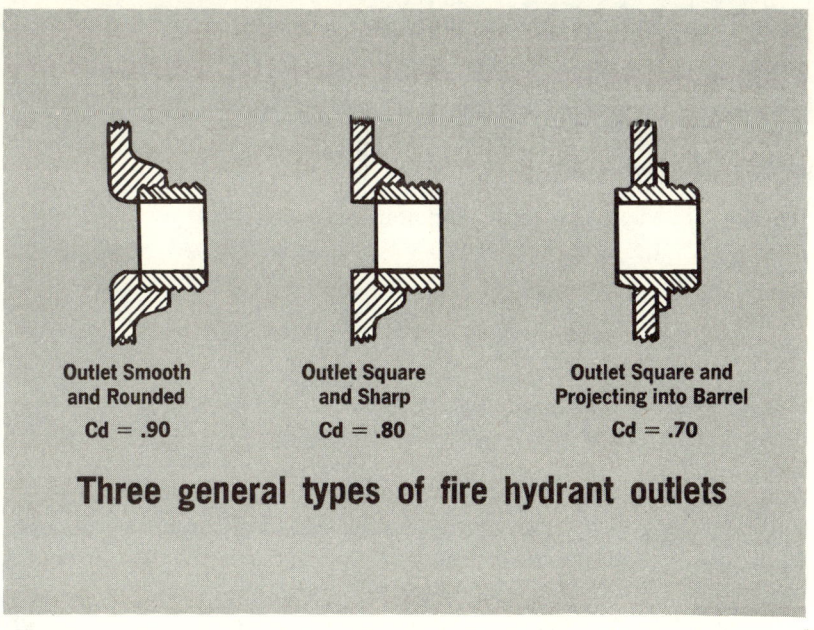

Figure 6

slight reduction in flow and apply the Freeman or AIA formulas without using loss coefficients.

Figure 5 shows the impact of orifice design on the resultant coefficient of discharge. Figure 6 shows three general types of fire hydrant outlets and their corresponding coefficients.

NOZZLE REACTION

The phenomenon known as *nozzle reaction* is familiar to every fireman who has held a hose nozzle that is discharging water. The backward thrust of the nozzle is governed by the same principles involved in modern jet propulsion. A flowing jet of water creates a considerable amount of force due to the mass of the water and the increase in velocity as the water discharges from the nozzle. The backward force exerted on the nozzle illustrates a basic law of physics known as Newton's Third Law of Motion: "For every action there is an equal and opposite reaction." Figure 7 shows the similarity between nozzle reaction and other common examples of "action and reaction" or more appropriately, "equal and opposite force."

Nozzle reaction can be calculated by using the following formula:

$$R = 2PA$$

where R is nozzle reaction in pounds of backward thrust, P is nozzle pressure in psi, and A is area of the nozzle discharge opening in square inches. By substituting the formula $A = \dfrac{\pi d^2}{4}$, we can reduce the reaction equation to a more common form:

$$R = 2PA$$

$$R = 2P\left(\frac{\pi d^2}{4}\right)$$

$$R = \frac{P\pi d^2}{2}$$

$$R = 1.57 d^2 P$$

When using $R = 1.57 d^2 P$, remember that d is the nozzle diameter in inches, P is the nozzle pressure in psi, and R is reaction in pounds of force.

It should be pointed out that if a hose was laid perfectly straight from hydrant to nozzle, the reaction force would be exerted against

VELOCITY AND DISCHARGE

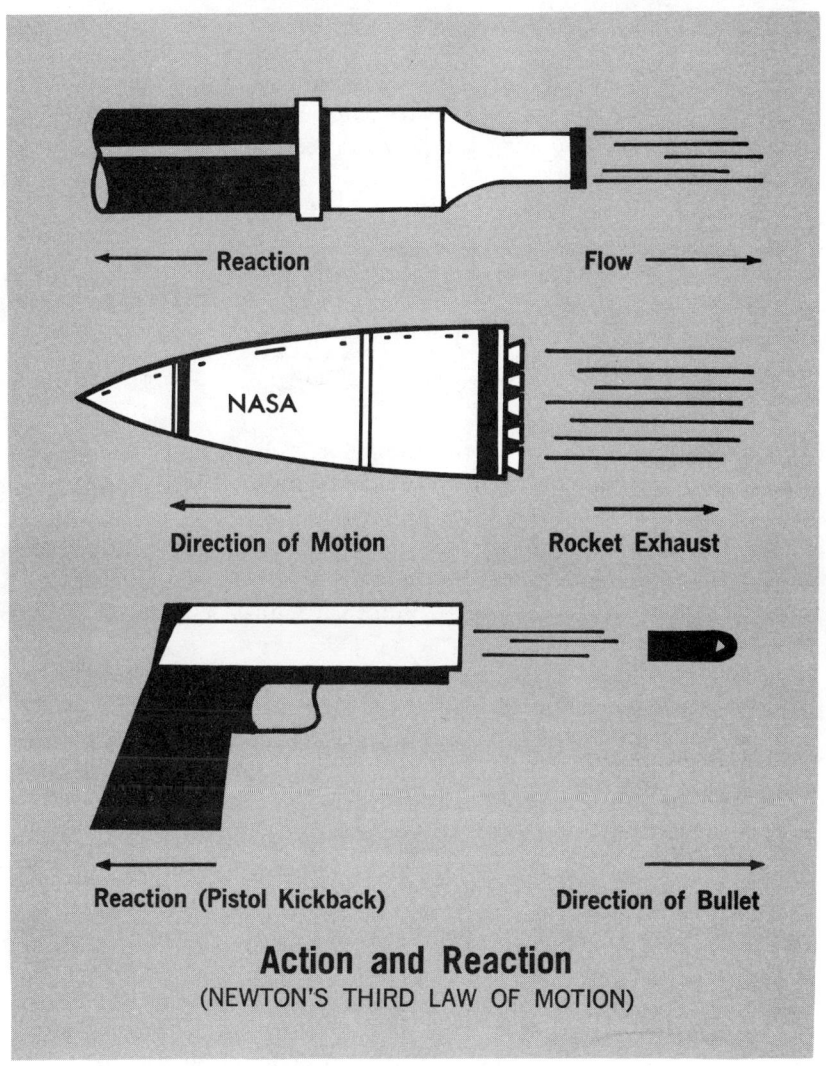

Figure 7

the hydrant. Since a perfectly straight run of hose is virtually impossible, the fireman holding the nozzle will nearly always have to cope with nozzle reaction. Fortunately, the fireman is usually exposed to forces less than shown in the table since the hose is usually in contact with the ground or otherwise restrained from movement. The

FIRE SERVICE HYDRAULICS

nozzle reaction formula neglects the effect of jet contraction and friction loss. However, these losses are minor and the formula is considered sufficiently accurate for our purposes.

A close review of the table will reveal that for a given flow, the larger the nozzle producing the flow the less the reaction. This fact is illustrated below:

Nozzle Diameter	Nozzle Pressure	gpm	Reaction
1 inch	80 psi	265	125 pounds
1⅛ inch	50 psi	265	100 pounds
1¼ inch	32 psi	265	80 pounds

Obviously the slight increase in diameter is more than compensated for by the considerable reduction in nozzle pressure needed with larger nozzles to create the same discharge rate.

We should also emphasize the importance of using caution when employing large discharge ladder pipes. Since ladder pipes may generate reactions in excess of 400 pounds, care must be taken to prevent damage to the ladder or ladder supports.

Another point to be considered when calculating reaction is that the momentary reaction that occurs when a nozzle is opened suddenly may be much greater than the reaction that is normally present during water flow. When a nozzle is shut off, the pressure at the nozzle is equal to the pressure at the source of supply (assuming equal elevation). This static pressure is always greater than the residual flowing pressure. If opened suddenly, the momentary reaction force will result from the static pressure. For example, if a 1-inch nozzle has a flowing nozzle pressure of 50 psi and a friction loss of 50 psi in the supply hose, the reaction would be 78 pounds. When the nozzle is closed, the static pressure at the nozzle would be 100 psi. If opened suddenly, the momentary reaction force would be 157 pounds. To prevent this sudden backward thrust, nozzles should be opened slowly.

In general, the reaction of spray nozzles is less than with the straight stream type since the small impinging streams tend to diffuse the reaction forces within the nozzle.

Water hammer

While nozzle reaction dictates that nozzles should be *opened* slowly, another hydraulic phenomenon dictates that they should also be

VELOCITY AND DISCHARGE

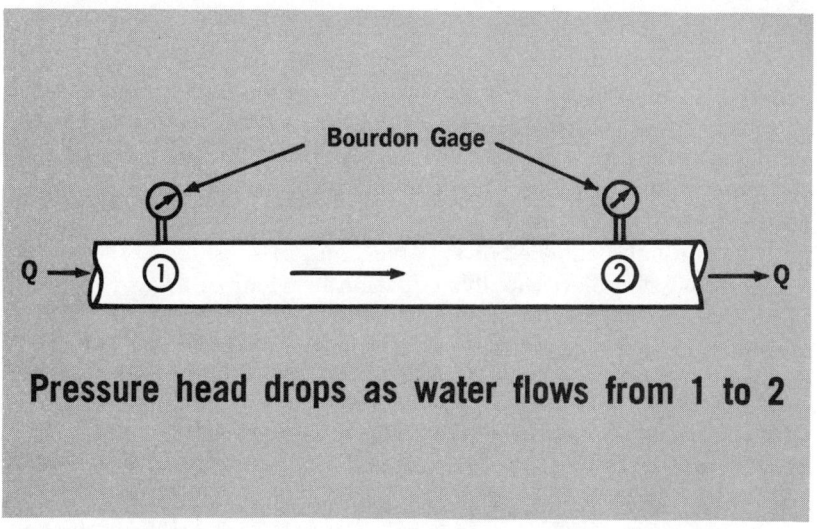

Figure 8

closed slowly. The sudden stoppage of water flow within a pipe or hose can result in the formation of pressure surges known as *water hammer*. Just as the sudden stopping of any moving object such as a bullet or speeding automobile exerts a considerable force, the sudden closing of a valve or fire hose nozzle likewise creates a potentially destructive force.

As discussed in Chapter 1, energy can neither be created nor destroyed. Hence, the kinetic energy possessed by moving water must be suddenly converted to another form of energy if water flow ceases. As a result, water hammer is created. The force of water hammer is sufficient in many cases to rupture pipe lines. Fortunately, the elasticity of fire hose tends to reduce the effect of water hammer but the sudden closing of shut-off nozzles can rupture hose or damage pumps.

FRICTION LOSS IN WATER CONDUCTORS

If we were to cause water to flow through a section of pipe as shown in Figure 8, we would notice that the Bourdon gage at point 1 records a slightly higher pressure than the gage at point 2. Since there is no difference in elevation between points 1 and 2, and since the pipe diameter is constant, the difference in gage readings must indicate a

FIRE SERVICE HYDRAULICS

drop in pressure head as water flows from point 1 to point 2. This loss in pressure head is due to numerous hydraulic energy losses commonly referred to as *friction loss*. Friction loss occurs to some extent in every flowing system. The loss reflects a loss in energy resulting from internal water friction, friction between the water and the pipe lining, and miscellaneous energy losses due to such factors as water contraction and expansion, change in flow direction, and passage through bends, elbows, valves, and fittings.

Our discussion of Bernoulli's equation in Chapter 1 was theoretical in nature and did not take into account hydraulic losses in a flowing system. In actuality, Bernoulli's equation should be refined as follows:

$$PH_① + EH_① + VH_① = PH_② + EH_② + VH_② + H_L$$

or

$$\frac{P_①}{D} + Z_① + \frac{V_①{}^2}{2g} = \frac{P_②}{D} + Z_② + \frac{V_②{}^2}{2g} + H_L$$

where H_L denotes *head loss* (or friction loss as it is commonly called). Head loss is the loss of head resulting from hydraulic energy losses (primarily due to friction) as water flows from point 1 to point 2.

It should be stressed that this revised form of Bernoulli's equation does not violate the basic law of physics stating that energy can neither be created nor destroyed. Actually, the total amount of energy still remains constant; it has merely been converted from one form to another. For the sake of simplicity, H_L represents the amount of energy that is converted from pressure energy to heat energy as a result of friction.

Problem No. 1

Problem: The gage at reference point 1 in Figure 9 reads 20 psi, while the gage at reference point 2 reads 15 psi. Reference point 1 is 10 feet above reference point 2. The pipe is flowing full and the diameter is the same at both points. Determine the friction loss (H_L) between the two points.

Answer:

$$PH_① + EH_① + VH_① = PH_② + EH_② + VH_② + H_L$$

Since pipe diameter is constant and $Q_1 = Q_2$, $V_① = V_②$

$$\frac{V_①{}^2}{2g} = \frac{V_②{}^2}{2g}$$

VELOCITY AND DISCHARGE

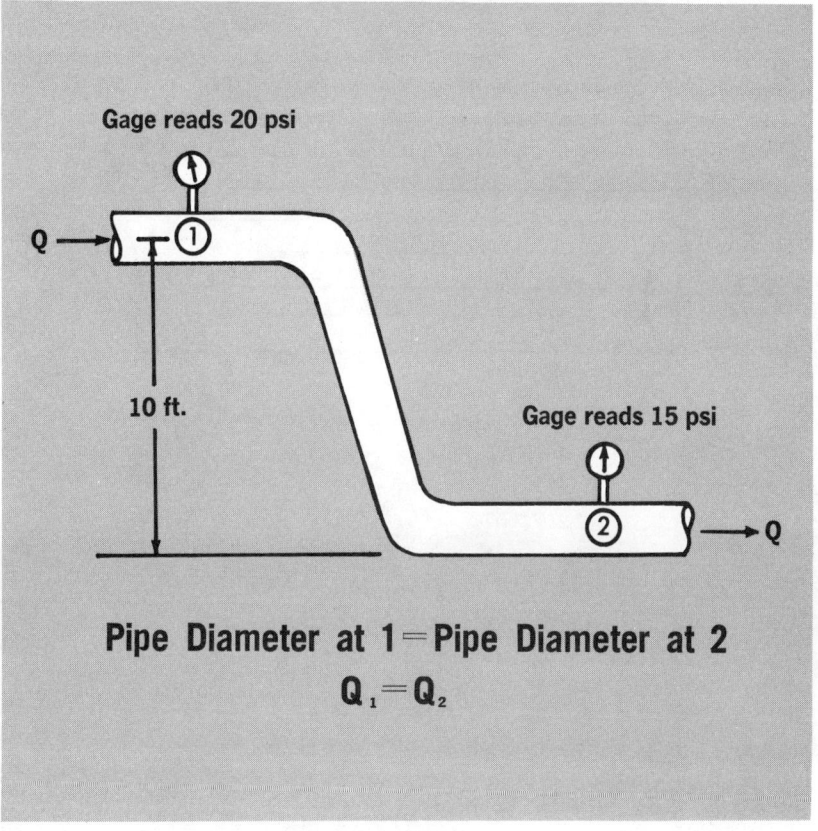

Figure 9

or
$$VH_① = VH_②$$
$$PH_① = 20 \text{ psi} \times 2.31 = 46.2 \text{ ft}$$
$$EH_① = 10 \text{ ft}$$
$$PH_② = 15 \text{ psi} \times 2.31 = 34.6 \text{ ft}$$
$$EH_② = 0$$
$$H_L = ?$$

By substituting and rearranging we have
$$H_L = (PH_① + EH_①) - PH_②$$
$$H_L = (46.2 \text{ ft} + 10 \text{ ft}) - 34.6 \text{ ft}$$
$$H_L = 56.2 \text{ ft} - 34.6 \text{ ft}$$
$$H_L = 21.6 \text{ ft of head}$$

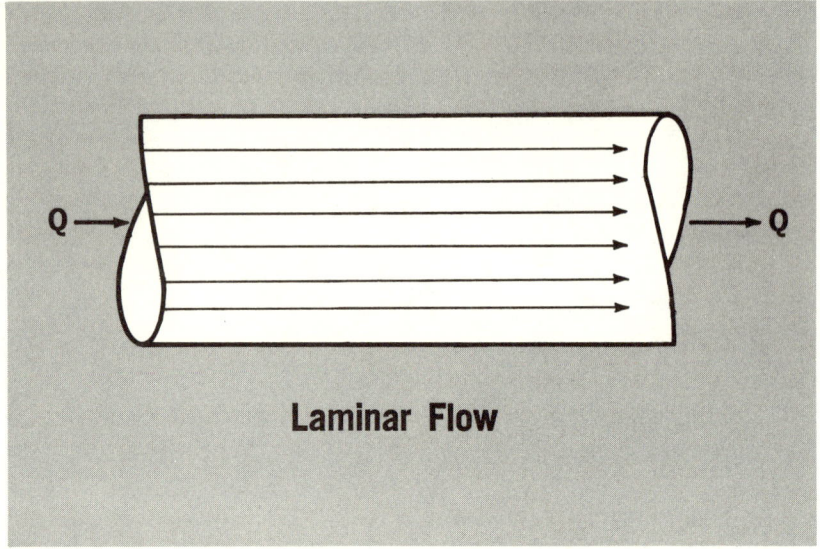

Figure 10

or
$$H_L = 21.6 \times .434 = 9.3 \text{ psi}$$

Effect of flow pattern on friction loss

There are two types of flow patterns that water can assume as it flows through a conductor: *laminar flow* or *turbulent flow*. The type of flow pattern, in turn, determines the friction loss in straight pipe (or hose) runs.

Figure 10 illustrates laminar flow.

During laminar flow, which occurs at relatively low velocities, the water actually flows in even parallel layers. If different colored dyes were injected into a glass pipe during laminar flow, each color would flow in a distinct straight line and it would appear that the flow consists of separate sheets of water with the sheets flowing on top of each other. In laminar flow the velocity of flow is greater at the center of the conductor and decreases toward the conductor walls. Figure 11 is a longitudinal section of a pipe containing laminar flow and shows the velocity distribution within the pattern. In reality there is no flow at the pipe wall, so surface roughness has little if any bearing on friction loss.

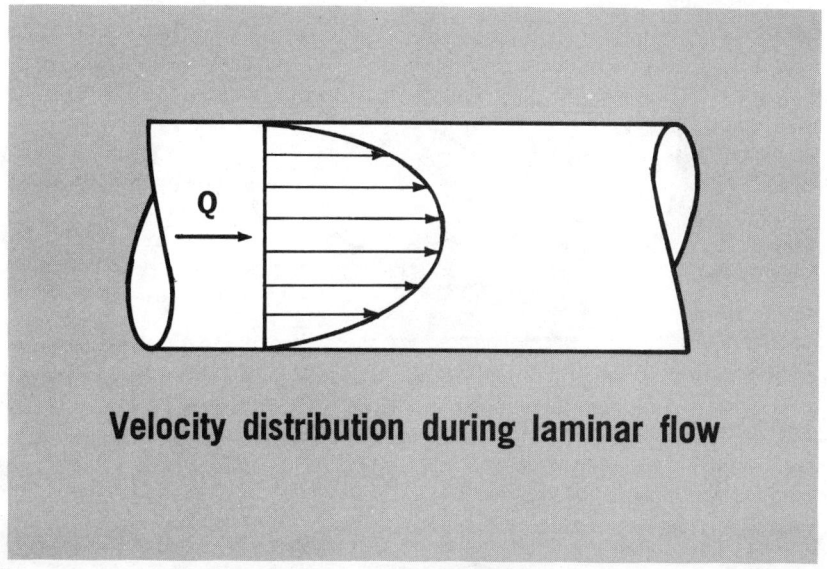

Figure 11

Viscosity effects

What is commonly called friction loss during laminar flow is really a result of a characteristic of all liquids known as *viscosity*. Viscosity is the tendency of a liquid to possess internal resistance to flow. During laminar flow, viscosity may be thought of as the friction between liquid layers. Molasses is a good example of a liquid having high viscosity. For any given liquid, viscosity varies inversely with temperature, meaning that the higher the temperature the lower the viscosity and vice versa. The inverse viscosity-temperature relationship is the reason that we use a higher "weight" (or viscosity) oil in our cars in the summer and a lower "weight" (or viscosity) oil in the winter.

Velocity effects

The velocity of water in fire mains or fire hose is usually too great for laminar flow to exist. Consequently, the flow in fire mains and fire hose is usually *turbulent*. Figure 12 shows the irregular flow pattern typical of turbulent flow conditions. During turbulent flow, friction loss is a function of the relative roughness of the conductor lining as well as the liquid viscosity.

FIRE SERVICE HYDRAULICS

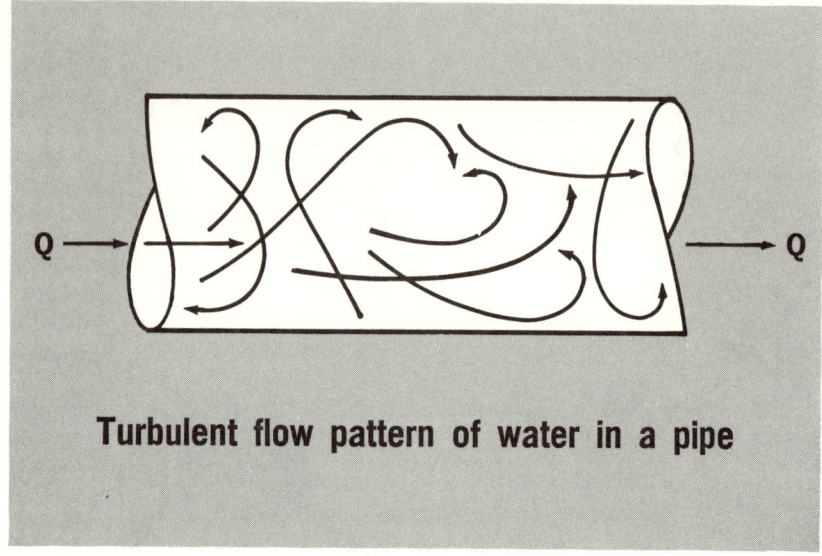

Figure 12

The effects of flow pattern on friction loss may be summarized as follows:
1. Friction loss under laminar flow in straight pipe or hose is strictly due to viscosity.
2. Friction loss under turbulent flow is a result of viscosity as well as the friction between the water and the interior surface of the pipe or hose.

Since there are a few minor differences between friction loss in hose and friction loss in pipe, each will be discussed separately.

Friction loss in fire hose

Friction loss in fire hose is governed by five fundamental rules:
1. Friction loss varies with the relative quality of the hose.
2. Friction loss varies directly with length.
3. Friction loss varies approximately with the square of the velocity (or with the square of Q, since $Q = AV$).
4. For a given velocity, friction loss varies inversely as the fifth power of the hose diameter.
5. For a given velocity, friction loss is nearly independent of pressure.

VELOCITY AND DISCHARGE

The primary characteristics of hose that affect friction loss are smoothness of the lining, age of hose, thickness of the lining and the weave of the jacket. Although rubber lining is comparatively smooth, surface friction continues to exist. If it were possible to make a lining smooth as glass, the friction loss would be materially reduced. Generally, older hose creates greater friction loss due to deterioration of the rubber lining.

The direct dependence of friction loss on length means simply that *if the length were doubled, then the friction loss would be doubled.* Four times the length means four times the friction loss.

However, since friction loss varies with the square of the velocity, *if the flow (or quantity flowing) is doubled, friction loss is quadrupled. If the velocity is increased by four times, the friction loss is increased by sixteen times.*

The most important factor of friction loss in hose is the inverse relationship between friction loss and the fifth power of the diameter. This relationship emphasizes that the best way to reduce friction loss is to increase the hose size. If the hose size is doubled, and the flow is unchanged, the friction loss is only $(1/2)^5$, or $1/32$ times the friction loss in the small size. Figure 13 shows a comparison of friction loss in $2\frac{1}{2}$-inch and $1\frac{1}{2}$-inch hose.

If hose diameter remained constant, friction loss would be independent of pressure. However, since hose expands under pressure, friction loss tends to decrease with higher pressures due to the increased hose diameter. Likewise, because of its greater expansion under pressure, single-jacketed hose shows less friction loss than double-jacketed hose. According to the 13th edition of the NFPA "Fire Protection Handbook," recent tests have shown that synthetic hose causes a higher friction loss than cotton-jacketed hose because of the lower expansion of synthetic fibers.

In addition to the losses due to viscosity and friction between water and hose, there are other sources of hydraulic losses in fire hose. Couplings, bends, valves and nozzle energy losses are minor, relative to viscosity and friction losses.

Methods of calculating friction loss in hose will be covered in a later chapter. However, for illustration purposes a few tables are included here to clarify some of the principles discussed.

Table 1 shows friction loss for hose ranging from $2\frac{1}{2}$-inch to 3-inch with various size couplings. Table 2 shows friction loss in smaller diameter rubber-lined hose. Table 3 shows allowances for friction loss in suction hose which are important for drafting.

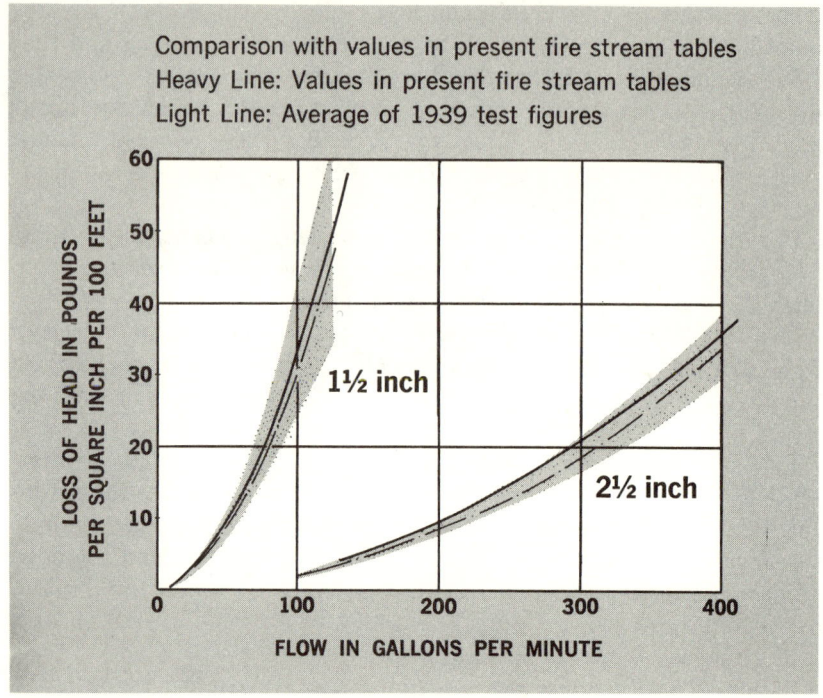

Figure 13

Common friction loss formulas

Methods of calculating friction loss in hose will be covered in detail in a later chapter. However, a few of the more common friction loss formulas are included here to emphasize some of the principles discussed.

The relationship between friction loss and the various factors affecting it can be summarized by the following general equation:

$$FL = \frac{KLQ^2}{d^5}$$

where FL = friction loss in psi
L = length of hose expressed as number of 50-foot sections
Q = quantity flowing ($Q = AV$) expressed in hundreds of gallons per minute
d = hose diameter in inches
K = a constant depending upon the material of the hose and hose roughness

VELOCITY AND DISCHARGE

Table 1 Friction Losses in Rubber-Lined Fire Hose
(Pounds per square inch per 100 feet of hose)

Flow in gpm	2½-In.*	2¾-In.† (3-In. couplings)	3-In.‡ (2½-In. couplings)	3-In.* (3-In. couplings)	Flow in gpm	2½-In.*	2¾-In.† (3-In. couplings)	3-In.‡ (2½-In. couplings)	3-In.* (3-In. couplings)
100	2.5	1.7	1.2	1.2	290	19.9	11.9	8.4	7.7
110	3.2	2.1	1.4	1.4	300	21.2	12.7	9.0	8.2
120	3.9	2.4	1.6	1.6	310	22.5	13.5	9.7	8.7
130	4.5	2.8	1.9	1.8	320	23.8	14.3	10.3	9.3
140	5.2	3.1	2.1	2.0	330	25.3	15.2	10.9	9.9
150	5.8	3.6	2.5	2.3	340	26.9	16.2	11.6	10.5
160	6.6	4.0	2.9	2.6	350	28.4	17.1	12.3	11.0
170	7.4	4.5	3.2	2.9	360	30.0	18.0	13.0	11.5
180	8.3	5.0	3.6	3.2	370	31.5	18.9	13.7	12.2
190	9.2	5.6	3.8	3.5	380	33.0	19.8	14.4	12.8
200	10.1	6.1	4.0	3.8	390	34.6	20.7	15.2	13.4
210	11.1	6.7	4.4	4.2	400	36.2	21.7	16.0	14.1
220	12.0	7.2	4.5	4.6	420	39.9	24.0	17.7	15.4
230	13.0	7.8	5.3	5.0	440	43.2	25.9	19.4	16.8
240	14.1	8.5	5.8	5.4	460	46.8	28.1	21.3	18.2
250	15.3	9.2	6.2	5.9	480	50.8	30.5	23.1	19.7
260	16.4	9.9	6.8	6.3	500	55.1	33.1	25.0	21.2
270	17.5	10.5	7.3	6.7					
280	18.7	11.2	7.8	7.2					

* The losses shown for nominal 2½-in. hose (actual 2⁹⁄₁₆-in. waterway) and for 3-in. hose are based upon tests conducted in 1909 by the Natinal Board of Fire Underwriters.

† Losses for 2¾-in. hose with 3-in. couplings are based upon tests by the Oakland, California, Fire Department. The 2¾-in. hose with 3-in. couplings, while widely used in the San Francisco Bay area, is not a nationally recognized standard fire hose. Three-inch hose should be equipped with 2½-in. couplings to be considered a standard size.

‡ Losses for 3-in. hose with standard 2½-in. couplings are based upon tests by the NFPA Fire Service Department in cooperation with the Boston Fire Department.

Reproduced by permission from the Fire Protection Handbook, 13th edition, Copyright 1969 National Fire Protection Association, Boston, Mass.

Note that the equation above agrees precisely with our previous discussion of the factors governing friction loss in hose: friction loss varies directly with length, directly with the square of Q, inversely as the fifth power of the hose diameter, and directly with relative hose quality.

Actual field experiments have shown that the friction loss in 100 feet of 2½-inch rubber-lined hose with 250 gpm flowing is equal to

FIRE SERVICE HYDRAULICS

Table 2 Friction Losses in Small Diameter Rubber or Rubber-Lined Fire Hose

(Pounds per square in. per 100 feet of hose)

Flow in gpm	¾-In. (Booster)	1-In. (Booster)	1½-In. (Lined standpipe & lined forestry)	1½-In. (Good quality fire dept. hose)	2-In.*
10	13.5	3.5	0.5	0.3	0.1
15	29.0	7.2	1.0	0.7	0.3
20	50.0	12.3	1.7	1.2	0.4
25	75.0	18.5	2.6	1.9	0.6
30	105.0	26.0	3.6	2.5	0.9
35	140.0	35.0	4.8	3.4	1.2
40	180.0	44.0	6.1	4.3	1.5
45	—	55.0	7.6	5.4	1.9
50	—	67.0	9.2	7.1	2.3
60	—	—	13.0	9.2	3.2
70	—	—	17.3	12.3	4.3
80	—	—	22.0	15.6	5.4
90	—	—	27.3	19.5	6.9
100	—	—	33.0	25.5	8.3
120	—	—	47.0	33.0	11.7
150	—	—	70.0	50.0	17.5
200	—	—	—	—	29.9

*The 2-inch hose is not recognized as a standard size but is used in some fire departments as a leader line because it is relatively easy to handle and carries more water at a given pressure than 1½-linch hose.

Reproduced by permission from the Fire Protection Handbook, 13th Edition Copyright 1969 National Fire Protection Association, Boston, Mass.

15 psi. By using the measured quantities, we can solve for K in the above formula.

$$FL = \frac{KLQ^2}{d^5}$$

or

$$K = \frac{FL(d^5)}{LQ^2}$$

$$K = \frac{15(2.5)^5}{2(2.5)^2}$$

$$K = 120$$

VELOCITY AND DISCHARGE

Table 3 Allowances for Friction Loss in Suction Hose

Rated Capacity of Pumper (gpm)	Diameter of Suction Hose (inches)	Allowance (feet)	
		For 10 Feet of Suction Hose	For Each Additional 10 Feet of Suction Hose
500	4 4½	6 3½	+1 +½
750	4½ 5	7 4½	+1½ +1
1000	5 6	8 4	+1½ +½
1250	6	6½	+½
1500	6	9	+1

Courtesy of American Insurance Association.

Hence, if considering only 2½-inch rubber-lined hose, the general friction loss formula can be simplified as follows:

$$FL = \frac{120LQ^2}{(2.5)^5}$$

$$FL = 1.2Q^2L$$

where FL = loss in psi
Q = flow in hundreds of gpm $(Q/100)$
L = length of 2½-inch rubber-lined hose expressed as number of 50-foot sections

Problem No. 2

Problem: A nozzle at the end of a 600-foot stretch of 2½-inch rubber-lined hose is discharging 250 gpm. What is the friction loss in the hose?

Answer:

$$FL = 1.2Q^2L$$
$$= 1.2(2.5)^2(12)$$
$$= 90 \text{ psi}$$

FIRE SERVICE HYDRAULICS

By a simple adjustment the above formula can be developed into the most commonly used fire service formula for determining friction loss per 100 feet of hose:

$$FL = 1.2Q^2L$$

or

$$FL = 1.2Q^2 \times 2, \quad \text{since } L = 2 \text{ for 100 feet}$$
$$= 2.4Q^2$$

which can be rewritten as:

$$FL = 2Q^2 + .4Q^2$$

For flows ranging from 100 gpm to 600 gpm, $.4Q^2$ can be considered equal to Q with only a negligible loss in accuracy. Hence, $FL = 2Q^2 + .4Q$ can be rewritten more simply as:

$$FL = 2Q^2 + Q$$

where: FL = loss in psi per 100 feet of 2½-inch rubber-lined hose
Q = flow in hundreds of gpm

(Editor's note: This formula lends itself to quick calculation in the field and is used exclusively in Part 3, "Fireground Hydraulics.")

Problem No. 3

Problem: What is the friction loss in 100 feet of 2½-inch rubber-lined hose when flowing 300 gpm?
Answer:

$$FL = 2Q^2 + Q$$
$$= 2(3)^2 + 3$$
$$= 21 \text{ psi}$$

For flows in 2½-inch hose that are less than 100 gpm, the above equation is modified to the following form:

$$FL = 2Q^2 + \tfrac{1}{2}Q$$

The friction loss in other than 2½-inch rubber-lined hose can also be determined from the above equations. The friction loss in other sizes or other types of hose for the same quantity flowing can be determined by dividing the friction loss in 2½-inch rubber-lined hose by the appropriate factors as explained in Part 3.

The nozzle pressure equation

Another common friction loss equation is the nozzle pressure equation. This equation relates engine pressure, nozzle pressure, friction

VELOCITY AND DISCHARGE

loss and hose diameter. It can be developed by starting with the formula previously developed:

$$FL = 1.2Q^2L \quad \text{(where } Q = \text{gpm}/100\text{)}$$

which can be rewritten

$$FL = 1.2(Q/100)^2L$$
$$FL = \frac{1.2Q^2L}{(100)^2}$$
$$FL = \frac{1.2Q^2L}{10,000}$$

Earlier in this chapter, we discussed two equations relating nozzle pressure to nozzle discharge: $Q = 29.71d^2\sqrt{P}$, which is known as Freeman's Formula; and $Q = 29.83d^2\sqrt{P}$ which is the AIA Flow Formula. These equations may be simplified to

$$Q = 30d^2\sqrt{P}$$
$$Q = 30d^2\sqrt{NP} \quad \text{(where } NP \text{ is used to denote nozzle pressure)}$$

Substituting this value of Q into the previous equation yields

$$FL = \frac{1.2Q^2L}{10,000}$$
$$FL = \frac{1.2(30d^2\sqrt{NP})^2L}{10,000}$$
$$= \frac{1.2(900d^4NP)L}{10,000}$$
$$= \frac{1.08d^4(NP)L}{10}$$

or more simply

$$FL = \frac{1.1d^4(NP)L}{10}$$

which can be rewritten as

$$FL = 1.1K(NP)L, \quad \text{where } K = \frac{d^4}{10}$$

By applying Bernoulli's Equation, it can be shown that pump discharge pressure equals nozzle pressure plus friction loss between pump and nozzle (if the pump and nozzle are at the same elevation). The reader may wish to prove this to himself to test comprehension of

Bernoulli's Equation. This relationship between pump discharge pressure, nozzle pressure, and friction loss may be expressed in equation form as:

Pump Discharge Pressure = Nozzle Pressure + Friction Loss

or

$$EP = NP + FL$$

where EP stands for "engine pressure" which is commonly accepted fire department terminology for pump discharge pressure.

By inserting the previous equation for friction loss, we get

$$EP = NP + FL$$
$$= NP + 1.1K(NP)L, \text{ where } K = \frac{d^4}{10}$$

or

$$EP = NP(1 + 1.1KL)$$

For actual application, the equation should be simplified to:

$$EP = NP(1.1 + KL)$$

It should be emphasized that there are other sources of hydraulic losses in fire hose in addition to viscosity and friction losses. Flow through couplings, bends, valves, fittings and nozzles always results in loss of energy. However, these losses are minor relative to viscosity and friction losses.

Summary of Chapter 2

(1) Important terms:

Lift	Reaction
Barometric pressure	Water hammer
Drafting	Friction loss
Hydraulic losses	Head loss
Discharge coefficient	Laminar flow
Standard orifice	Turbulent flow
Coefficient of velocity	Viscosity
Coefficient of contraction	Minor losses

(2) Important Equations:
  (a) $Q = 29.71d^2 \sqrt{P}$ (Freeman's Formula)
  (b) $Q = 29.83d^2 \sqrt{P}$ (AIA Formula)
  (c) $\quad\quad Q = 29.83C_d d^2 \sqrt{P}$
  (d) $\quad\quad\quad R = 2PA$, or
  $\quad\quad\quad\quad R = 1.57d^2P$

(e) $PH_{①} + EH_{①} + VH_{①} = PH_{②} + EH_{②} + VH_{②} + H_L$

(f) $FL = \dfrac{KLQ^2}{d^5}$

(g) $FL = 1.2Q^2L$

(h) $FL = 2Q^2 + Q$

(i) $EP = NP(1.1 + KL)$

SELECTED REFERENCE LIST

Fire Flow Tests, National Board of Fire Underwriters, New York, 1963.
Fire Department Pumper Tests and Fire Stream Tables, National Board of Fire Underwriters, New York, 1959.
Water Supply Engineering, Fifth Edition, Harold E. Babbitt and James J. Doland, McGraw-Hill, New York, 1955.
Introduction to Fluid Mechanics, R. W. Henke, Addison-Wesley Publishing Co., Reading, Mass., 1966.
Fire Protection Handbook, Thirteenth Edition, National Fire Protection Association, Boston, Mass., 1969.
Fluid Mechanics for Engineers, M. L. Albertson, J. R. Barton, and D. B. Simons, Prentice Hall Inc., Englewood Cliffs, N.J., 1960.
Flow of Fluids, Technical Paper No. 410, The Crane Co., Chicago, Ill., 1957.
The Fire Chief's Handbook, First Edition, Fred Shepperd, Case-Shepperd-Mann Publishing Corporation, New York, 1932.

PART TWO: **water**

CHAPTER THREE

Water distribution systems

Public water supply systems have two principal functions: to provide water for domestic, commercial, and industrial use, and to provide water for fire protection. Most of the water used for fire protection is delivered from water distribution mains through fire hydrants to fire department pumpers. These in turn increase the pressure to develop the fire streams needed. In addition to providing water for hose streams, pumpers can augment the water supply to sprinkler and standpipe systems in buildings through fire department connections installed on such systems.

When water distribution systems began to appear on the municipal scene and their possible use as a source of water for fire fighting was considered, no special devices such as hydrants were available. In those days water mains were made of bored out wooden logs. When a fire occurred, the main was excavated and a hole bored into it. The water which was under low pressure filled the excavation and provided a supply for the pumpers and buckets of that day. After the fire a wood plug was driven into the hole in the main and the excavation filled in. Hopefully, fire fighters remembered the location of the plug for future use. These wooden plugs were called "fire plugs" (Figure 1).

It is quite evident that this means of obtaining water for fire protection was not very satisfactory. Time was lost in excavating the hole to reach the plug, dirt from the excavation fouled the pump valves,

WATER DISTRIBUTION SYSTEMS

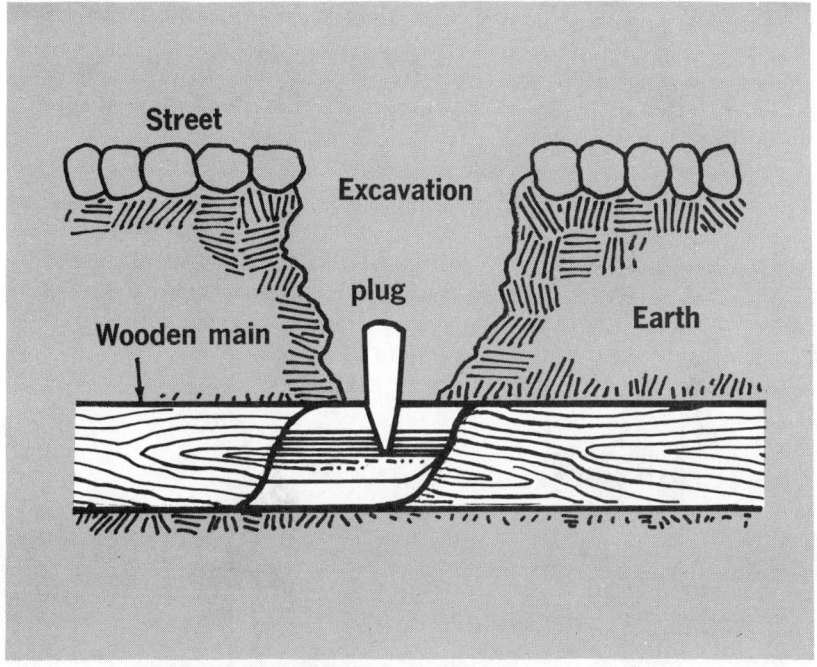

Figure 1

and streets were left in poor condition. And as streets began to be paved, it became more difficult to excavate to reach the main.

When iron mains appeared, a special T-shaped fitting was placed at intervals in the main and a wooden plug driven into the open leg of the tee. A metal or wooden shield that extended to the street surface was placed around the top of the tee. This shield protected the wooden plug from damage and prevented soil from washing away when the plug was pulled for a fire. Portable canvas cisterns that formed a reservoir were sometimes used with this type of fire plug to provide a source of suction for the pumper. Later "metal standpipes" were inserted into the tee after the plug was pulled. The standpipes were tapered and had a hose connection at the upper end for supplying suction for the pumper. However, it was a wet job to insert such a standpipe into the tee with water discharging from the main.

As water supply engineering developed and water systems improved, the need arose for a more satisfactory device than the fire plug for fire department use. This need led to the development of the fire hydrant.

FIRE SERVICE HYDRAULICS

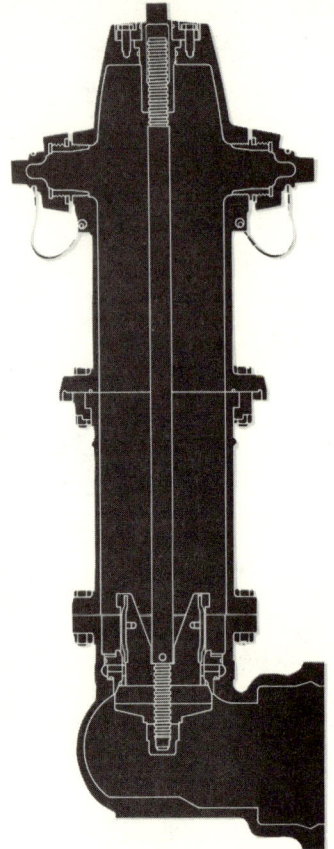
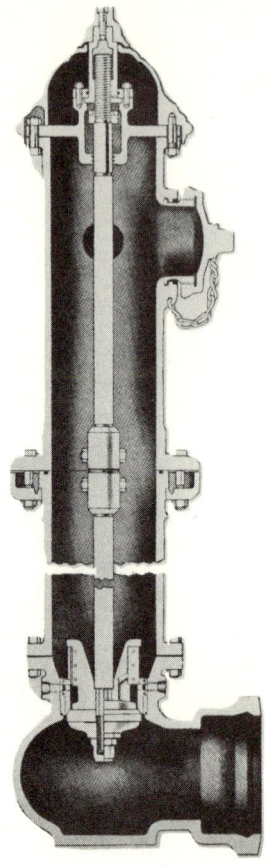

Figure 2a   Figure 2b

In spite of the marked difference between a modern hydrant and the fire plug of the early days, the old term is still used in many places to designate hydrants.

Two principal types of hydrants are in use today—the dry barrel and the wet barrel. The dry barrel type is by far the more common because it can be used in all climates. The wet barrel type can be used only in mild climates where freezing temperatures do not occur. The American Water Works Association has prepared standards for both types: AWWA Standard C502 for Fire Hydrants for Ordinary Water Works Service and AWWA Standard C503 for Wet Barrel Fire Hydrants for Ordinary Water Works Service. These standards cover hydrants designed for a working pressure of 150 psi and provide for a

WATER DISTRIBUTION SYSTEMS

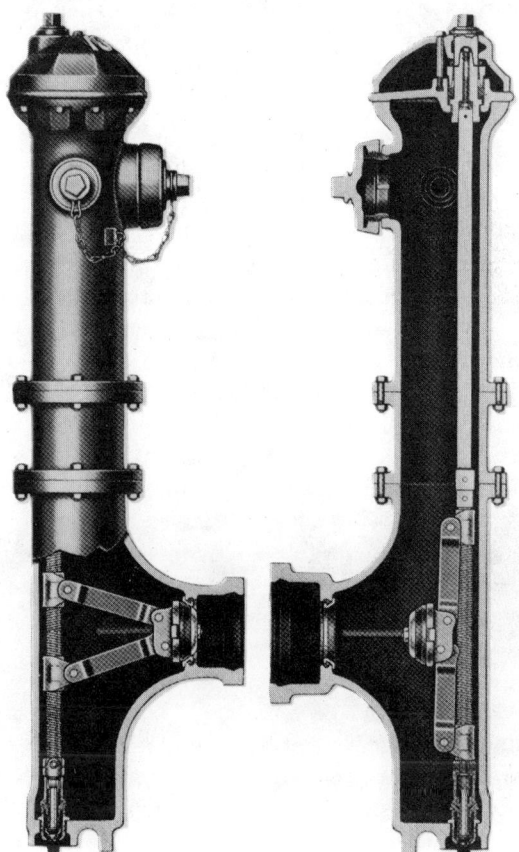

Figure 2c

hydrostatic shop test of each hydrant at 300 psi. The use of these standards enables a municipality to purchase hydrants that are well designed and constructed and well adapted to regular municipal service. As a result, most hydrants being installed by municipalities today conform to one of these standards as a minimum requirement.

Dry-barrel hydrants

The dry barrel hydrant, although it actually has many parts, consists essentially of a footpiece, a barrel, a bonnet, an operating stem, a main valve and a drain. This, of course, is an oversimplification but should prove helpful in becoming familiar with the basic parts and their functions (Figures 2a, 2b, and 2c).

FIRE SERVICE HYDRAULICS

The footpiece, sometimes called the elbow or shoe, provides the inlet for the hydrant from the branch connection that extends from the main. It also contains the seat for the main valve and the drain outlet.

The barrel or riser, which may have more than one section, extends vertically from the footpiece and continues above the ground surface. It contains the outlet nozzles and houses the operating stem.

The bonnet, or top, is fastened to the barrel at its upper end to form a protecting cover. The mechanism for turning the stem is housed in the bonnet and is controlled by the operating nut which extends out of the top.

The operating stem extends through the length of the barrel and usually carries the valve at its lower end. Its upper end is connected to the mechanism in the bonnet.

The main valve, which is attached to the stem, moves away from the seat in the footpiece as the hydrant is opened and moves against the seat as the hydrant is closed. The valves on most hydrants manufactured today are arranged to open against the pressure and to close with the pressure. An obvious advantage of this arrangement is that if the stem should be broken, the pressure in the footpiece on the supply side of the valve would hold it in the closed position.

Dry barrel hydrants require a means by which the water remaining in the barrel after use can be removed. Drain holes, together with a suitable drain mechanism, are provided in the footpiece for this purpose. When a hydrant is closed, the drain opens. As the hydrant is opened, the drain closes (during the first few turns) and remains closed while the hydrant discharges. As the hydrant is shut down, the drain opens (during the last few turns) and remains open until the hydrant is operated again.

At the time a dry barrel hydrant is installed, it is necessary to make certain that it will drain properly into the ground. For this purpose a sufficient quantity of clean stone or coarse gravel is placed around the lower portion of the footpiece and extended up to several inches above the drain holes. This practice will usually enable hydrants to drain properly even where soils may have a low permeability. It also helps to prevent the clogging of drain holes with dirt or sand.

In some locations the ground water level may normally be higher than the hydrant drain holes. When this happens, water enters the hydrant barrel and rises to the level of the ground water on the outside. The water is then subject to freezing in winter, which could render the hydrant inoperative. It is common practice, therefore, to plug the drains of hydrants installed in areas where the ground water table is high. Plugging the drain makes it necessary to pump out the water remaining in the barrel after each use. In some business districts where normal drainage from hydrants has leaked into basements, it also has been found necessary to plug the drains to prevent damage. It is essential that hydrants with plugged drains be properly identified so that fire department and water department personnel will be on notice to pump them out after use.

Wet barrel hydrants

A wet barrel hydrant (Figure 3) consists essentially of a *bury* section and a *top* section, and has fewer parts than a dry barrel hydrant. The bury section provides the inlet for the hydrant from the branch connection that extends from the main. It has a 90° bend just like the footpiece of a dry barrel hydrant but extends vertically to a point above the ground surface. It is essentially a piece of cast iron pipe with a 90° bend at one end, and there are no moving parts within it.

The top section, also called the body, extends vertically from the bury section and contains the outlet nozzles and the independent valves that control the flow from each nozzle. Each valve is mounted on a threaded stem which is supported by a housing on the opposite side (180° apart) of the body. This housing contains the threads that enable the stem to move the valve on and off its seat which is on the inside end of the outlet nozzle. The operating nut is on the exterior end of the stem that protrudes from its housing (Figure 4).

Outlet nozzles

Every hydrant should have at least two outlet nozzles so that a damaged outlet nozzle will not prevent the hydrant from being used. Further, because pumpers are needed to provide satisfactory fire streams in most municipalities, one of these outlet nozzles should be a pumper outlet nozzle. An exception to this can be made in areas in which the fire flow needed can be obtained from the water distribution system at residual pressures (not less than 75 psi) suitable for the use of direct hydrant hose streams and hydrants are spaced 250 to 300 feet apart so that hose lines will not be too long for such streams to be effective.

For most municipalities, hydrants with a pumper outlet nozzle and one or two 2½-inch outlet nozzles will provide satisfactory service. In some large cities hydrants with two pumper outlet nozzles are installed in districts where large flows are needed. In order for such hydrants to be used effectively, the distribution system should be capable of supplying 2000 gpm at each hydrant. Fire department

FIRE SERVICE HYDRAULICS

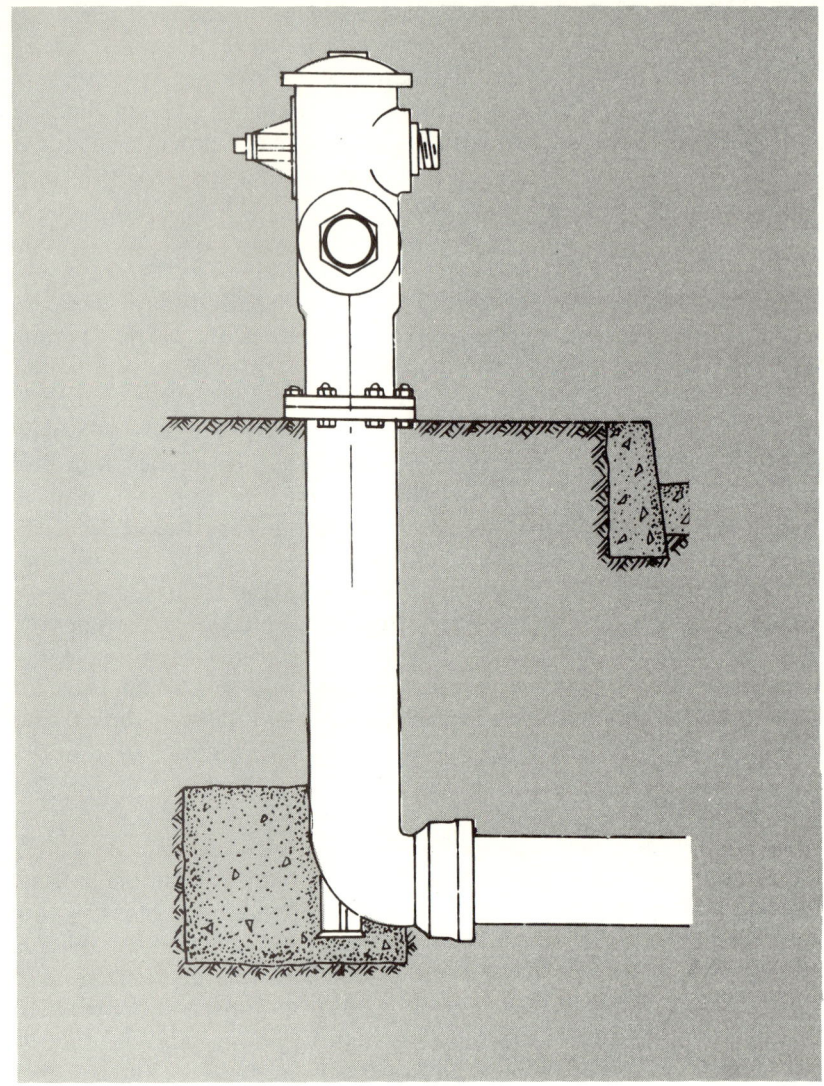

Figure 3

standard operating procedures should include appropriate methods for connecting two pumpers to a single hydrant.

The threads on outlet nozzles should conform to American National Standard Institute Standard B 26, "National (American) Standard Fire Hose Coupling Screw Threads," except where the threads in use

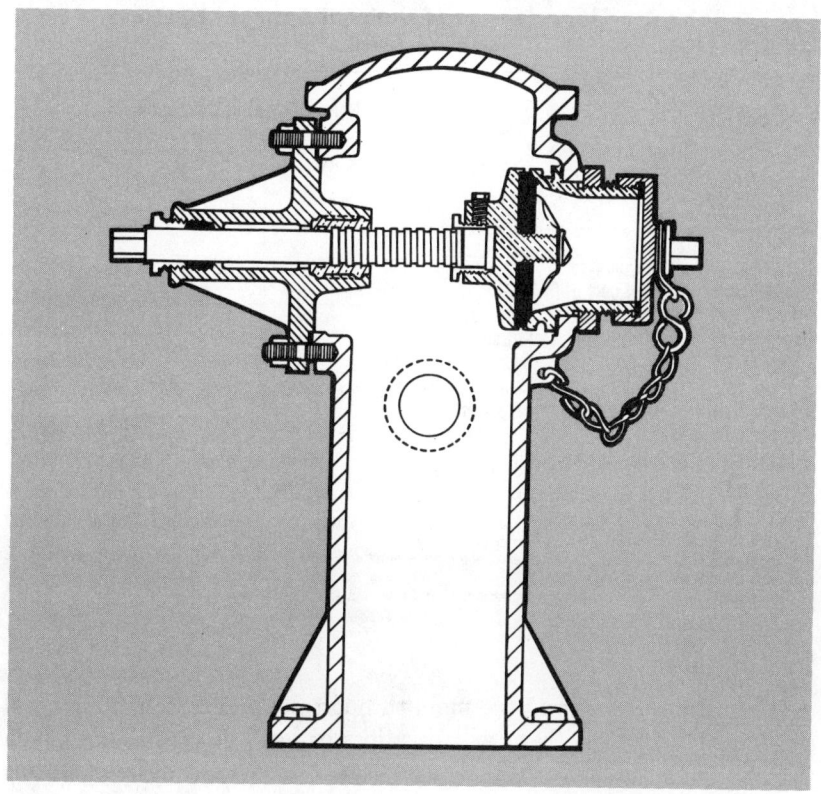

Figure 4

in the municipality installing the hydrant are of different dimensions. The characteristics of the National Standard threads are shown in Table 1.

The outlet nozzles for both dry and wet barrel hydrants are provided with caps. In the case of the dry barrel type, the caps are needed to protect the threads from damage, to prevent foreign material from entering the barrel, and to prevent discharge from the outlet nozzles not in use during hydrant operation. In the wet-barrel type, the caps protect the nozzle threads and also the portion of the valve that would be otherwise exposed. Outlet nozzle caps are often provided with chains connected to the hydrant barrel to prevent the caps from being stolen or lost at a fire. The installation of chains is optional with the municipality or utility purchasing the hydrant.

FIRE SERVICE HYDRAULICS

Table 1 Characteristics of National Standard Fire Hose Coupling Thread*

Characteristic**	Nominal ID of Hose Coupling (in.)				
	2½	3	3½	4†	4½
Number of threads per inch	7½	6	6	4	4
Total length of threaded part of hydrant nipple, external (minimum)—in.	1	1⅛	1⅛	1¼	1¼
Distance from face of nipple to start of second turn—in.	¼	5⁄16	5⁄16	7⁄16	7⁄16
Depth of coupling swivel to washer seat—in.	15⁄16	1 1⁄16	1 1⁄16	1 3⁄16	1 3⁄16
Distance from face of coupling swivel to start of second turn—in.	3⁄16	¼	¼	⅜	⅜
Depth of thread of coupling swivel—in.	11⁄16	13⁄16	13⁄16	15⁄16	15⁄16

* [ANSI B26-1925 (R1953)] The complete document may be obtained from the American Society of Mechanical Engineers, 345 East 47th Street, New York, N.Y. 10017.
** Outer ends of external and internal thread should be terminated by the "Higbee Cut" on full thread to avoid crossing and mutilation of thread.
† Dimensions in this column approved by National Fire Protection Assn., 1956. (Not ANSI.)

Loss of head

When water flows from the main through the branch connection and through the hydrant there will be a loss of head or pressure between the main and the hydrant outlet nozzles. This loss will be small for small flows, but may increase to excessive amounts for large flows in poorly designed hydrants.

The loss of head in the hydrant itself is limited by the AWWA standards as shown in Table 2, Maximum Permissible Loss of Head for Hydrants. A study of this table will show that for a 1000-gpm delivery through the pumper nozzle (irrespective of size), the loss cannot exceed 5 psi. The standards also specify that if a hydrant (such as

Table 2 Maximum Permissible Loss of Head for Hydrants

No. of Outlet Nozzles	Nom. Diameter of Outlet Nozzles (in.)	Total Flow From Outlet Nozzles (gpm)	Max. Permissible Head Loss (psi)
1	2½	250	1.0
2	2½	500*	2.0
3	2½	750*	3.0
4	2½	1000*	4.0
1	4½	1000	5.0†

* 250 gpm, approximately from each outlet nozzle.
† Also to apply to pumper outlet nozzles of other sizes.

WATER DISTRIBUTION SYSTEMS

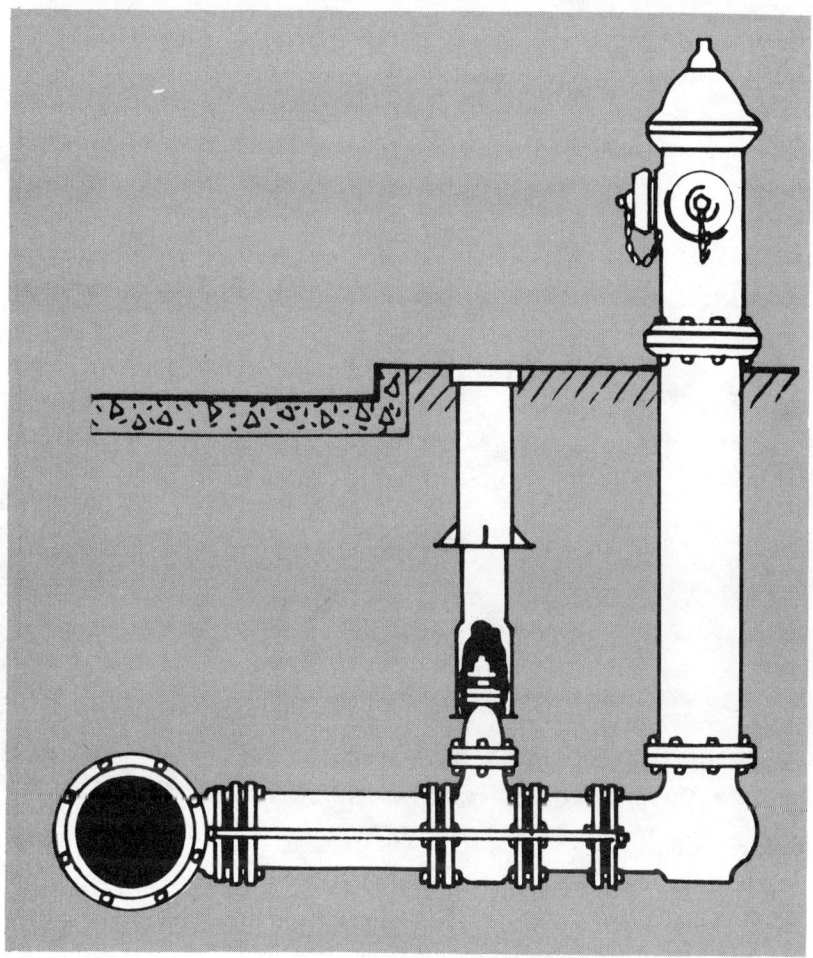

Figure 5

those with two pumper outlets) is designed to deliver more than 1000 gpm, the total loss of head cannot exceed 5 psi no matter how large the discharge. Therefore, it is quite evident that hydrants purchased under AWWA standards should not present any problems arising out of excessive head loss. Although a great many hydrants installed at present conform to AWWA standards, there are hydrants in service in many municipalities that do not meet the standards and in which there will be excessive head losses with large flows. In some of these hydrants the loss may be so high that an adequate discharge

FIRE SERVICE HYDRAULICS

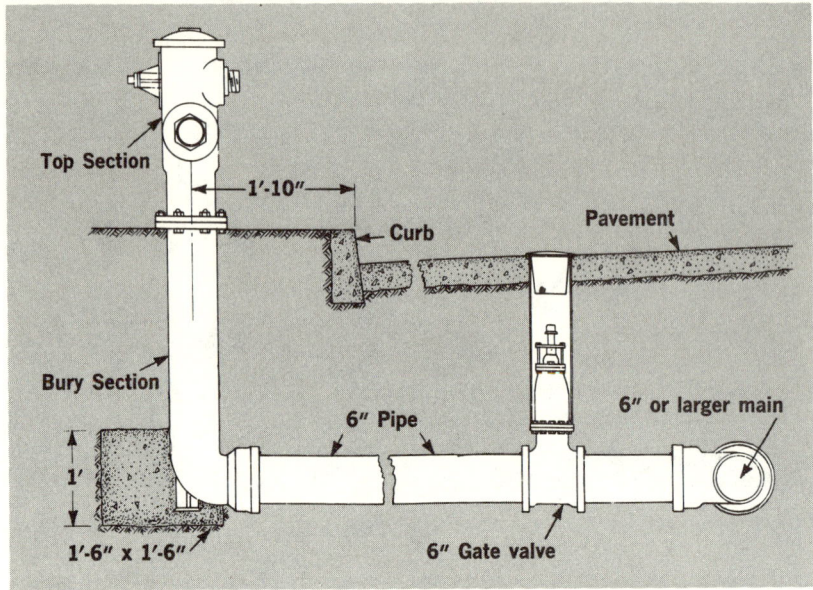

Figure 6

cannot be obtained. Many such hydrants are old or of makes no longer available. They are gradually being replaced.

Branch connection

Part of the loss of head between the main and the outlet nozzle takes place in the branch connection (Figures 5 and 6). The minimum size considered satisfactory for this connection is 6-inch, and this is the size generally being installed. Hydrants which have two pumper outlets, and may be required to deliver 2000 gpm or more, need an 8-inch branch connection. There are, however, many water distribution systems in which supply to hydrants is restricted by 4-inch branch connections. And in some cases the originally inadequate size has been further reduced in capacity by tuberculation.

It is good practice to install a gate valve in the branch connection so that a hydrant can be replaced or repaired without shutting down a portion of the distribution system. The gate valve is of special importance for wet barrel hydrants because if the barrel is broken off due to an accident, the water will discharge geyser-like until a shut-off can be made. The gate valve also has been found to be very helpful when

WATER DISTRIBUTION SYSTEMS

after a long period of use at a fire, a hydrant cannot be closed because of a defect in one or more of its parts.

Breakable hydrants

Most hydrant manufacturers now produce at least one dry barrel model which has replaceable sections or components that are designed to have lower breaking strength than the rest of the unit. This design permits the upper portions of the barrel and operating stem to be broken off (accidentally) at ground level without damaging the buried portion and therefore without any discharge of water. Figure 7 shows a hydrant of this type. If the barrel is struck by a vehicle, the safety flange breaks and the upper part of the barrel falls over on its side. At the same time a special coupling on the operating stem breaks, separates the upper section of the stem from the lower section, and prevents the stem from being bent. The main hydrant valve remains closed, since it is held in place against its seat by the pressure in the main. If the upper portion of the barrel has not been damaged by the vehicle, the hydrant can be reassembled using a new stem coupling and safety flange and placed back in service in a relatively short time.

It can be readily understood that this type of arrangement would not be suitable for wet barrel hydrants, but there is a type manufactured that is designed to give a "dry break" (Figure 8). A spring-loaded check valve, located in a recessed portion of the bury section, is held in the "open" position by a special break-off bar fastened to the inner side of the top section. A special break-off flanged riser section with localized break-off points near each end is placed between the bury section and the top section. If the hydrant top section is struck by a vehicle, the break-off section is broken off at one of its two weak points. The top section falls on its side. The break-off bar then releases the check valve which is forced by its spring into the waterway. Here it is forced against a seat by the pressure of the water and prevents any further flow. The only water that is discharged is that which was held in the upper break-off sections plus a small amount that flows by the check valve until it is fully seated. If the top section has not been damaged by the vehicle, the hydrant can be placed back in service using a new break-off section and bar. It is necessary to close the gate valve in the branch connection when the hydrant is repaired.

Flush hydrants

Post-type hydrants are, of course, not suitable for all installations. Airport runways and loading areas, for example, call for a hydrant that

FIRE SERVICE HYDRAULICS

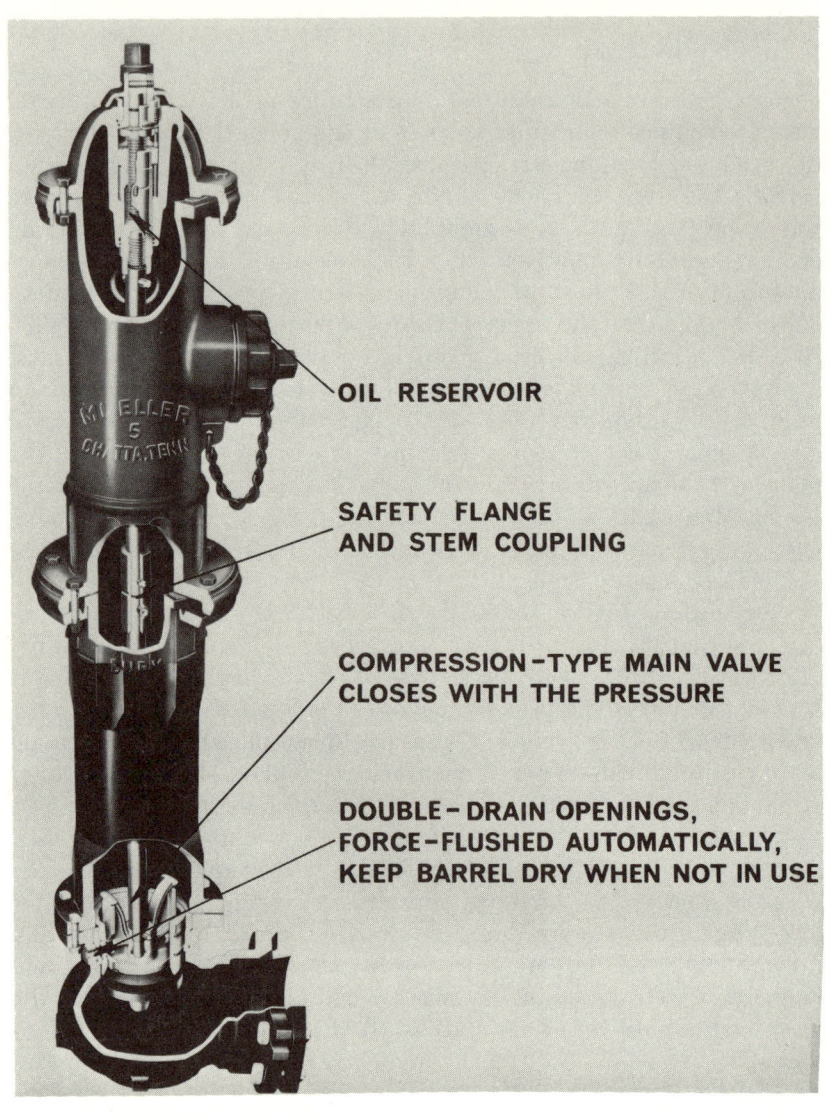

Figure 7

WATER DISTRIBUTION SYSTEMS

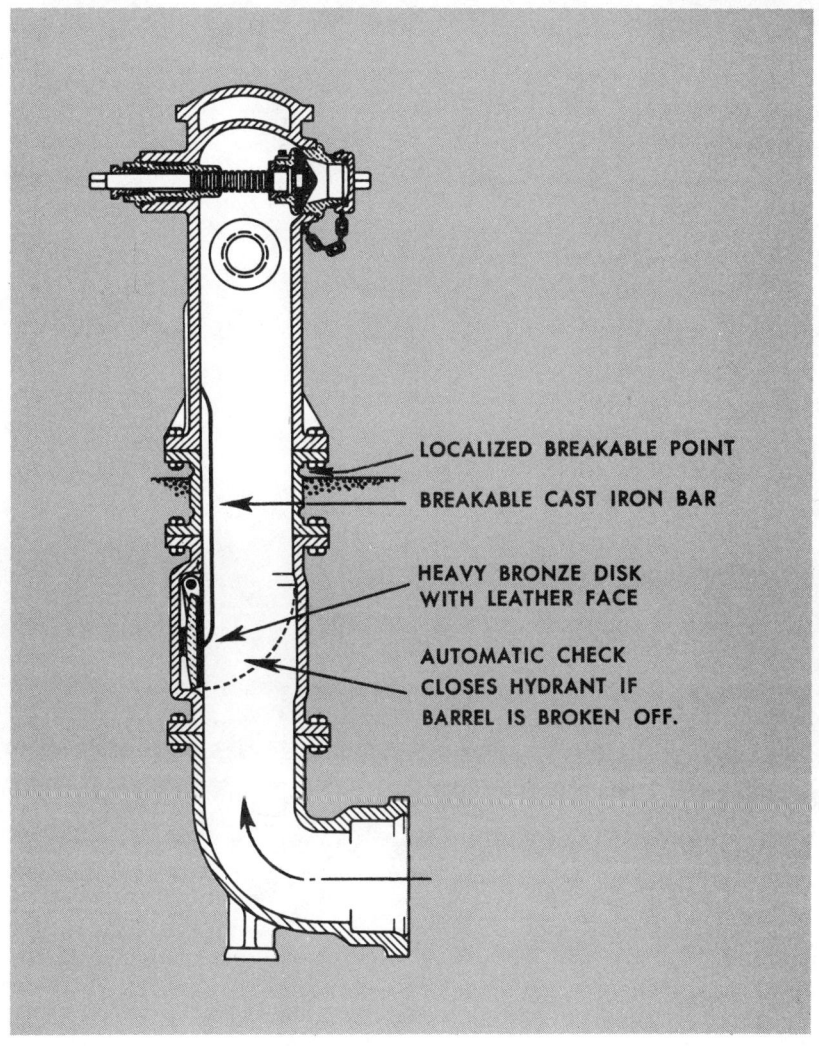

Figure 8

is flush with the ground. These flush-type hydrants are also made in wet and dry barrel types (Figures 9 and 10).

The outlet nozzles of a flush hydrant are located just below the surface of the ground and are enclosed in a special cast iron box or a manhole. The cover for the box must be strong enough to carry the weight of vehicles and, at the same time, be light enough to be easily removed for fire use. It also should be clearly marked so that it can be

FIRE SERVICE HYDRAULICS

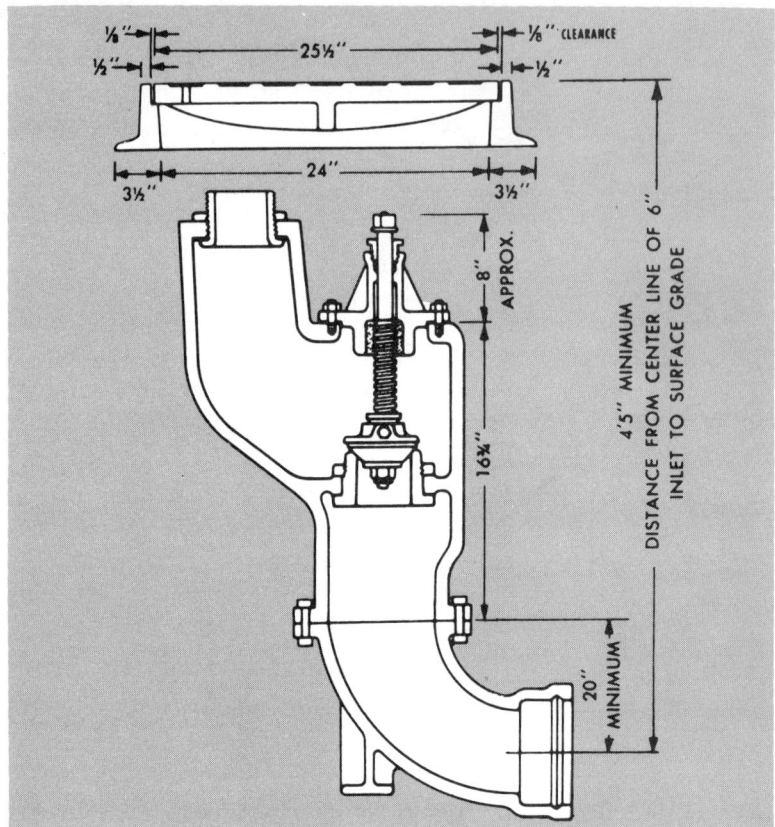

Figure 9

readily identified. Flush hydrants were used for regular municipal service in many cities in the United States, particularly in New England, during the period of transition from the fire plug to the modern hydrant. Many of these consisted essentially of a covered barrel that had a valve at the bottom. Fire apparatus carried a portable "chuck" which had a number of independently gated outlet nozzles, and which was designed to be quickly connected to the top of a hydrant. Some of these hydrants are still in service today. In fact, the hydrants on the Baltimore high pressure fire system are of this type.

It is quite obvious that there are a number of disadvantages to the use of flush hydrants. They are difficult to locate particularly in a street which may have other manhole or box covers. Motor vehicles may be parked directly over them. And in winter they may be cov-

WATER DISTRIBUTION SYSTEMS

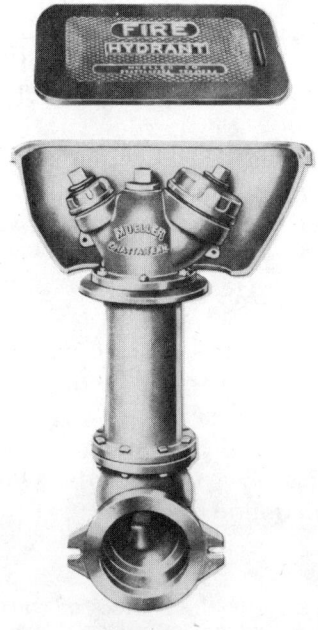

Figure 10

ered with snow or ice. Further, there is a delay involved in getting a fire stream into operation as compared with a post-type hydrant. This delay is greater where it is necessary to use a portable chuck.

Private yard hydrants

Many industrial plants maintain a private yard main system for fire protection purposes only. These systems are supplied by special fire pumps and are usually designed to supply the yard hydrants at pressures suitable for direct hose streams. The hydrants generally have from two to four hose outlets, each provided with an independent gate. Such hydrants may be located in hose houses and have hose connected to one of the outlets for immediate use. Standards for private yard systems are published by the National Fire Protection Association in NFPA Pamphlet 24, "Outside Protection."

High pressure hydrants

In some of the larger cities special high pressure fire systems are provided to augment the supply available for fire fighting from the

FIRE SERVICE HYDRAULICS

regular public water supply system. These systems were designed to operate at pressures ranging from 150 to 300 psi. The use of pressures of such magnitude require specially designed mains, gate valves and hydrants. The hydrants are of the post type (except in Baltimore) and are of extra heavy design. They are usually provided with two, three or four outlet nozzles with independent gates. The branch connections are mainly 8-inch in size. The portable chucks used with the flush hydrants on the Baltimore system have four outlet nozzles with independent pressure-regulating valves.

WATER SUPPLY SYSTEMS

Up to this point, consideration has been given mainly to hydrants. No matter how well hydrants may be designed, installed and maintained, their effectiveness depends upon the amount of water that is available from the water supply system of which they are a part. Water supply systems have two basic components: (1) the supply works and (2) the distribution system.

Supply works

The supply works include the sources of water such as lakes, rivers, springs and wells, and also the facilities necessary to take the water from the source, treat and purify it as necessary, and then deliver it to the distribution system. These facilities include impounding reservoirs, intakes, pumping stations, treatment and purification plants, and supply mains.

Distribution system

The distribution system consists of a network of pipes usually provided with local storage at certain points in the form of elevated tanks. The mains of the distribution systems are of three general types: arteries, secondary feeders and minor distributors.

Arterial mains are the largest and extend from supply mains or other supply works to all portions of the system. They are generally looped and supply the secondary feeders. These are generally smaller than the arteries and in turn supply the minor distributors.

The three terms are relative, and there is not always a sharp line of demarcation between them. Further, it is quite obvious that the size of an arterial main may range from a 12-inch in a small community to a 60-inch in a large city. Although smaller, there will be a similar variation in size for secondary feeders. But the minor distributors will generally be 6-inch or 8-inch, except in areas where there is a

WATER DISTRIBUTION SYSTEMS

high required flow or high consumption, which may require 12 to 16-inch. Sizes smaller than 6-inch are not suitable for providing fire protection.

The minor distributors generally form a pattern which is called a gridiron. The gridiron of pipes has service connections to provide water to individual buildings for domestic use and in some cases for private fire protection systems (sprinklers and standpipes). It is also equipped with hydrants for fire protection use. In order to shut down portions of the system to control breaks in mains or when it is necessary to make repairs, valves are installed at certain locations, usually near street intersections.

Main capacity

Assuming that an adequate supply at good pressure is available from the supply works (reservoirs or pumping stations), the amount of water available for fire protection and general consumption depends upon the carrying capacity of the mains—those extending to the distribution system and those that comprise the distribution system.

When water flows in a pipe, hose or other conduit, there is a loss of pressure or head. This loss is attributed to what is commonly called pipe friction, but this term in connection with flow in a pipe is actually a misnomer because the resistance to flow is not due to friction in the true sense of the word. As a result of many studies, it is generally agreed that the resistance to flow is due to shearing forces between particles or layers of water. Turbulence in the flow due to roughness of pipe walls or high velocities causes an increase in the effects of the shearing forces and greater resistance to flow. The greater the resistance to flow, the larger will be the friction loss. Some studies have resulted in the development of formulas for flow in pipes that give the relationships between the various factors involved.

The best information available on flow in pipes indicates that the loss of head, H, is:

1. Directly proportional to the length (l) of pipe
2. Variable with some power, n, of the velocity, V, of the water flowing
3. Variable with the roughness of the pipe
4. Inversely proportional to some power, x, of the pipe diameter, d

This can be written in equation form as follows:

$$H = K \frac{lV^n}{d^x}$$

where K is a constant of proportionality which takes into account the roughness of the pipe. This particular formula is difficult to use because n varies with the roughness of the pipe and x varies with the velocity. Therefore, in solving pipe flow problems more convenient formulas which are modifications of the above equation based on actual studies of pipe flow are used. One such empirical formula which has been employed by hydraulic engineers for many years is the Hazen-Williams equation which gives the velocity V as follows:

$$V = CR^{0.63}h^{0.54} \times 0.001^{-0.04}$$

where C is the Hazen-Williams Coefficient, R is the hydraulic radius, and h is the loss of head per foot of pipe. In this formula, V is expressed in feet per second and h in feet per foot of pipe. The hydraulic radius R is equal to the cross-sectional area of a conduit divided by its wetted perimeter. For a circular pipe flowing full, R is equal to the area of a circle divided by the circumference (wetted perimeter) or $\frac{\pi d^2}{4} \div \pi d = \frac{d}{4}$, where d is the diameter of the pipe in feet.

Discharge

The discharge Q through a pipe is equal to the product of the cross-sectional area A of the pipe and the velocity V. Q is expressed in cubic feet per second, A in square feet, and V in feet per second.

$$Q = AV$$

Substituting the value of V from the Hazen-Williams equation,

$$Q = ACR^{0.63}h^{0.54} \times 0.001^{-0.04}$$

This is a more useful form of the Hazen-Williams equation because it gives the discharge if the size of pipe, loss of head per foot, and coefficient C are known.

The coefficient C varies with the roughness of the pipe wall. It can be seen from the formula that for a given size pipe, which fixes the cross-sectional area A and the hydraulic radius R, and for a given loss of head h, the discharge Q will vary directly with the coefficient. In other words, the higher the coefficient C, the greater will be the discharge. It follows that the smoother pipes will have the higher coefficients and the rougher pipes the lower coefficients.

Pipe materials

Supply mains extending from reservoirs or pumping stations are of large size and may be made of reinforced concrete, steel, cast iron, or asbestos cement. In recent years ductile iron has also been used.

WATER DISTRIBUTION SYSTEMS

Although mains of any of these materials extend into and form parts of distribution systems, the majority of the distributing mains are cast iron or asbestos cement. Cast iron mains were first installed in Versailles, France, in the year 1664 at the order of King Louis XIV to provide water for the city and the fountains on the palace grounds. Some of these mains are still in service. Cast iron pipe was introduced in the United States about 1816. Asbestos cement pipe was developed in Europe about 1913 and was introduced in this country in 1929.

Hazen-Williams coefficients

For new bituminous-coated cast iron pipe the coefficient C is equal to 130. However, the value of C will usually decrease after a period of use with most waters due to tuberculation or corrosion. Tuberculation occurs after the coating wears off or deteriorates and the water comes in contact with the iron. It is characterized by growths of tubercles on the interior surface of the pipe. This makes for a rough surface that results in increased turbulence during flow and consequently, additional head loss (Figure 11). The formation of tubercles and of corrosion can only occur where the water comes in contact with the metal of the pipe. The remedy for this situation is to use cement-mortar linings wherever aggressive waters are to be carried by the pipe. The coefficient C for new cement-mortar lined cast iron pipe will range from 140–150. A coefficient of 140 is recommended for design of supply lines and 130 for distributing mains of smaller size and shorter length.

Although most new cast iron pipe installed in water distribution systems in recent years is cement-mortar lined, there are many miles of unlined cast iron water pipe in service, and in many instances the coefficient C is 100 or less. Where C is 100, the carrying capacity of the mains will usually be satisfactory since this value of C was used for many years in designing water distribution systems. However, if the values fall to any appreciable extent below 100, it is quite likely that the mains will be found to be of inadequate capacity. As an example, a pipe found to have a coefficient of 50 would have only one half the carrying capacity for which it was designed (assuming a design coefficient of 100) and only about 40 percent of its actual original capacity (based on $C = 130$).

Restoration of capacity

The capacity of tuberculated mains or mains otherwise incrusted as a result of corrosion can be temporarily restored by cleaning the interior with special tools designed for the purpose (Figure 12).

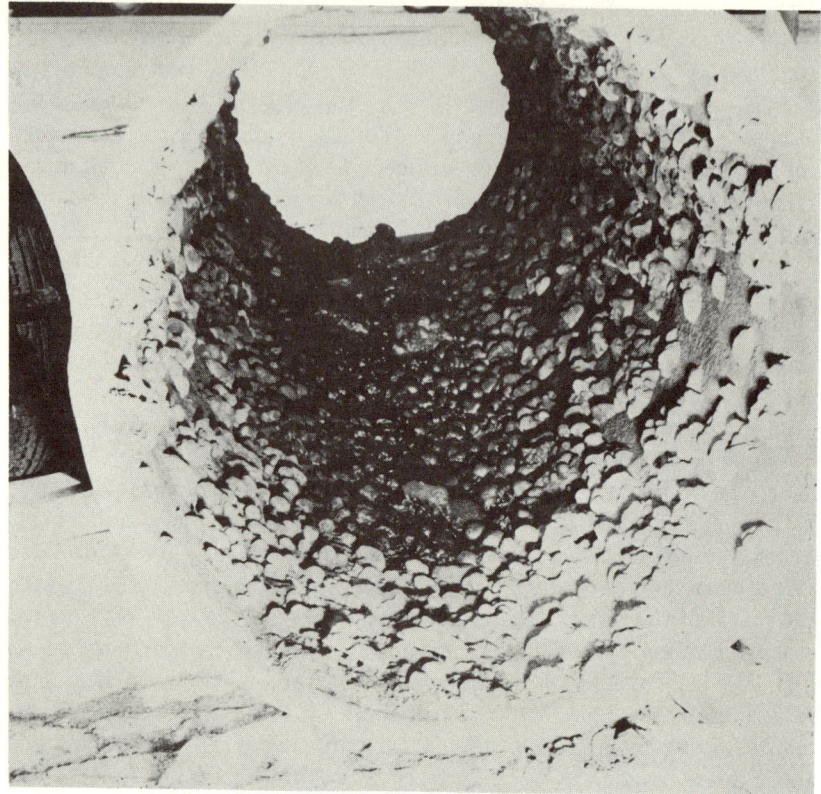

Figure 11

These cleaning tools have a series of spring steel scrapers that maintain contact with the interior walls of the pipe as the tool is pushed through by water pressure or pulled through by a cable attached to a power winch. A tight-fitting squeegee is forced or pulled through after the cleaning tool to remove any loose material or sludge remaining in the pipe.

Unless some means is provided for protecting the metal in the pipe wall from the action of the water, tubercles or incrustation will again begin to accumulate and usually at a more rapid rate than when the pipe was new. Therefore, a protective coating should be applied after cleaning. Several processes have been developed for lining buried mains with cement mortar or other materials. However, the cement mortar linings have proven to be most satisfactory and are the type generally used in the United States.

WATER DISTRIBUTION SYSTEMS

Figure 12

Whenever a main is cleaned or lined, it must be placed out of service. If the main being cleaned and lined has service connections to buildings, temporary lines are usually run along the surface of the ground to supply the connections. The temporary lines are too small to supply hydrants so it is necessary to plan the work so that some mains and hydrants in the area remain in service.

Large supply mains and arterial mains can also be cleaned and lined if necessary. This work is often simplified by the absence of service connections and hydrants. But care must be taken to make certain that the supply for the municipality is not interrupted or impaired during the cleaning and lining process.

The coefficient C for asbestos cement pipe is 140, and the original value is maintained throughout the life of the pipe just as in the case of cement-mortar lined cast iron. Although steel and reinforced concrete pipe are used for supply lines and arterial mains, they generally do not comprise an appreciable portion of most distribution systems. Steel pipe behaves very much like cast iron with respect to tuberculation and interior corrosion, so that a cement-mortar lining or other appropriate lining must be employed if the original coefficient is to be maintained. The coefficient for reinforced concrete pipe is about the same as for asbestos cement, and it does not materially change during the life of the pipe.

Pipe standards

Standards for cast iron, steel, asbestos cement, and reinforced concrete pipe have been developed by the American Water Works Association, and most pipe manufactured and installed in municipal water supply systems in the United States conforms to these standards as a minimum.

Plastic pipe has been used in some water distribution systems, but most of this pipe is less than 6 inches in size and has not been installed with the intent of supplying hydrants. A good field for this pipe has been found in areas where soil conditions are such that the use of the more generally accepted materials previously discussed would not be suitable, but its use has not been limited to such applications. Six-inch and larger sizes are available, but only a relatively few installations have been made in municipal water distribution systems. Until suitable standards are prepared and sufficient experience has been developed, plastic pipe is not likely to be used to any great extent in distribution systems providing fire protection. At present a committee of the American Water Works Association is preparing standards for plastic pipe for use in distribution systems.

Relative capacity of pipes

On numerous occasions it is important to know or be able to determine the relative carrying capacities of different sizes of pipes. These capacities are based upon the same loss of head for a given length of pipe of the same material and interior surface characteristics. Refer-

WATER DISTRIBUTION SYSTEMS

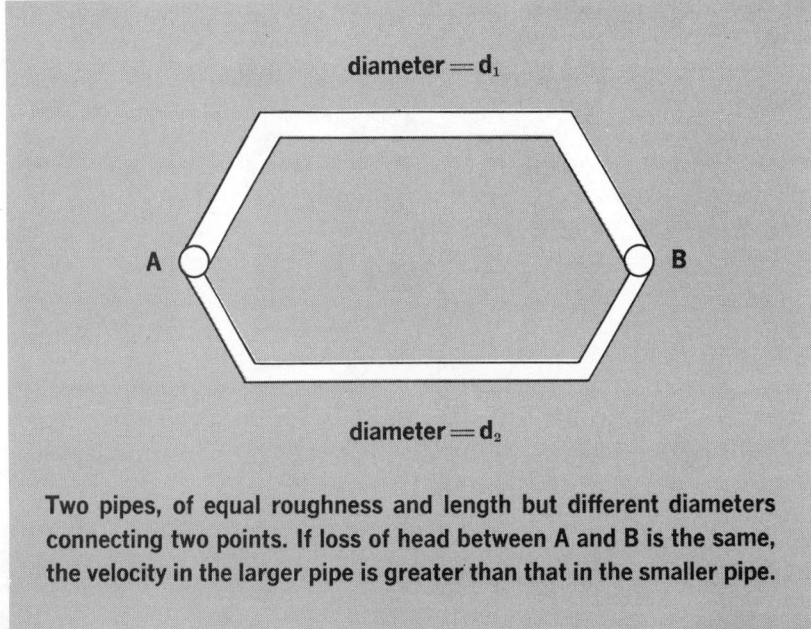

Two pipes, of equal roughness and length but different diameters connecting two points. If loss of head between A and B is the same, the velocity in the larger pipe is greater than that in the smaller pipe.

Figure 13

ring to the Hazen-Williams equation for velocity, it is evident that with the coefficient C and the loss of head per foot of pipe constant, the velocity will change with the diameter. That is, the greater the diameter the greater the velocity. Hence the velocity in the larger of the two pipes being compared will be greater, resulting in an increased carrying capacity considerably greater than that due to the relative difference in the cross-sectional areas of the pipes.

This can be shown as follows: Referring to Figure 13, which shows two pipes of equal roughness and length but different in size (diameters d_1 and d_2) extending between two points A and B, the Hazen-Williams equation can be written for each pipe.

$$Q_1 = A_1 C_1 R_1^{0.63} h^{0.54} \times 0.001^{-0.04}$$
$$Q_2 = A_2 C_2 R_2^{0.63} h^{0.54} \times 0.001^{-0.04}$$

Since the pipes are of the same roughness, $C_1 = C_2$. Further, the loss of head between points A and B has to be the same irrespective of the route (pipe 1 or pipe 2) followed. Since the pipes are of the same length, the loss of head per foot of length h_1 for pipe 1 equals

FIRE SERVICE HYDRAULICS

the loss of head per foot of length h_2 for pipe 2. Cancelling out the equal terms, the two equations become

$$Q_1 = A_1 R_1^{0.63}$$
$$Q_2 = A_2 R_2^{0.63}$$

Since the area $A = \dfrac{\pi d^2}{4}$ and the hydraulic radius $R = \dfrac{d}{4}$,

$$Q_1 = \frac{\pi d_1^2}{4} \times \left(\frac{d_1}{4}\right)^{0.63}$$

$$Q_2 = \frac{\pi d_2^2}{4} \times \left(\frac{d_2}{4}\right)^{0.63}$$

Dividing the first equation by the second,

$$\frac{Q_1}{Q_2} = \frac{\dfrac{\pi d_1^2}{4} \times \left(\dfrac{d_1}{4}\right)^{0.63}}{\dfrac{\pi d_2^2}{4} \times \left(\dfrac{d_2}{4}\right)^{0.63}} = \frac{d_1^2 \times d_1^{0.63}}{d_2^2 \times d_2^{0.63}}$$

$$\frac{Q_1}{Q_2} = \frac{d_1^{2.63}}{d_2^{2.63}}$$

This shows that the discharges for the two pipes vary as the 2.63 power of their respective diameters.

If the relative discharges in the pipes were in proportion to the respective cross-sectional areas, the relationship would be

$$\frac{Q_1}{Q_2} = \frac{A_1}{A_2} = \frac{\dfrac{\pi d_1^2}{4}}{\dfrac{\pi d_2^2}{4}} = \frac{d_1^2}{d_2^2}$$

This would indicate a smaller difference in the discharges than is actually the case, and this equation would apply only if the velocities in the two pipes were equal. This could not be the case if the loss of head in the two pipes were the same. However, quite frequently when a question is asked as to how much more water can be carried by an 8-inch pipe than a 4-inch, the answer will be given as four times, based on the fact that the cross-sectional area of the 8-inch is four times that of the 4-inch. This is a result of the use of the equation

$$\frac{Q_1}{Q_2} = \frac{d_1^2}{d_2^2} = \frac{(8)^2}{(4)^2} = \frac{64}{16} = 4$$

WATER DISTRIBUTION SYSTEMS

which is incorrect. The correct equation to use is

$$\frac{Q_1}{Q_2} = \frac{d_1^{2.63}}{d_2^{2.63}} = \frac{(8)^{2.63}}{(4)^{2.63}} = 6.2$$

As a rule of thumb in designing distribution systems, an 8-inch pipe is usually considered to have six times the carrying capacity of a 4-inch. Table 3 shows the relative carrying capacities for pipes ranging in size

Table 3   Relative Carrying Capacity of Pipes

Diameter (in.)	4	6	8	10	12	14	16	18	20	24	30	36	42	48	54	60
4	1															
6	2.9	1														
8	6.2	2.1	1													
10	11	3.8	1.8	1												
12	18	6.2	2.9	1.6	1											
14	27	9.3	4.4	2.4	1.5	1										
16	38	13	6.2	3.4	2.1	1.4	1									
18	52	18	8.4	4.7	2.9	1.9	1.4	1								
20	69	24	11	6.2	3.8	2.6	1.8	1.3	1							
24		38	18	10	6.2	4.1	2.9	2.1	1.6	1						
30		69	32	18	11	7.4	5.2	3.8	2.9	1.8	1					
36			52	29	18	12	8.4	6.2	4.7	2.9	1.6	1				
42				44	27	18	13	9.3	7.0	4.4	2.4	1.5	1			
48				62	38	26	18	13	10	6.2	3.4	2.1	1.4	1		
54					52	35	25	18	14	8.4	4.7	2.9	1.9	1.4	1	
60					69	46	32	24	18	11	6.2	3.8	2.6	1.8	1.3	1

This table shows the relative carrying capacity of pipes 4 to 60 inches in diameter for the same loss of head. Under such conditions of flow the velocity in the larger pipe is greater than in the smaller so that the larger pipe will discharge at a rate greater than that which would be expected by a comparison of their respective cross-sectional areas. The values in the table are based on the Hazen-Williams equation in which the discharges vary as the 2.63 power of the respective diameters.

To use the table find the larger pipe size in the left hand column, and read the value in the horizontal line to the right under the smaller pipe size. This value will give the number of times the carrying capacity of the larger pipe is greater than the smaller or the number of smaller pipes needed to deliver the same discharge as the larger.

FIRE SERVICE HYDRAULICS

from 4-inch to 60-inch based on the variation of discharge with the 2.63 power of the diameter. The use of this table obviates the necessity of solving the exponential equation

$$\frac{Q_1}{Q_2} = \frac{d_1{}^{2.63}}{d_2{}^{2.63}}$$

As an example, if it is desired to determine the carrying capacity of a 36-inch pipe with respect to a 12-inch, locate 36-inch in the left-hand column and read the figure in the horizontal line to the right under 12-inch, which is 18. This means that a 36-inch pipe has the carrying capacity of about eighteen 12-inch pipes. In similar fashion, it can be seen that a 6-inch has the capacity of about three (2.9) 4-inch and that an 8-inch has the capacity of about two (2.1) 6-inch.

A study of the table will show that:

If the diameter is	carrying capacity is increased
doubled	6 times
tripled	18 times
quadrupled	38 times
quintupled	69 times

Use of Hazen-Williams equation

It is quite obvious that the Hazen-Williams equation can be used to solve many engineering problems involving flow in pipes and conduits. In the design of a water supply system, pipe sizes can be determined to provide the needed discharge, Q, when the loss of head per foot, h, is limited to a specific figure. On the other hand, if a certain size pipe exists in a system, the loss of head per foot, h, can be determined for a specific discharge, Q. The determination of loss of head per foot, h, is also important in choosing a suitable pump for delivering a certain discharge, Q, through a pipe line of given size and length. Further, if a pipe exists in a system and it is supplied by a reservoir or pumping station that provides a specific head, H, the discharge, Q, available can be determined.

Because the solution of the equation requires the use of logarithms and it would be rather time-consuming to solve the equation for each individual application, special tables, charts, nomographs, and slide rules have been developed that greatly simplify the solution. If any two of three variables, Q, d, or h are known, the other can be determined. Of course, in any given situation it is necessary to know the value of the Hazen-Williams coefficient C.

Following are a few examples showing the application of the Hazen-

WATER DISTRIBUTION SYSTEMS

Williams equation to the solution of basic problems in pipe flow. The nomograph (Figure 14) is one of the means that are used to solve the Hazen-Williams equation and will be used in these examples. An examination of the nomograph will reveal that the values of four variables, flow or discharge (in cubic feet per second and also in gallons per minute), pipe diameter (in inches), head loss (in feet per thousand feet), and velocity (in feet per second) are shown on vertical lines. The units for pipe diameter and head loss are not the same as previously used in the discussion of the equation, but they are commonly used in water supply and hydraulic computations, and the nomograph has been laid out on this basis. In using the nomograph, if a straight edge is used to connect the values of any two variables that are known, the other variable will be found along the straight edge where it intersects the vertical line for the unknown variable. It should be noted that this nomograph is for a Hazen-Williams coefficient of $C = 100$.

Example A. It is desired to find the size of pipe needed to deliver 2000 gpm with a loss of head not to exceed 4 feet per thousand; a Hazen-Williams coefficient of $C = 100$ is to be used for design purposes.

A straight edge is placed on the nomograph intersecting the loss of head line at 4 feet per thousand feet and the flow line at 2000 gpm; the intersection of the straight edge with the pipe diameter line is a point between 14 and 16 inches. This indicates that a pipe larger than 14-inch is needed and that a 16-inch will provide a small amount of extra capacity. A 16-inch would be chosen.

Example B. It is desired to determine the loss of head in 2500 feet of 8-inch pipe ($C = 100$) when the flow is 1000 gpm.

The straight edge is placed on the nomograph intersecting the pipe diameter line at 8 inch and the flow line at 1000 gpm; the intersection of the straight edge with the loss of head line is at 30 feet per thousand feet. To obtain the loss for 2500 feet, this figure is multiplied by $2500/1000 = 2.5$.

$$2.5 \times 30 = 75.0 \text{ feet}$$

It is often convenient to have the loss in a pipe line in psi. The loss of head in feet can be converted to psi by multiplying by 0.434:

$$75.0 \times 0.434 = 32.6 \text{ psi}$$

Example C. It is desired to select the head at which a 7000-gpm pump should be designed for supplying its capacity at a pres-

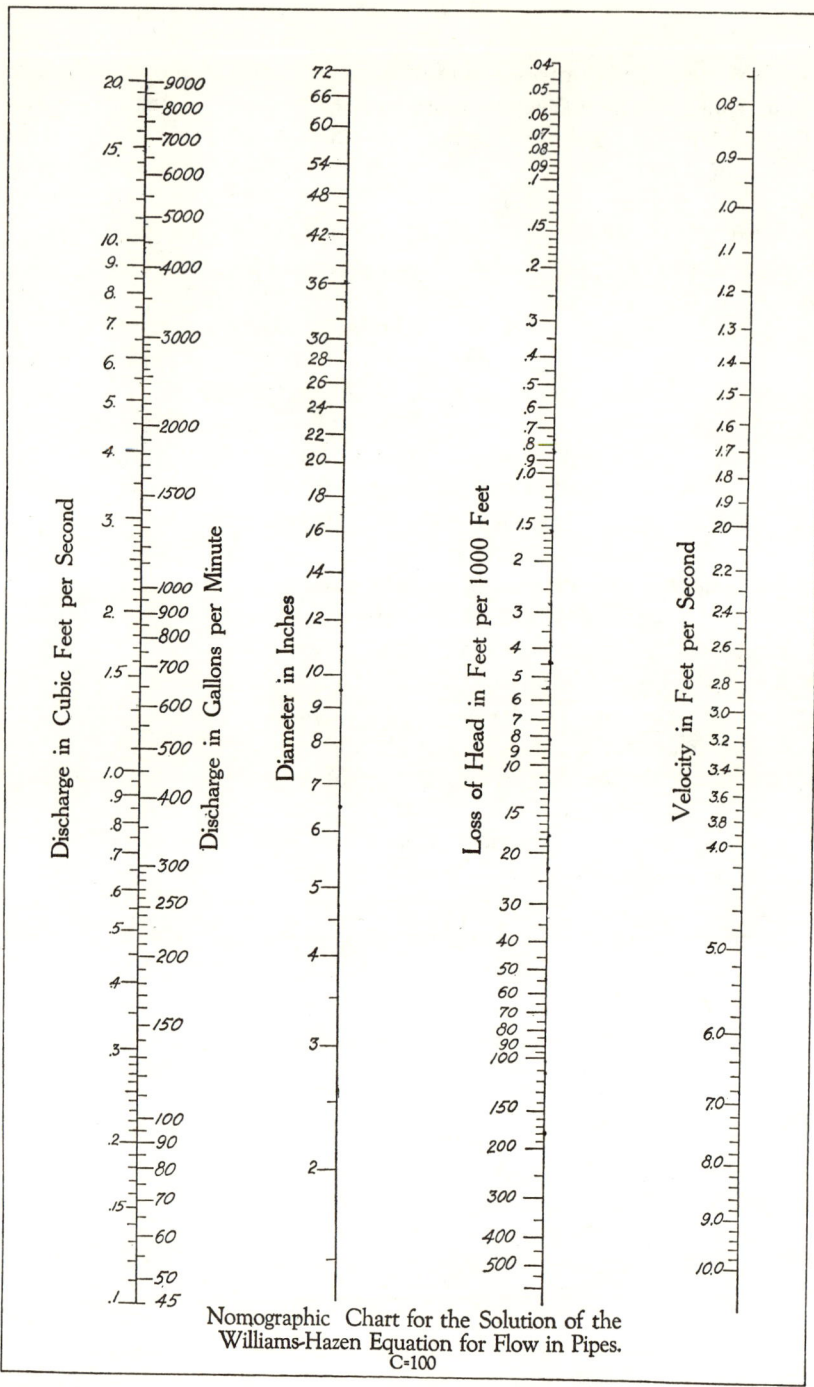

Figure 14

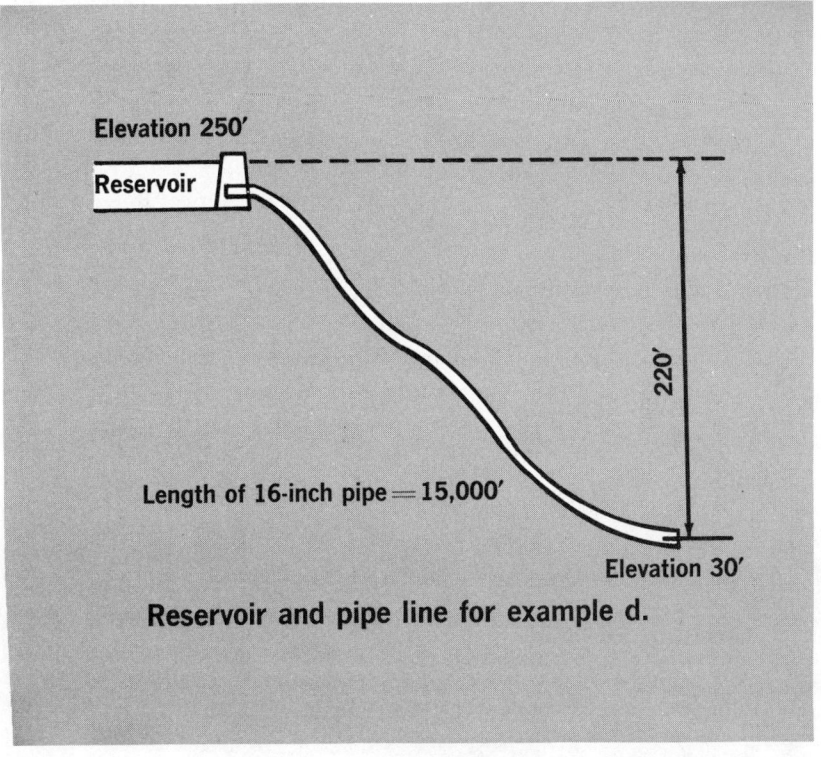

Figure 15

sure of 50 psi at the end of a level 24-inch pipe line ($C = 100$) 10,000 feet long.

The straight edge is placed on the nomograph intersecting the pipe diameter line at 24-inch and the flow line at 7000 gpm; the intersection of the straight edge with the loss of head line is at 5.2 feet per thousand feet. For 10,000 feet the total loss is $5.2 \times 10 = 52$ feet. Since the pressure required at the end of the line is 50 psi, this is converted to head in feet and added to the loss in the line to determine the head at which the pump must deliver. Pressure in psi can be converted to head in feet by dividing by 0.434 or multiplying by its reciprocal 2.31. The total head at which the pump must operate is $52 + (50 \times 2.31) = 52 + 115.5 = 167.5$ feet. The final head chosen for the pump would be somewhat higher than 167.5 feet to take care of the minor losses of head that exist in all pipe lines.

FIRE SERVICE HYDRAULICS

Example D. A reservoir with a water surface elevation of 250 feet supplies a 16-inch pipe line ($C = 100$) 15,000 feet in length with its lower end at elevation 30 feet. It is desired to find the discharge that is available with a pressure of 60 psi at the lower end and also the velocity of the water flowing in the line (Figure 15).

Referring to the figure, it can be seen that there is a difference in elevation between the reservoir surface and the pipe line terminus of 220 feet. This would be the head that would be available to be used in delivering water through the pipeline if there was no pressure required at the end of the line. However, since a pressure of 60 psi is required at the end of the line, its equivalent in feet must be subtracted from 220 feet to obtain the head available to produce discharge in the line. Since 60 psi is equal to $60 \times 2.31 = 138.6$ feet, the head available is $220 - 138.6 = 81.4$ feet. The line is 15,000 feet long so that $81.4 \div 15 = 5.43$ feet is the head available per thousand feet. The straight edge is placed on the nomograph intersecting the pipe diameter line at 16-inch and the head loss line at 5.4 feet per thousand feet; the intersection of the straight edge with the flow line is at 2450 gpm and with the velocity line is at 3.9 feet per second. Therefore, the pipeline would deliver at a rate of 2450 gpm from the reservoir to its terminus where the pressure would be 60 psi, and the velocity of the water flowing in the pipe would be 3.9 feet per second.

In Examples C and D, factors were introduced for converting head in feet to pressure in pounds per square inch and vice versa. Table 4 gives values of head in feet for pressures in pounds per square inch up to 260 psi.

FIRE FLOW TESTS

Fire flow tests are made on water distribution systems to determine the rate of flow available at various locations for fire fighting purposes. Depending upon the method to be employed in combating fires at a particular location, a certain residual pressure in the main is specified. The rate of flow must be available, at this residual pressure if the water is to be used effectively. As an example, if fire department pumpers are to be used to supply hose streams, a residual pressure of 20 psi is usually specified. This residual pressure in the mains is necessary to overcome the friction loss in the hydrant branch, in the hydrant, in the suction hose and fittings of the pumper, and to provide the water velocity needed at the pumper suction inlet. If hose streams

WATER DISTRIBUTION SYSTEMS

Table 4 Water Pressure Conversion Table

psi	ft.	psi	ft.	psi	ft.	psi	ft.	psi	ft.
1	2.31	53	122.43	105	242.55	157	362.67	209	482.79
2	4.62	54	124.74	106	244.86	158	364.98	210	485.10
3	6.93	55	127.05	107	247.17	159	367.29	211	487.41
4	9.23	56	129.36	108	249.48	160	369.60	212	489.72
5	11.55	57	131.67	109	251.79	161	371.91	213	492.03
6	13.86	58	133.98	110	254.10	162	374.22	214	494.34
7	16.17	59	136.29	111	256.41	163	375.53	215	496.65
8	18.48	60	138.60	112	258.72	164	378.84	216	498.96
9	20.79	61	140.91	113	261.03	165	381.15	217	501.27
10	23.10	62	143.22	114	263.34	166	383.46	218	503.58
11	25.41	63	145.53	115	265.65	167	385.77	219	505.89
12	27.72	64	147.84	116	267.96	168	388.08	220	508.20
13	30.03	65	150.15	117	270.27	169	390.39	221	510.51
14	32.34	66	152.46	118	272.58	170	392.70	222	512.82
15	34.65	67	154.77	119	274.89	171	395.01	223	515.13
16	36.96	68	157.08	120	277.20	172	397.32	224	517.44
17	39.27	69	159.39	121	279.51	173	399.63	225	519.75
18	41.58	70	161.70	122	281.82	174	401.94	226	522.06
19	43.89	71	164.01	123	284.13	175	404.25	227	524.37
20	46.20	72	166.32	124	286.44	176	406.56	228	526.68
21	48.51	73	168.63	125	288.75	177	408.87	229	528.99
22	50.82	74	170.94	126	291.06	178	411.18	230	531.30
23	53.13	75	173.25	127	293.37	179	413.49	231	533.61
24	55.44	76	175.56	128	295.68	180	415.80	232	535.92
25	57.75	77	177.87	129	297.99	181	418.11	233	538.23
26	60.06	78	180.18	130	300.30	182	420.42	234	540.54
27	62.37	79	182.49	131	302.61	183	422.73	235	542.85
28	64.68	80	184.80	132	304.92	184	425.04	236	545.16
29	66.99	81	187.11	133	307.23	185	427.35	237	547.47
30	69.30	82	189.42	134	309.54	186	429.66	238	549.78
31	71.61	83	191.73	135	311.85	187	431.97	239	552.09
32	73.92	84	194.04	136	314.16	188	434.28	240	554.40
33	76.23	85	196.35	137	316.47	189	436.59	241	556.71
34	78.54	86	198.66	138	318.78	190	438.90	242	559.02
35	80.85	87	200.97	139	321.09	191	441.21	243	561.33
36	83.16	88	203.28	140	323.40	192	443.52	244	563.64
37	85.47	89	205.59	141	325.71	193	445.83	245	565.95
38	87.78	90	207.90	142	328.02	194	448.14	246	568.26
39	90.09	91	210.21	143	330.33	195	450.45	247	570.57
40	92.40	92	212.52	144	332.64	196	452.76	248	572.88
41	94.71	93	214.83	145	334.95	197	455.07	249	575.19
42	97.02	94	217.14	146	337.26	198	457.38	250	577.50
43	99.33	95	219.45	147	339.57	199	459.69	251	579.81
44	101.64	96	221.76	148	341.88	200	462.00	252	582.12
45	103.95	97	224.07	149	344.19	201	464.31	253	584.43
46	106.26	98	226.38	150	346.50	202	466.62	254	586.74
47	108.57	99	228.69	151	348.81	203	468.93	255	589.05
48	110.88	100	231.00	152	351.12	204	471.24	256	591.36
49	113.19	101	233.31	153	353.43	205	473.55	257	593.67
50	115.50	102	235.62	154	355.74	206	475.86	258	595.98
51	117.81	103	237.93	155	358.05	207	478.17	259	598.29
52	120.12	104	240.24	156	360.36	208	480.48	260	600.60

are to be used directly from hydrants, a much higher residual pressure is required. This is because it will be necessary to have sufficient pressure to overcome the friction loss between the main and hydrant outlet and in the hose line and nozzle attached, and still provide a nozzle pressure which will give an effective stream. A good average residual pressure for direct hydrant streams is 75 psi, but this should be increased for congested areas, particularly those with tall buildings.

In some cases, as when the flow test is made for the purpose of determining the quantity available for supply to automatic sprinkler or other private fire protection installations, the rate of flow must be available at a specified residual pressure at a certain elevation. In the case of tests for automatic sprinkler supply, the residual pressure is usually specified as 15 psi, and the elevation is that of the top line of sprinklers.

The procedure followed in the field consists of discharging water at a measured rate of flow from the system at a given location and observing the corresponding pressure drop in the mains. From the data thus obtained, it is possible to compute the rate of flow available at any residual pressure.

Fire flow test hydraulics

In order to understand exactly what occurs when a fire flow test is run, a knowledge of a few fundamental principles of hydraulics is necessary.

Consider any pipe at any grade or series of grades supplied by a reservoir (Figure 16). The level of the water in the reservoir is shown by the line designated by WS. When there is no flow in the pipe and a pressure reading is taken on a gage located at any point T, the value of the pressure converted to feet of water can be represented graphically on a vertical line drawn through point T by the length PT. The point P will be at the same elevation as that of the water surface in the reservoir. If a vertical tube were inserted in the top of the pipe at point T, the water would rise to the same level as that in the reservoir. However, with a certain rate of flow, Q, through the pipe from the reservoir, a pressure reading at point T will give a value which can be represented by the length FT. This indicates that the rate of flow in the pipe has caused a drop in pressure at point T which is represented by the length PF. The line WF joining the reservoir level with the value of the pressure at point T is termed the hydraulic grade line. For each different value of Q there will be a different hydraulic grade line and corresponding drop in pressure at T.

To simplify the solution of problems involving systems of pipes of

WATER DISTRIBUTION SYSTEMS

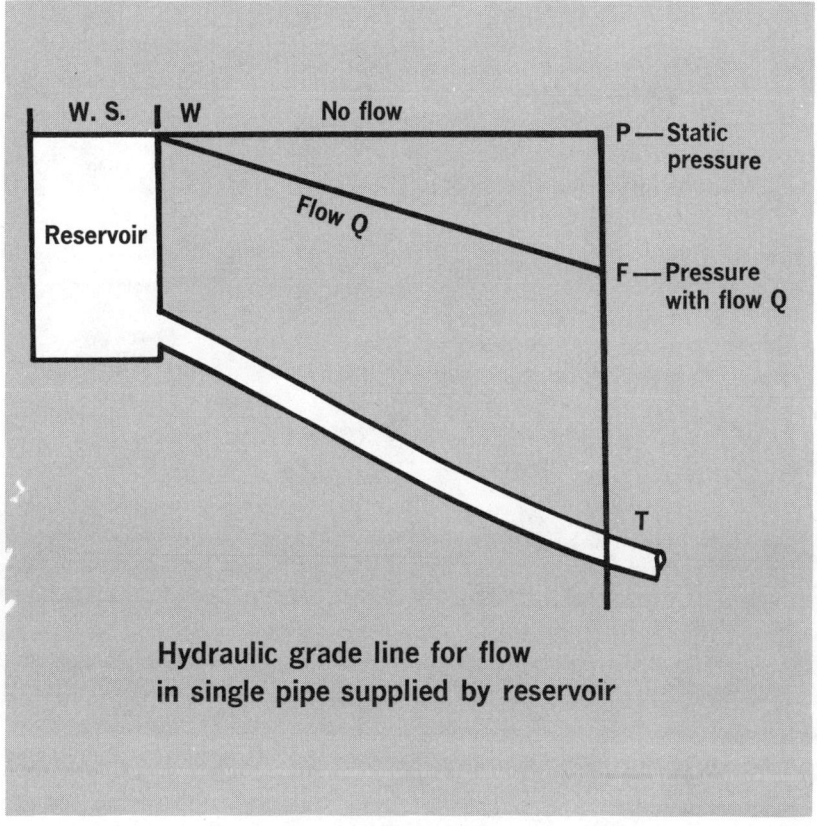

Figure 16

various sizes, it is possible to substitute an equivalent pipe for any given system of pipes. An equivalent pipe is one of such size and length that the pressure drop for any given rate of flow through it will be the same as that for the original system of pipes. Therefore, an entire distribution system between its source of supply and the point where a flow test is to be conducted can be considered a single pipe and the hydraulic grade line drawn as for the single pipe in Figure 16. In this case point T will be taken as any point on a distribution system with supply from a reservoir (Figure 17). If supply is from pumps, the same principles will apply if the pressure at the pumps can be maintained. With no flow in the system, the hydraulic grade line is horizontal, and the pressure at point T will be equal to the full head of the reservoir at point T. This is represented by the vertical

FIRE SERVICE HYDRAULICS

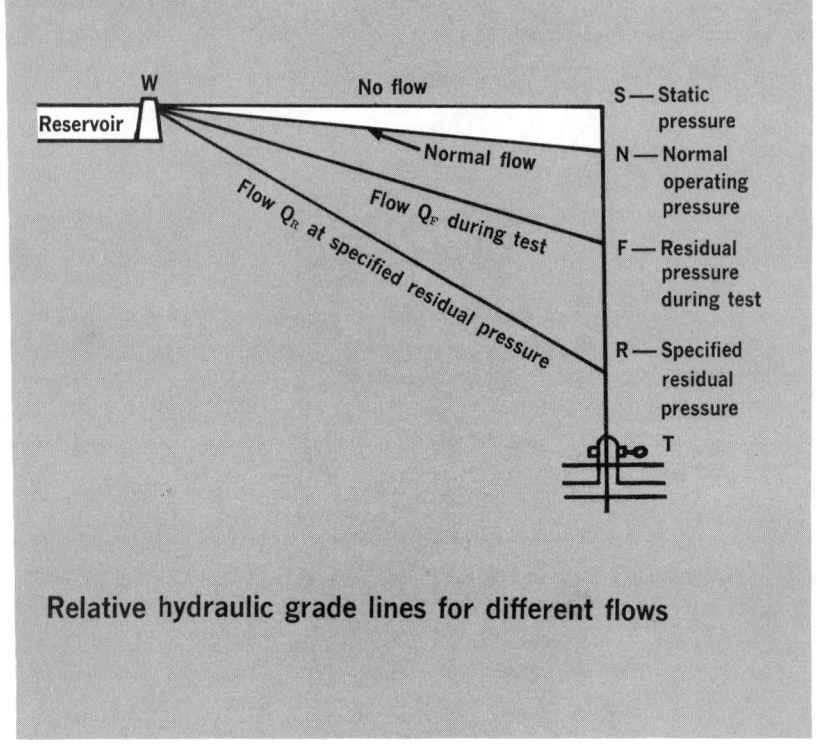

Figure 17

distance ST in Figure 17 and is the true static pressure. Since there is always some water being drawn from a water system or being wasted through leaks, this true static pressure at point T will never actually be reached. But during periods of low consumption—late at night or very early morning—this pressure may be closely approached. Under normal operating conditions there will be a drop in pressure, SN, due to the flow of normal consumption through the mains. This drop increases with a high consumption rate and decreases with a small rate. The normal operating pressure is represented by NT and the hydraulic grade line for normal operation by WN.

When a flow test is run at point T and a relatively high rate of flow Q_F is discharged, the pressure in the mains will be reduced so that the pressure at T will be represented by FT and the hydraulic grade line by WF. The pressure FT will be the residual pressure during the test, and the line NF will represent the pressure drop during the test. The specified residual pressure at which the quantity of water Q_R

must be available for effective fire fighting is represented by RT and the corresponding hydraulic grade line by WR. The test observations provide the value of NT (the normal pressure at the test location), of Q_F (the rate of flow during the test) and of FT (the residual pressure during the test). The pressure RT (the residual pressure at which the fire flow must be delivered) is known for each test. From these values, the fire flow Q_R available at the specified residual pressure RT can be computed.

Test locations

The locations at which tests should be made depend upon the purpose of the tests. When tests are made to analyze the distribution system of a municipality, it is customary to conduct them on the arteries and secondary feeders to determine their adequacy and also on the minor distributors to determine the efficiency of local distribution. Tests are also run to determine the fire flow available in high-value or high-hazard districts. These include major and minor business districts, industrial, warehouse, institutional, and apartment districts, and shopping centers. Amusement park and resort hotel areas would also be included. In addition, tests are made in a number of representative residential districts. From the results of such a group of tests and a study of the distribution map, it is possible to analyze the ability of the system to deliver fire flow at most locations.

In some cases, tests are not made for the purpose of analyzing a whole distribution system, but to ascertain the supply available at a given location. The purpose of such tests may be to determine the fire flow available at a certain group of buildings, the supply available for automatic sprinkler protection in a given building, or the ability of the system to supply a proposed extension into a new area.

Test layout

Once the test site has been chosen, the engineer or inspector in charge selects a group of test hydrants. Among these hydrants he picks one that he designates as the *residual hydrant*. This hydrant is so chosen that the other hydrants are between it and the larger mains which constitute the immediate sources of supply in the area. Figure 18 shows several test layouts; a circle encloses each residual hydrant.

The number of hydrants to be used in any test depends upon the strength of the distribution system in the vicinity of the test location. To obtain satisfactory results, the engineer flows a sufficient number of hydrants to cause a pressure drop in the residual hydrant of not less than 10 psi. If the mains are small and the system weak, only one

FIRE SERVICE HYDRAULICS

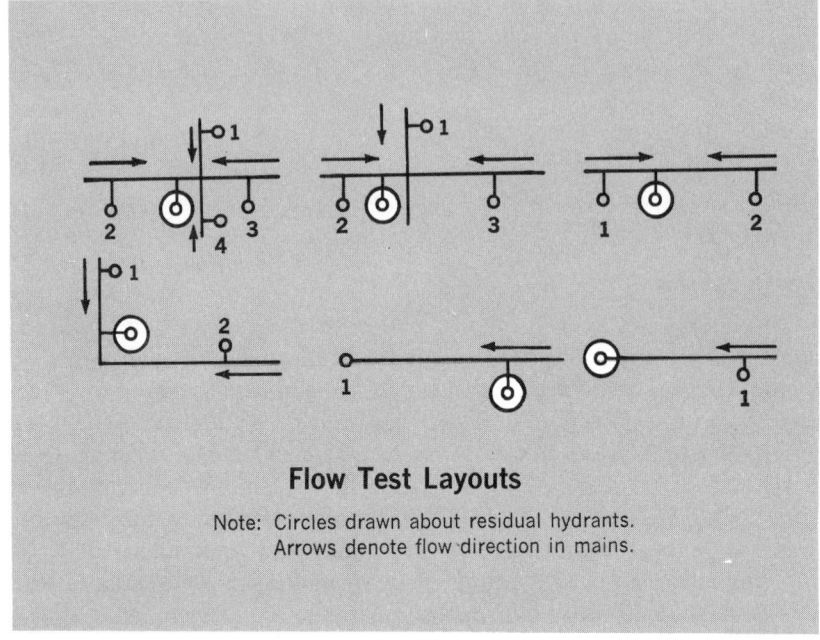

Flow Test Layouts

Note: Circles drawn about residual hydrants.
Arrows denote flow direction in mains.

Figure 18

or two hydrants need to be flowed. If, on the other hand, the mains are larger and the system strong, it may be necessary to flow as many as seven or eight hydrants. Using a map of the distribution system, he makes a study of each test location to select the hydrants for each test. He then makes a sketch that shows the mains, hydrants, and valves in the vicinity of the test. A circle is drawn around the residual hydrant, and each other hydrant to be used in the test is numbered for identification. This preliminary choice of hydrants may be modified as to number, location, or both when the test is made. In making the final selection of the hydrants to be flowed, it is necessary to make certain that the discharging streams of water from these hydrants will not cause damage to nearby property.

To make a test on a dead-end main, either the dead-end hydrant or the next hydrant may be used as the residual hydrant. In both cases, the rate of flow and the residual pressure observed will be that available at the next to the dead-end hydrant. This assumes the hydrants to be at the same elevation, but if they are not and the dead-end hydrant is used as the residual hydrant, the residual pressure should be corrected by an amount equal to the difference in elevation.

From a practical standpoint it is actually better to flow the dead-end hydrant. First, this method offers an opportunity to flush the dead end. Second, it can indicate if any obstructions exist in the main between the two hydrants.

Sometimes the distance between the hydrants can be great or other conditions exist which could make the result of a test run as described above of little value in evaluating the supply available. However, if better data is needed (for example, in connection with the installation of a new sprinkler system), it is sometimes possible to flow the dead-end hydrant and observe the residual pressure on a service connection to the main near the hydrant. When this procedure is followed, all water use on the service should be discontinued during the test.

If there is no convenient place to observe the residual pressure on a dead-end main, a test can be made by opening the dead-end hydrant wide and measuring the discharge. Such a test will only indicate the rate of flow available. The residual pressure in the main is equal to the observed velocity pressure at the hydrant outlet nozzle plus the friction loss between the main and hydrant outlet nozzle and can be estimated fairly well. Tests of this type are not usually made, but may be of value in indicating the supply available for a pumper at the dead-end hydrant. With a short suction line a pumper should be able to draw from the hydrant a rate of flow at least equal to that indicated by the test. It should be noted, however, that the residual pressure in the main may be less than the desired 20 psi.

Equipment needed

The equipment necessary for the field work consists of a single 200-psi Bourdon pressure gage with 1-psi gradations, a number of pitot blades and 50 or 60-psi Bourdon pressure gages with $\frac{1}{2}$-psi gradations (one blade and gage for each hydrant to be flowed), and a hydrant cap tapped with a $\frac{1}{4}$-inch hole into which a short length of $\frac{1}{4}$-inch brass pipe is fitted. This pipe has a tee for the 200-psi gage and a cock at the end for bleeding air from the hydrant. In some cases a valve or a cock may be inserted between the gage and the tee. A coupling in the $\frac{1}{4}$-inch pipe is sometimes provided between the hydrant cap and tee to permit adjusting the gage to a convenient position after the hydrant cap has been tightened. Figure 19 shows equipment of this type. A number of small scales with $\frac{1}{16}$-inch gradations for measuring the diameters of hydrant outlet nozzles and suitable forms for recording the field data are also necessary.

FIRE SERVICE HYDRAULICS

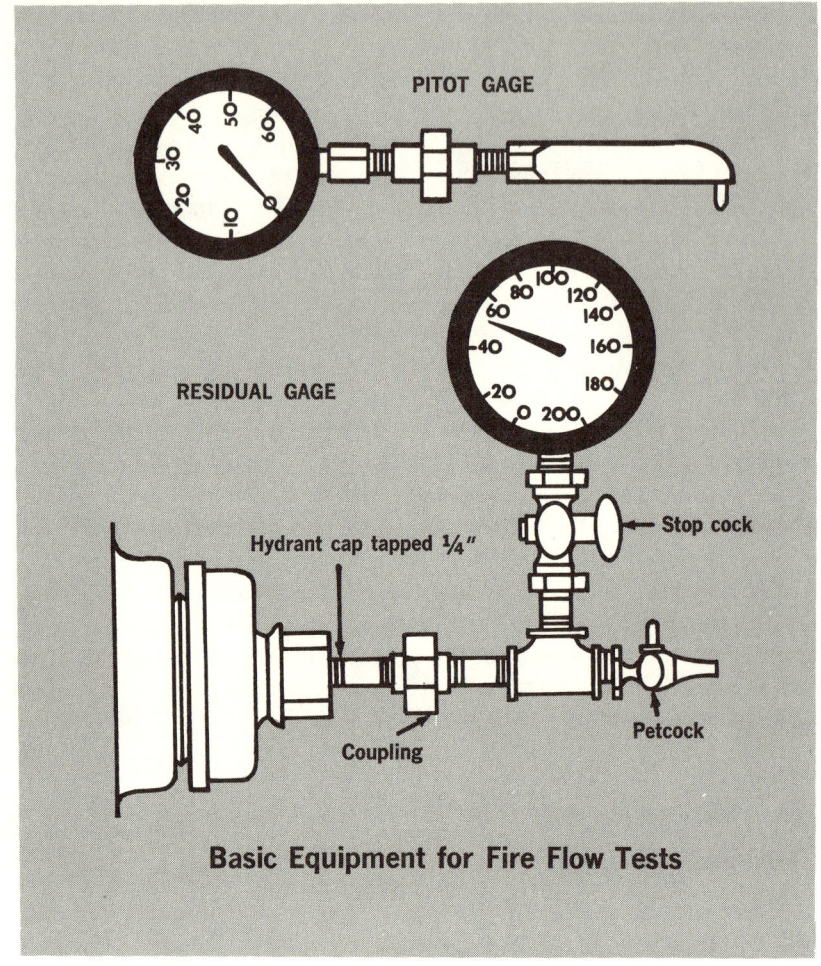

Figure 19

Test procedure

In conducting a typical test, the chief of the field party assigns a man to each hydrant to be flowed and takes his own station at the residual hydrant. Here he attaches the 200-psi gage to one of the 2½-inch outlet nozzles using the special cap and opens the cock on the gage piping assembly. After all outlet nozzle caps have been checked to see that they are secure, he opens the hydrant valve fully. When the air is exhausted from the barrel (indicated by a flow of water), the cock is closed. In some cases the needle on the gage may oscillate. It is best to wait until it stops before taking a reading, but if the needle continues to fluctuate, he makes his reading at the center of the swing. He then gives the signal to open one of the hydrants. When this hydrant is flowing, he checks the residual gage to see if the pressure has been affected by the flow. Each of the other hydrants is opened in succession with the chief's residual gage similarly checked.

When all hydrants are flowing, he gives a signal to the men at the hydrants to take a pitot gage reading in the streams. This signal should not be given until the residual gage needle is at rest. Simultaneously the chief reads the residual pressure at his hydrant.

The drop in the residual gage reading from that taken with no flow to that taken with all hydrants flowing should be not less than 10 psi. It is also good practice to prevent the pressure within the distribution system from reaching too low a figure, and hence residual gage readings generally should not be permitted to fall below 20 psi. The magnitude of the pressure drop can be controlled by the number of hydrants used, the number of outlet nozzles opened on each, and the extent to which the valve on each hydrant is opened.

After the readings have been taken, each hydrant is shut down. Hydrants should be shut down slowly and one at a time to prevent undue surges in the system. When all the flowing hydrants are closed, the residual gage should read the same as at the start of the test. Before closing the valve on the residual hydrant, the cock on the gage piping should be opened to prevent a vacuum being placed on the gage as the hydrant drains. If this is not done the gage may be damaged.

At the hydrants used for flow during the test, the discharges from the outlet nozzles are determined from measurements of the diameter of the outlet nozzles flowed and of the velocity pressures of the streams as indicated by the pitot gage reading. The formula used to compute the discharge, Q, from these measurements is:

$$Q = 29.83cd^2 \sqrt{P} \qquad \text{(Equation } A\text{)}$$

where c is a coefficient, d the diameter of the outlet nozzle in inches and P the velocity pressure in psi. The application of a coefficient is necessary due to the existence of a velocity at the outer edge of the stream that is somewhat less than that in the major portion of the cross section.

Discharges from various size hydrant outlet nozzles when flowing full, as determined by pitot readings, are compared with the flows in the main supplying the hydrants by venturi meter, pitometer and current meter. These comparative measurements indicated that c, the coefficient of discharge, averaged 0.91, and a value of 0.90 has been adopted for general use. Discharges for circular outlets from $2\frac{1}{4}$ to $5\frac{3}{16}$ inches in diameter for velocity pressures of $\frac{1}{4}$ to 36 psi are included in the table at the end of this chapter. This table is based upon the use of a coefficient of 0.90 which assumes a full and relatively smooth flow from the hydrant outlet nozzle.

If hydrants are of wet barrel type, it is not usually possible to obtain discharges that can be measured with sufficient accuracy. This is because of the variation in velocity over the cross sections of the outlet nozzles due to the effect of the independent valves. In such cases discharge tubes are placed on the outlet nozzles to produce streams that can be measured with sufficient accuracy. Figure 20 shows such a tube. Tests have shown that a coefficient of 0.90 should be used with the type of tube shown.

When measuring discharges from hydrant outlet nozzles, it is always preferable for accuracy to use $2\frac{1}{2}$-inch outlet nozzles rather than pumper outlet nozzles. Experience has shown that the $2\frac{1}{2}$-inch outlet nozzles are generally filled across the entire cross section during flow, while with the larger outlet nozzles there is frequently a void near the bottom. When measuring the pitot pressure of a stream of practically uniform velocity, the orifice in the pitot blade is held at the center of the stream and approximately 1 inch from the face of the hydrant outlet nozzle. The center line of the orifice should be at right angles to the plane of the face of the outlet nozzle. Where a stream is not uniform in velocity over the major portion of its cross section, the pitot orifice is moved across the cross section to obtain an average. However, in doing this, the area close to the edge should not be explored because the lower velocity there is considered in applying the coefficient c. For $2\frac{1}{2}$-inch outlets the center reading is all that is necessary unless the stream has a ragged appearance. For ragged streams a smaller coefficient should be applied or a discharge tube (Figure 20) used.

The number of outlet nozzles to flow depends upon the number of hydrants in the test and the strength of the system at the point of test. It is not necessary to fully open the valves on the hydrants flowed. In fact, this procedure is undesirable for most tests because the high discharges can cause an excessive drop in system pressure. A flow of 500 gpm is considered reasonable on a $2\frac{1}{2}$-inch outlet. By referring to the discharge tables, it is seen that for an outlet exactly $2\frac{1}{2}$ inches in diameter, a velocity pressure of 9 psi is required for a flow of 500 gpm.

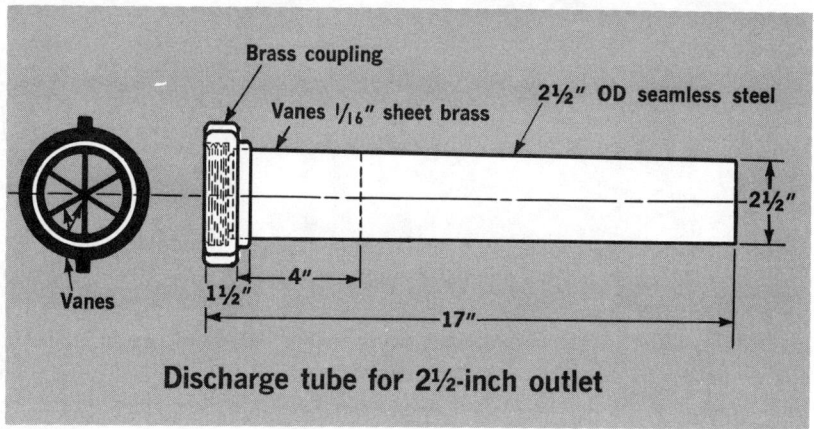

Discharge tube for 2½-inch outlet

Figure 20

With this discharge the pitot blade can be held in the stream without difficulty, and the gage can be read easily since there will be little vibration of the needle. If two outlet nozzles are flowed, 1000 gpm can be obtained. For good results it is desirable to limit the maximum flow from a given hydrant to about 1000 gpm in most tests. Pitot gage pressures in excess of 10 psi should be avoided wherever possible because such pressures cause considerable vibration of the needle and make reading difficult. Higher pressure and consequent higher flows can also damage roads and lawns.

Before flowing a hydrant, the caps on those outlet nozzles not being used should be checked to make certain that they are tight. The diameters of the outlet nozzles to be flowed should be measured to the nearest 16th of an inch. Gage readings should be taken to the nearest ¼ psi for readings up to 5 psi and to the nearest ½ psi for higher readings. Where a low reading is taken, the blade should be held horizontally because if the blade is long, a difference of ¼ psi will result from the gage being held 7 inches above or below the center of the stream.

If it is necessary to use a pumper outlet nozzle on a test, the best results can be obtained with a velocity pressure held between 5 and 10 psi. For pumper outlets the approximate discharge can be computed from Equation A using the velocity pressure at the center of the stream and multiplying the result by one of the following coefficients related to the velocity pressure. These coefficients are applied in

FIRE SERVICE HYDRAULICS

addition to the coefficient in Equation A and are for average type hydrants.

Velocity Pressure (psi)	Coefficients
2	0.97
3	0.92
4	0.89
5	0.86
6	0.84
7 and over	0.83

These coefficients are for average conditions and their application in some instances may lead to incorrect results. The best policy to follow is to avoid the use of the pumper outlet nozzle for measuring discharge. However, if pumper outlet nozzles must be used and there is reason to believe that the above coefficient will not give satisfactory results, the pitot tube should be moved over the cross section to obtain an average as previously described. In cases of this kind, however, particularly where the outlet nozzles have independent valves, the best results can be obtained by using discharge tubes.

If a pitot blade is not available for measuring the hydrant discharge, a 50 or 60-psi gage tapped into a hydrant cap may be used. The hydrant cap with gage attached is placed on one outlet nozzle and the flow allowed to take place through the other outlet nozzle at the same elevation. The readings obtained on a gage so located and on a pitot gage held in the stream are practically the same. The reasons for this are that with a freely discharging stream, practically all the pressure head in the hydrant barrel during flow is changed into velocity in the stream, and the velocity head in the hydrant barrel is so small in relation to that in the stream that it may be ignored.

An advantage in measuring hydrant discharge in this manner results from the fact that the gage needle will not vibrate as much as in the case of a pitot gage needle, and hence it is easier to obtain a more accurate reading. This method is also advantageous when trained observers are not available because a certain degree of skill is required to hold a pitot blade in a flowing stream and obtain a correct reading. However, since it is more convenient, because both nozzles can be used for discharge, and since the personnel making the flow tests are usually trained and experienced, the pitot blade is normally employed in preference to the fixed gage.

Calculation of results

The formula used to calculate the discharge available at the desired residual pressure can be developed from the Hazen-Williams equation

written in the form:
$$Q = ACR^{0.63}h^{0.54} \times 0.001^{-0.04}$$

For a given length of pipe L, if the loss of head is H, the loss of head per foot h will be $\frac{H}{L}$. Substituting,

$$Q = ACR^{0.63}\left(\frac{H}{L}\right)^{0.54} \times 0.001^{-0.04}$$

Since for any given pipe the value of the area A, the coefficient C, the hydraulic radius R, the length L, and the constant $0.001^{-0.04}$ do not change with a change in the discharge Q, it is evident that discharge Q is proportional to the 0.54 power of the loss of head H. The equation for this condition may be expressed as follows:

$$Q = KH^{0.54}$$

where K is a constant.

As previously explained, the distribution system between the source and test may be considered as a single equivalent pipe. For the flow during the test the equation takes the form

$$Q_F = KH_F^{0.54}$$

where Q_F is the discharge and H_F the loss of head which is equal to the pressure drop during the test (Figure 17). The flow Q_R at the desired residual pressure will be:

$$Q_R = KH_R^{0.54}$$

where H_R represents the drop from normal pressure to the desired residual pressure. Dividing equation Q_F by Q_R

$$\frac{Q_F}{Q_R} = \frac{KH_F^{0.54}}{KH_R^{0.54}} = \frac{H_F^{0.54}}{H_R^{0.54}}$$

Solving for Q_R

$$Q_R = Q_F \frac{H_R^{0.54}}{H_F^{0.54}} \qquad \text{(Equation B)}$$

In this equation any units of discharge or pressure drop may be used as long as the same units are used for each value of the same variable. In other words, if Q_R is expressed in gpm, Q_F must be in gpm, and if H_R is expressed in psi, H_F must be expressed in psi. These are the units which are normally used in applying this formula to fire flow test computations.

Referring to Figure 17, the pressure read on the gage before the hydrants were opened is the normal pressure N, and the pressure read

FIRE SERVICE HYDRAULICS

FORMULA AND TABLE FOR COMPUTING FIRE FLOW TEST RESULTS

$$Q_R = Q_F \times \frac{H_R^{0.54}}{H_F^{0.54}}$$

Q_R = Flow available at desired residual pressure.
Q_F = Flow during test.
H_R = Pressure drop to desired residual pressure.
H_F = Pressure drop during test.

VALUES OF "H" TO THE 0.54 POWER

H	$H^{0.54}$	H	$H^{0.54}$	H	$H^{0.54}$	H	$H^{0.54}$	H	$H^{0.54}$	H	$H^{0.54}$	H	$H^{0.54}$
1	1.00	26	5.81	51	8.36	76	10.37	101	12.09	126	13.62	151	15.02
2	1.45	27	5.93	52	8.44	77	10.44	102	12.15	127	13.68	152	15.07
3	1.81	28	6.05	53	8.53	78	10.51	103	12.22	128	13.74	153	15.13
4	2.11	29	6.16	54	8.62	79	10.59	104	12.28	129	13.80	154	15.18
5	2.39	30	6.28	55	8.71	80	10.66	105	12.34	130	13.85	155	15.23
6	2.63	31	6.39	56	8.79	81	10.73	106	12.41	131	13.91	156	15.29
7	2.86	32	6.50	57	8.88	82	10.80	107	12.47	132	13.97	157	15.34
8	3.07	33	6.61	58	8.96	83	10.87	108	12.53	133	14.02	158	15.39
9	3.28	34	6.71	59	9.04	84	10.94	109	12.60	134	14.08	159	15.44
10	3.47	35	6.82	60	9.12	85	11.01	110	12.66	135	14.14	160	15.50
11	3.65	36	6.93	61	9.21	86	11.08	111	12.72	136	14.19	161	15.55
12	3.83	37	7.03	62	9.29	87	11.15	112	12.78	137	14.25	162	15.60
13	4.00	38	7.13	63	9.37	88	11.22	113	12.84	138	14.31	163	15.65
14	4.16	39	7.23	64	9.45	89	11.29	114	12.90	139	14.36	164	15.70
15	4.32	40	7.33	65	9.53	90	11.36	115	12.96	140	14.42	165	15.76
16	4.47	41	7.43	66	9.61	91	11.43	116	13.03	141	14.47	166	15.81
17	4.62	42	7.53	67	9.69	92	11.49	117	13.09	142	14.53	167	15.86
18	4.76	43	7.62	68	9.76	93	11.56	118	13.15	143	14.58	168	15.91
19	4.90	44	7.72	69	9.84	94	11.63	119	13.21	144	14.64	169	15.96
20	5.04	45	7.81	70	9.92	95	11.69	120	13.27	145	14.69	170	16.01
21	5.18	46	7.91	71	9.99	96	11.76	121	13.33	146	14.75	171	16.06
22	5.31	47	8.00	72	10.07	97	11.83	122	13.39	147	14.80	172	16.11
23	5.44	48	8.09	73	10.14	98	11.89	123	13.44	148	14.86	173	16.16
24	5.56	49	8.18	74	10.22	99	11.96	124	13.50	149	14.91	174	16.21
25	5.69	50	8.27	75	10.29	100	12.02	125	13.56	150	14.97	175	16.26

METHOD OF USE: Insert in the formula the values of $H_R^{0.54}$ and $H_F^{0.54}$ determined from the table, and the value of Q_F, and solve the equation for Q_R.

with all hydrants flowing is the residual F during the test. The difference between these two pressures is the pressure drop H_F during the test and is represented by NF in Figure 17. The first step in computing the results is to find H_F by subtracting the residual pressure F from the normal pressure N.

The next step is to determine the flow Q_F during the test. The

WATER DISTRIBUTION SYSTEMS

velocity pressure observed at each flowing outlet nozzle and the diameter of the outlet nozzle are used to find the discharge by substituting in Equation A or referring to the table "Discharge for Circular Outlets" at the end of this chapter. The sum of the discharges from all the outlet nozzles flowed is the total flow Q_F for the test.

H_R, the drop from the normal pressure N to the specified residual pressure R is represented by NR in Figure 17. It is found by subtracting the specified residual pressure R from the pressure reading N taken before the hydrants were opened.

By substituting the values of H_F, Q_F, and H_R determined from the test data in Equation B, the flow Q_R at the specified residual pressure can be found. The solution of Equation B requires the use of logarithms, but the process can be simplified by the use of the special "Table for Computing Fire Flow Test Results." This table gives the values of the 0.54 power of numbers from 1 to 175, so that for any value of H from 1 to 175, the corresponding value of $H^{0.54}$ can be read from the table.

The procedure for calculating test results can be shown best by examples.

Example 1: The test layout is shown on the accompanying sketch. Four hydrants numbered 1, 2, 3, and 4 were flowed, and the residual pressure was observed at the hydrant at Raynor Avenue and Evelyn Street. The discharge from each outlet nozzle flowed is determined from "Discharge Table for Circular Outlets" (at end of chapter), using the diameter and the velocity pressure. The data obtained during the test and from the table follows:

Pressure before hydrants were flowed			82 psi
Pressure with four hydrants flowing			53 psi
Pressure drop during tests = H_F			29 psi

Hydrant No.	Diameter of Outlet Nozzle (In.)	Velocity Pressure (psi)	Discharge (gpm)
1	2½	9¼	510
	2½	9¼	510
2	2⅜	10½	490
3	2⅝	5¼	420
	2⅝	5¼	420
4	2½	7½	460
	2½	7½	460
		Total flow during test = Q_F =	3270

FIRE SERVICE HYDRAULICS

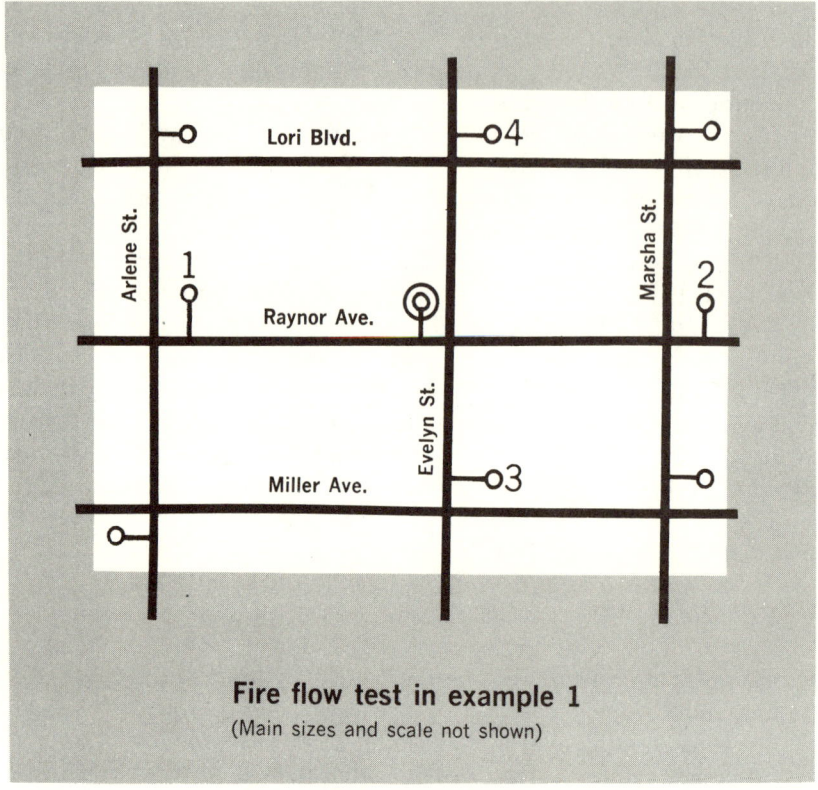

Fire flow test in example 1
(Main sizes and scale not shown)

Since it is desired to find the discharge Q_R that would be available with a residual pressure of 20 psi, $H_R = 82 - 20 = 62$ psi. Substituting in Equation B:

$$Q_R = Q_F \frac{H_R^{0.54}}{H_F^{0.54}} = 3270 \times \frac{62^{0.54}}{29^{0.54}}$$

From the "Table for Computing Fire Flow Test Results" $62^{0.54} = 9.29$ and $29^{0.54} = 6.16$. Substituting these values,

$$Q_R = 3270 \times \frac{9.29}{6.16} = 4930 \text{ gpm}$$

Results are usually rounded off to the nearest 1000 gpm for discharges of more than 100 gpm and to the nearest 50 gpm for smaller discharges. In this case, the result would be reported as 4900 gpm.

WATER DISTRIBUTION SYSTEMS

If it is desired to find out how much water could be delivered at a residual pressure of 75 psi for streams directly from hydrants, it is only necessary to substitute a new value of H_R, which would be $82 - 75 = 7$ psi, in the equation

$$Q_R = 3270 \times \frac{7^{0.54}}{29^{0.54}} = 3270 \times \frac{2.86}{6.16} = 1520 \text{ gpm}$$

This would be rounded off to 1500 gpm.

If it is desired to determine the discharge that would be available with a residual pressure of 15 psi on the top line of an automatic sprinkler system in a 14-story building, it is again necessary to make the appropriate change in the value of H_R. The height of a story is generally estimated at 10 feet for modern buildings and 12 feet for older buildings. Assuming that the building is a modern one, the residual pressure of 15 psi is required at an elevation of approximately (14×10) 140 feet above the ground. This difference in elevation between the point (top line of sprinklers) at which the residual pressure is required and the point (hydrant) at which the residual pressure is being observed must be taken into account in determining the new value of H_R. Therefore, the approximate difference in elevation of 140 feet is converted to $140 \times 0.434 = 61$ psi and added to the required residual pressure of 15 psi for the sprinklers. The required residual pressure at the hydrant thus equals $15 + 61 = 76$ psi. For this residual pressure H_R will be $82 - 76 = 6$ psi, and

$$Q_R = 3270 \times \frac{6^{0.54}}{29^{0.54}} = 3270 \times \frac{2.63}{6.16} = 1400 \text{ gpm}$$

Suppose it was desired to supply an automatic sprinkler system in a modern 16-story building at or near the test location. The difference in elevation to be considered would be (16×10) 160 feet which is equivalent to $160 \times 0.434 = 69$ psi. Adding this to the 15 psi residual pressure needed for sprinklers gives $69 + 15 = 84$ psi. Since the normal pressure at the site is 82 psi, there would not be sufficient pressure available from the water distribution system to provide a satisfactory supply for sprinklers.

If the building were 15 stories in height, the difference in elevation would be (15×10) 150 feet which is equivalent to $150 \times 0.434 = 65$ psi. This added to 15 psi gives $65 + 15 = 80$ psi so

FIRE SERVICE HYDRAULICS

that H_R would be $82 - 80 = 2$ psi. Substituting in the formula

$$Q_R = 3270 \times \frac{2^{0.54}}{29^{0.54}} = 3270 \times \frac{1.45}{6.16} = 770 \text{ gpm}$$

This would be rounded off to 750 gpm.

Although 750 gpm is indicated to be available at 15 psi on the top line of sprinklers, it should be noted that the normal pressure is only 2 psi higher than that required for satisfactory operation of the sprinklers. Since the pressure in a water distribution system varies with the rate of water use, it would be possible for the pressure to drop below normal under heavy demand, introducing the possibility of unsatisfactory sprinkler operation under such conditions. If the variations in distribution system pressure exceeded 6 psi, the supply for the sprinklers in the 14-story building could be similarly adversely affected.

Example 2: The test layout is shown on the accompanying sketch. Two hydrants numbered 1 and 2 were flowed, and the residual pressure was observed at Hiller Road and Karen Street. The test data is as follows:

Pressure before hydrants were flowed			54 psi
Pressure with two hydrants flowing			31 psi
Pressure drop during test = H_F =			23 psi

Hydrant No.	Diameter of Outlet Nozzle (in.)	Velocity Pressure (psi)	Discharge (gpm)
1	2¼	5	300
	2¼	5	300
2	2⁷⁄₁₆	3	280
	2⁷⁄₁₆	3	280
		Total flow during test = Q_F =	1160

In order to find the discharge Q_R at 20 psi, H_R will be $54 - 20 = 34$ psi. Substituting in equation (b),

$$Q_R = 1160 \times \frac{34^{0.54}}{23^{0.54}} = 1160 \times \frac{6.71}{5.44} = 1430 \text{ gpm}$$

This is rounded off to 1400 gpm.

Because the pressure with no flow from hydrants is only 54 psi, there is no discharge available at a pressure (75 psi) suitable for direct hydrant hose streams.

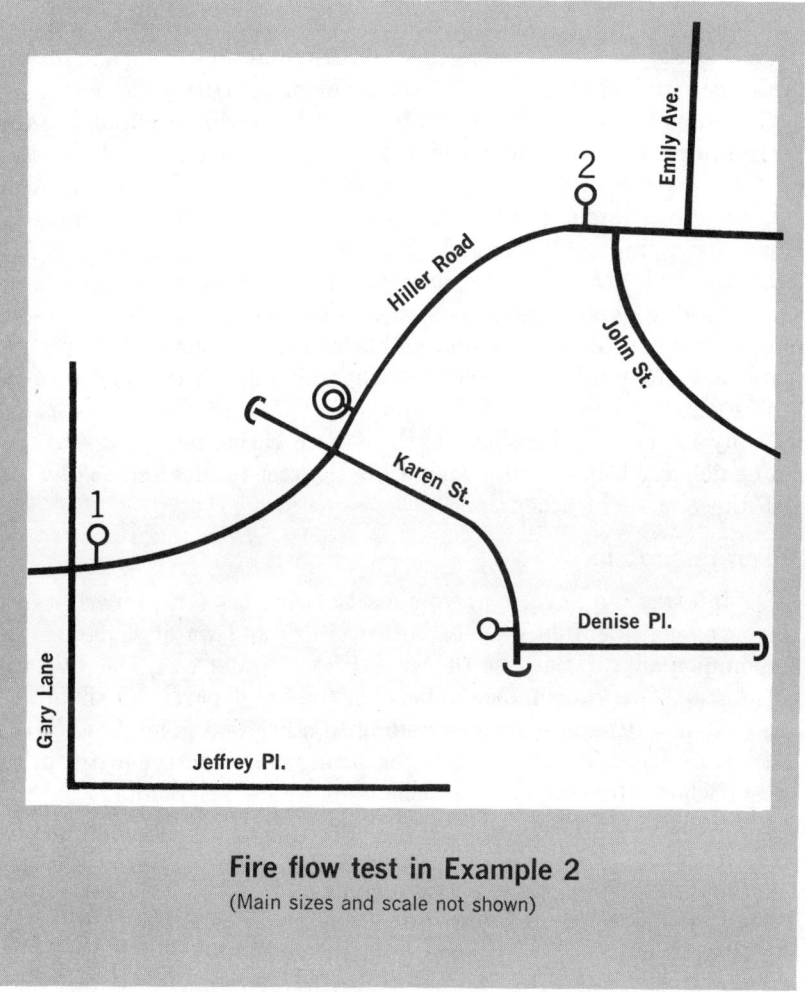

Fire flow test in Example 2
(Main sizes and scale not shown)

In order to determine the discharge available at 15 psi residual pressure for sprinklers in an old four-story building, the difference in elevation between the top line of sprinklers and the hydrant will be (4 × 12) 48 feet or the equivalent of 48 × 0.434 = 21 psi. The required residual pressure is 21 + 15 = 36 psi, so that $H_R = 54 - 36 = 18$ psi. Substituting in Equation B,

$$Q_R = 1160 \times \frac{18^{0.54}}{23^{0.54}} = 1160 \times \frac{4.76}{5.44} = 1000 \text{ gpm}$$

FIRE SERVICE HYDRAULICS

Reporting test results

In interpreting the results of the fire flow tests, it should be remembered that they show the strength of the distribution system and do not necessarily indicate the degree of adequacy of the entire water system. In this connection it is important to check the facilities supplying the distribution system and determine the length of time the discharges indicated by the tests can be maintained. In reporting test results, the conditions under which the tests were run should be indicated. This information should include the water consumption for the day and, if possible, the rate of consumption during the tests, the elevations of water levels in reservoirs and tanks before and after the tests, and the number, discharge rates, and discharge pressures of the pumps operating. Recording gage charts, where available, are good sources of this type of data. It will be found that this information is useful not only in determining the length of time the indicated discharges can be delivered, but also in comparing the test results and conditions with those made at other times.

Use of test results

The information developed from a set of flow tests, if properly used, can be of considerable help to both the fire and water departments of a municipality. Because the quantities of water available at various locations are known, chief officers of the fire department should be able to deploy the pumpers responding to a fire in a given area to the best advantage. They will know the strong points on the distribution system where the supply is sufficient for numerous pumpers as well as the weak spots where not more than one or two pumpers can be used effectively. To illustrate, in Example 1 above, where the test indicated that 4900 gpm was available at 20 psi, five 1000-gpm or six 750-gpm pumpers could be used at or near full capacity. With the six 750-gpm units, an additional pumper could also be operated but not at full capacity. In Example 2, with 1400 gpm available, only one 1000-gpm unit at full capacity and another pumper at partial capacity could be used; two 750-gpm pumpers operating at 700 gpm each could be used as an alternative.

Since test results will reveal the weak points in a water distribution system, they can be used by the water department to determine where and to what extent improvements are needed. They are also helpful in providing data which can be used in estimating the supply available for extensions into newly developing areas.

When tests are repeated after a number of years at the same location and under similar conditions, a comparison of the results may show decreased quantities available because of tuberculation of mains, increased consumption, or both. If such warnings are heeded, the necessary steps can be taken to provide additional distribution capacity before the available quantities become seriously inadequate. In a similar manner, repeated tests can be used to determine the increased quantities that may be available in areas where improvements have been made.

In reviewing flow tests results and comparing them with results obtained by other methods of analysis, it is often found that certain sections of a distribution system are not delivering their full capacity. This indicates that there are either restrictions somewhere in the mains or that one or more valves may be inadvertently closed. It points out the need for a physical check on the mains and valves in the vicinity of the test. Many closed valves in distribution systems have been located as a result of fire flow tests.

REFERENCE LIST

AWWA Standard for Fire Hydrants for Ordinary Water Works Service—AWWA C-502-64, American Water Works Association, New York.

AWWA Standard for Wet-Barrel Hydrants for Ordinary Water Works Service—AWWA C-503-70, New York.

Fire Flow Tests, Discharge Tables for Circular Outlets, Friction Losses in Pipes, American Insurance Association, New York, 1963 (1967 reprint).

Historical Development of the Fire Hydrant, L. A. Jackson, Journal of the American Water Works Association, Vol. 36, No. 9, September 1944.

Hydraulic Tables, Gardner S. Williams and Allen Hazen, John Wiley & Sons, New York, 3rd Edition, 1952.

FIRE SERVICE HYDRAULICS

DISCHARGE TABLE FOR CIRCULAR OUTLETS
Outlet pressure measured by pitot gage.

Outlet Pres. in lbs. per sq. in.	OUTLET DIAMETER IN INCHES							
	2¼	2⁵⁄₁₆	2⅜	2⁷⁄₁₆	2½	2⁹⁄₁₆	2⅝	2¹¹⁄₁₆
	Gallons per Minute							
¼	70	70	80	80	80	90	90	100
½	100	100	110	110	120	120	130	140
¾	120	120	130	140	150	150	160	170
1	140	140	150	160	170	180	180	190
¼	150	160	170	180	190	200	210	220
½	170	180	190	200	210	220	230	240
¾	180	190	200	210	220	230	240	260
2	190	200	210	230	240	250	260	270
¼	200	220	230	240	250	260	280	290
½	220	230	240	250	270	280	290	310
¾	230	240	250	260	280	290	310	320
3	240	250	260	280	290	310	320	340
¼	250	260	270	290	300	320	330	350
½	250	270	280	300	310	330	350	360
¾	260	280	290	310	330	340	360	380
4	270	290	300	320	340	350	370	390
¼	280	300	310	330	350	360	380	400
½	290	300	320	340	360	370	390	410
¾	300	310	330	350	370	380	400	420
5	300	320	340	360	380	390	410	430
¼	310	330	350	370	390	400	420	440
½	320	340	350	370	390	410	430	450
¾	330	340	360	380	400	420	440	460
6	330	350	370	390	410	430	450	470
¼	340	360	380	400	420	440	460	480
½	350	370	390	410	430	450	470	490
¾	350	370	390	410	440	460	480	500
7	360	380	400	420	440	470	490	510
¼	370	390	410	430	450	480	500	520
½	370	390	410	440	460	480	510	530
¾	380	400	420	440	470	490	510	540
8	380	410	430	450	480	500	520	550
¼	390	410	440	460	480	510	530	560
½	400	420	440	460	490	510	540	560
¾	400	420	450	470	500	520	550	570
9	410	430	450	480	500	530	550	580
¼	410	440	460	480	510	540	560	590
½	420	440	470	490	520	540	570	600
¾	420	450	470	500	520	550	580	600
10	430	450	480	500	530	560	580	610

WATER DISTRIBUTION SYSTEMS

DISCHARGE TABLE FOR CIRCULAR OUTLETS
Outlet pressure measured by pitot gage.

Outlet Pres. in lbs. per sq. in.	OUTLET DIAMETER IN INCHES							
	$2\tfrac{1}{4}$	$2\tfrac{5}{16}$	$2\tfrac{3}{8}$	$2\tfrac{7}{16}$	$2\tfrac{1}{2}$	$2\tfrac{9}{16}$	$2\tfrac{5}{8}$	$2\tfrac{11}{16}$
	Gallons per Minute							
$10\tfrac{1}{4}$	440	460	480	510	540	570	590	620
$\tfrac{1}{2}$	440	470	490	520	540	570	600	630
$\tfrac{3}{4}$	450	470	500	520	550	580	610	640
11	450	480	500	530	560	590	610	640
$\tfrac{1}{4}$	460	480	510	530	560	590	620	650
$\tfrac{1}{2}$	460	490	510	540	570	600	630	660
$\tfrac{3}{4}$	470	490	520	550	580	600	630	660
12	470	500	520	550	580	610	640	670
$\tfrac{1}{2}$	480	510	540	560	590	620	650	690
13	490	520	550	570	610	640	670	700
$\tfrac{1}{2}$	500	530	560	590	620	650	680	710
14	510	540	570	600	630	660	690	730
$\tfrac{1}{2}$	520	550	580	610	640	670	700	740
15	530	560	590	620	650	680	720	750
$\tfrac{1}{2}$	540	570	600	630	660	700	730	760
16	540	570	610	640	670	710	740	780
$\tfrac{1}{2}$	550	580	620	650	680	720	750	790
17	560	590	620	660	690	730	760	800
$\tfrac{1}{2}$	570	600	630	670	700	740	770	810
18	580	610	640	680	710	750	780	820
$\tfrac{1}{2}$	590	620	650	690	720	760	800	830
19	590	630	660	700	730	770	810	840
$\tfrac{1}{2}$	600	640	670	700	740	780	820	860
20	610	640	680	710	750	790	830	870
21	620	660	690	730	770	810	850	890
22	640	670	710	750	790	830	870	910
23	650	690	730	770	810	850	890	930
24	670	700	740	780	820	860	910	950
25	680	720	760	800	840	880	920	970
26	690	730	770	810	860	900	940	990
27	710	750	790	830	870	920	960	1010
28	720	760	800	840	890	930	980	1020
29	730	770	820	860	910	950	1000	1040
30	750	790	830	870	920	970	1010	1060
31	760	800	840	890	940	980	1030	1080
32	770	810	860	900	950	1000	1050	1100
33	780	830	870	920	970	1010	1060	1110
34	790	840	880	930	980	1030	1080	1130
35	810	850	900	940	990	1040	1090	1140
36	820	860	910	960	1010	1060	1110	1160

Coefficient .90

FIRE SERVICE HYDRAULICS

DISCHARGE TABLE FOR CIRCULAR OUTLETS
Outlet pressure measured by pitot gage.

Outlet Pres. in lbs. per sq. in.	OUTLET DIAMETER IN INCHES							
	2¾	2¹³⁄₁₆	2⅞	2¹⁵⁄₁₆	3	3¹⁄₁₆	3⅛	3³⁄₁₆
	Gallons per Minute							
¼	100	110	110	120	120	130	130	140
½	140	150	160	160	170	180	190	190
¾	180	180	190	200	210	220	230	240
1	200	210	220	230	240	250	260	270
¼	230	240	250	260	270	280	290	300
½	250	260	270	280	300	310	320	330
¾	270	280	290	310	320	330	350	360
2	290	300	310	330	340	360	370	390
¼	310	320	330	350	360	380	390	410
½	320	340	350	370	380	400	420	430
¾	340	350	370	380	400	420	440	450
3	350	370	380	400	420	440	450	470
¼	370	380	400	420	440	460	470	490
½	380	400	420	430	450	470	490	510
¾	390	410	430	450	470	490	510	530
4	410	430	440	460	480	500	530	550
¼	420	440	460	480	500	520	540	560
½	430	450	470	490	510	530	560	580
¾	440	460	480	510	530	550	570	600
5	460	480	500	520	540	560	590	610
¼	470	490	510	530	550	580	600	630
½	480	500	520	540	570	590	620	640
¾	490	510	530	560	580	610	630	650
6	500	520	540	570	590	620	640	670
¼	510	530	550	580	600	630	660	680
½	520	540	570	590	620	640	670	700
¾	530	550	580	600	630	660	680	710
7	540	560	590	610	640	670	690	720
¼	550	570	600	620	650	680	710	740
½	560	580	610	630	660	690	720	750
¾	570	590	620	650	670	700	730	760
8	570	600	630	660	680	710	740	770
¼	580	610	640	670	690	720	750	780
½	590	620	650	680	700	740	770	800
¾	600	630	660	690	710	750	780	810
9	610	640	670	700	720	760	790	820
¼	620	650	670	700	730	770	800	830
½	630	660	680	710	740	780	810	840
¾	630	660	690	720	750	790	820	850
10	640	670	700	730	760	800	830	860

DISCHARGE TABLE FOR CIRCULAR OUTLETS

Outlet pressure measured by pitot gage.

Outlet Pres. in lbs. per sq. in.	OUTLET DIAMETER IN INCHES							
	2¾	2 13/16	2⅞	2 15/16	3	3 1/16	3⅛	3 3/16
	Gallons per Minute							
10¼	650	680	710	740	770	810	840	870
½	660	690	720	750	780	820	850	880
¾	670	700	730	760	790	830	860	890
11	670	700	740	770	800	840	870	900
¼	680	710	740	780	810	850	880	910
½	690	720	750	790	820	860	890	930
¾	700	730	760	790	830	870	900	940
12	710	740	770	800	840	870	910	950
½	720	750	780	820	850	890	930	960
13	730	770	800	840	870	910	950	980
½	750	780	810	850	890	930	960	1000
14	760	800	830	870	900	940	980	1020
½	770	810	850	880	920	960	1000	1040
15	790	820	860	900	940	980	1020	1060
½	800	840	870	920	950	990	1030	1070
16	820	850	890	930	970	1010	1050	1090
½	830	860	900	940	980	1020	1060	1110
17	840	870	920	960	1000	1040	1080	1130
½	850	890	930	970	1010	1050	1100	1140
18	860	900	940	980	1020	1070	1110	1160
½	870	910	950	1000	1040	1080	1130	1170
19	890	930	970	1010	1050	1100	1140	1190
½	900	940	980	1020	1070	1110	1160	1210
20	910	950	990	1040	1080	1130	1170	1220
21	930	970	1020	1060	1110	1150	1200	1250
22	950	1000	1040	1090	1140	1180	1230	1280
23	980	1020	1060	1110	1160	1210	1260	1310
24	1000	1040	1090	1140	1190	1230	1280	1340
25	1020	1060	1110	1160	1210	1260	1310	1370
26	1040	1080	1130	1180	1230	1280	1340	1390
27	1060	1100	1150	1200	1260	1310	1360	1420
28	1070	1120	1170	1230	1280	1330	1390	1450
29	1090	1140	1190	1250	1300	1360	1410	1470
30	1110	1160	1210	1270	1320	1380	1440	1500
31	1130	1180	1230	1290	1350	1400	1460	1520
32	1150	1200	1250	1310	1370	1420	1480	1540
33	1170	1220	1270	1330	1390	1450	1510	1570
34	1180	1240	1290	1350	1410	1470	1530	1590
35	1200	1260	1310	1370	1430	1490	1550	1620
36	1220	1280	1330	1390	1450	1510	1570	1640

Coefficient .90

FIRE SERVICE HYDRAULICS

DISCHARGE TABLE FOR CIRCULAR OUTLETS
Outlet pressure measured by pitot gage.

Outlet Pres. in lbs. per sq. in.	OUTLET DIAMETER IN INCHES							
	$3\frac{1}{4}$	$3\frac{5}{16}$	$3\frac{3}{8}$	$3\frac{7}{16}$	$3\frac{1}{2}$	$3\frac{9}{16}$	$3\frac{5}{8}$	$3\frac{11}{16}$
	Gallons per Minute							
$\frac{1}{4}$	140	150	150	160	160	170	180	180
$\frac{1}{2}$	200	210	220	220	230	240	250	260
$\frac{3}{4}$	250	250	260	270	280	290	310	320
1	280	290	310	320	330	340	350	360
$\frac{1}{4}$	320	330	340	350	370	380	390	410
$\frac{1}{2}$	350	360	370	390	400	420	430	450
$\frac{3}{4}$	380	390	400	420	430	450	470	480
2	400	420	430	450	460	480	500	520
$\frac{1}{4}$	430	440	460	480	490	510	530	550
$\frac{1}{2}$	450	470	480	500	520	540	560	580
$\frac{3}{4}$	470	490	510	530	550	560	580	610
3	490	510	530	550	570	590	610	630
$\frac{1}{4}$	510	530	550	570	590	610	640	660
$\frac{1}{2}$	530	550	570	590	620	640	660	680
$\frac{3}{4}$	550	570	590	610	640	660	680	710
4	570	590	610	630	660	680	710	730
$\frac{1}{4}$	590	610	630	650	680	700	730	750
$\frac{1}{2}$	600	630	650	670	700	720	750	770
$\frac{3}{4}$	620	640	670	690	720	740	770	800
5	630	660	680	710	730	760	790	820
$\frac{1}{4}$	650	680	700	730	750	780	810	840
$\frac{1}{2}$	670	690	720	740	770	800	830	860
$\frac{3}{4}$	680	710	730	760	790	820	850	880
6	700	720	750	780	810	830	870	890
$\frac{1}{4}$	710	740	760	790	820	850	880	910
$\frac{1}{2}$	720	750	780	810	840	870	900	930
$\frac{3}{4}$	740	770	800	820	850	890	920	950
7	750	780	810	840	870	900	930	970
$\frac{1}{4}$	760	790	820	850	890	920	950	980
$\frac{1}{2}$	780	810	840	870	900	930	970	1000
$\frac{3}{4}$	790	820	850	880	920	950	980	1020
8	800	830	860	900	930	960	1000	1030
$\frac{1}{4}$	820	850	880	910	940	980	1010	1050
$\frac{1}{2}$	830	860	890	920	960	990	1030	1060
$\frac{3}{4}$	840	870	900	940	970	1010	1040	1080
9	850	880	920	950	990	1020	1060	1100
$\frac{1}{4}$	860	900	930	960	1000	1040	1070	1110
$\frac{1}{2}$	870	910	940	980	1010	1050	1090	1120
$\frac{3}{4}$	890	920	960	990	1030	1060	1100	1140
10	900	930	970	1000	1040	1080	1120	1150

WATER DISTRIBUTION SYSTEMS

DISCHARGE TABLE FOR CIRCULAR OUTLETS
Outlet pressure measured by pitot gage.

Outlet Pres. in lbs. per sq. in.	OUTLET DIAMETER IN INCHES							
	3¼	3⁵⁄₁₆	3⅜	3⁷⁄₁₆	3½	3⁹⁄₁₆	3⅝	3¹¹⁄₁₆
	Gallons per Minute							
10¼	910	940	980	1020	1050	1090	1130	1170
½	920	960	990	1030	1070	1100	1140	1180
¾	930	970	1000	1040	1080	1120	1160	1200
11	940	980	1010	1050	1090	1130	1170	1210
¼	950	990	1030	1060	1100	1140	1180	1220
½	960	1000	1040	1070	1110	1150	1200	1240
¾	970	1010	1050	1090	1130	1170	1210	1250
12	980	1020	1060	1100	1140	1180	1220	1260
½	1000	1040	1080	1120	1160	1210	1250	1290
13	1020	1060	1100	1140	1190	1230	1270	1320
½	1040	1080	1120	1160	1210	1250	1300	1340
14	1060	1100	1140	1190	1230	1280	1320	1370
½	1080	1120	1160	1210	1250	1300	1340	1390
15	1100	1140	1180	1230	1270	1320	1370	1420
½	1120	1160	1200	1250	1300	1340	1390	1440
16	1140	1180	1220	1270	1320	1360	1410	1460
½	1150	1200	1240	1290	1340	1380	1430	1480
17	1170	1220	1260	1310	1360	1400	1460	1510
½	1190	1230	1280	1330	1380	1430	1480	1530
18	1200	1250	1300	1350	1400	1450	1500	1550
½	1220	1270	1320	1360	1410	1470	1520	1570
19	1240	1280	1330	1380	1430	1490	1540	1590
½	1260	1300	1350	1400	1450	1510	1560	1610
20	1270	1320	1370	1420	1470	1530	1580	1630
21	1300	1350	1400	1450	1510	1560	1620	1670
22	1330	1380	1430	1490	1540	1600	1660	1710
23	1360	1410	1470	1520	1580	1630	1690	1750
24	1390	1440	1500	1550	1610	1670	1730	1790
25	1420	1470	1530	1590	1640	1700	1760	1820
26	1450	1500	1560	1620	1680	1740	1800	1860
27	1480	1530	1590	1650	1710	1770	1830	1900
28	1500	1560	1620	1680	1740	1800	1870	1930
29	1530	1590	1650	1710	1770	1840	1900	1970
30	1550	1610	1670	1740	1800	1870	1930	2000
31	1580	1640	1700	1760	1830	1900	1970	2030
32	1610	1670	1730	1790	1860	1930	2000	2060
33	1630	1690	1760	1820	1890	1960	2030	2100
34	1660	1720	1780	1850	1920	1990	2060	2130
35	1680	1740	1810	1870	1950	2020	2090	2160
36	1700	1770	1840	1900	1970	2050	2120	2190

Coefficient .90

DISCHARGE TABLE FOR CIRCULAR OUTLETS
Outlet pressure measured by pitot gage.

Outlet Pres. in lbs. per sq. in.	OUTLET DIAMETER IN INCHES							
	$3\frac{3}{4}$	$3\frac{13}{16}$	$3\frac{7}{8}$	$3\frac{15}{16}$	4	$4\frac{1}{16}$	$4\frac{1}{8}$	$4\frac{3}{16}$
	Gallons per Minute							
$\frac{1}{4}$	190	200	200	210	220	220	230	240
$\frac{1}{2}$	270	280	290	290	300	310	320	330
$\frac{3}{4}$	330	340	350	360	370	380	400	410
1	380	390	400	420	430	440	460	470
$\frac{1}{4}$	420	440	450	470	480	500	510	530
$\frac{1}{2}$	460	480	490	510	530	540	560	580
$\frac{3}{4}$	500	520	530	550	570	590	610	620
2	530	550	570	590	610	630	650	670
$\frac{1}{4}$	570	590	610	630	650	670	690	710
$\frac{1}{2}$	600	620	640	660	680	700	720	750
$\frac{3}{4}$	630	650	670	690	710	740	760	780
3	660	680	700	720	750	770	790	820
$\frac{1}{4}$	680	710	730	750	780	800	830	850
$\frac{1}{2}$	710	730	760	780	810	830	860	880
$\frac{3}{4}$	730	760	780	810	830	860	890	910
4	760	780	810	830	860	890	920	940
$\frac{1}{4}$	780	810	830	860	890	920	940	970
$\frac{1}{2}$	800	830	860	890	910	940	970	1000
$\frac{3}{4}$	830	850	880	910	940	970	1000	1030
5	850	880	900	930	960	990	1020	1050
$\frac{1}{4}$	870	900	930	960	990	1020	1050	1080
$\frac{1}{2}$	890	920	950	980	1010	1040	1070	1110
$\frac{3}{4}$	910	940	970	1000	1030	1060	1100	1130
6	930	960	990	1020	1050	1090	1120	1150
$\frac{1}{4}$	950	980	1010	1040	1080	1110	1140	1180
$\frac{1}{2}$	960	1000	1030	1060	1100	1130	1170	1200
$\frac{3}{4}$	980	1020	1050	1080	1120	1150	1190	1230
7	1000	1030	1070	1100	1140	1170	1210	1250
$\frac{1}{4}$	1020	1050	1090	1120	1160	1190	1230	1270
$\frac{1}{2}$	1040	1070	1110	1140	1180	1210	1250	1290
$\frac{3}{4}$	1050	1090	1120	1160	1200	1230	1270	1310
8	1070	1110	1140	1180	1210	1250	1290	1340
$\frac{1}{4}$	1090	1120	1160	1200	1230	1270	1310	1360
$\frac{1}{2}$	1100	1140	1180	1220	1250	1290	1330	1380
$\frac{3}{4}$	1120	1160	1190	1230	1270	1310	1350	1400
9	1130	1170	1210	1250	1290	1330	1370	1420
$\frac{1}{4}$	1150	1190	1230	1270	1310	1350	1390	1440
$\frac{1}{2}$	1170	1210	1240	1280	1320	1370	1410	1460
$\frac{3}{4}$	1180	1220	1260	1300	1340	1380	1430	1470
10	1200	1240	1280	1320	1360	1400	1450	1490

WATER DISTRIBUTION SYSTEMS

DISCHARGE TABLE FOR CIRCULAR OUTLETS
Outlet pressure measured by pitot gage.

Outlet Pres. in lbs. per sq. in.	OUTLET DIAMETER IN INCHES							
	3¾	3¹³⁄₁₆	3⅞	3¹⁵⁄₁₆	4	4¹⁄₁₆	4⅛	4³⁄₁₆
	Gallons per Minute							
10¼	1210	1250	1290	1330	1380	1420	1460	1510
½	1230	1270	1310	1350	1390	1440	1480	1530
¾	1240	1280	1320	1370	1410	1450	1500	1550
11	1250	1300	1340	1380	1430	1470	1520	1570
¼	1270	1310	1350	1400	1440	1490	1530	1580
½	1280	1330	1370	1410	1460	1500	1550	1600
¾	1300	1340	1380	1430	1470	1520	1570	1620
12	1310	1350	1400	1440	1490	1540	1580	1640
½	1340	1380	1430	1470	1520	1570	1620	1670
13	1360	1410	1460	1500	1550	1600	1650	1700
½	1390	1440	1480	1530	1580	1630	1680	1730
14	1420	1460	1510	1560	1610	1660	1710	1770
½	1440	1490	1540	1590	1640	1690	1740	1800
15	1470	1510	1560	1610	1660	1720	1770	1830
½	1490	1540	1590	1640	1690	1750	1800	1860
16	1510	1560	1610	1670	1720	1770	1830	1890
½	1540	1590	1640	1690	1750	1800	1860	1920
17	1560	1610	1660	1720	1770	1830	1890	1950
½	1580	1640	1690	1740	1800	1850	1910	1980
18	1600	1660	1710	1770	1820	1880	1940	2000
½	1630	1680	1740	1790	1850	1910	1970	2030
19	1650	1700	1760	1820	1870	1930	1990	2060
½	1670	1730	1780	1840	1900	1960	2020	2090
20	1690	1750	1810	1860	1920	1980	2050	2110
21	1730	1790	1850	1910	1970	2030	2100	2160
22	1770	1830	1890	1960	2020	2080	2150	2220
23	1810	1870	1940	2000	2060	2120	2190	2270
24	1850	1910	1980	2040	2110	2170	2240	2310
25	1890	1950	2020	2080	2150	2220	2280	2360
26	1930	1990	2060	2130	2190	2260	2330	2410
27	1970	2030	2100	2170	2230	2300	2380	2450
28	2000	2070	2140	2210	2280	2350	2420	2500
29	2040	2110	2170	2240	2320	2390	2460	2540
30	2070	2140	2210	2280	2350	2430	2510	2580
31	2110	2180	2250	2320	2390	2470	2550	2620
32	2140	2210	2280	2360	2430	2500	2590	2670
33	2170	2250	2320	2390	2470	2550	2630	2710
34	2210	2280	2350	2430	2510	2580	2670	2750
35	2240	2310	2390	2460	2540	2620	2710	2790
36	2270	2340	2420	2500	2580	2660	2740	2830

Coefficient .90

FIRE SERVICE HYDRAULICS

DISCHARGE TABLE FOR CIRCULAR OUTLETS
Outlet pressure measured by pitot gage.

Outlet Pres. in lbs. per sq. in.	OUTLET DIAMETER IN INCHES							
	4¼	4⁵⁄₁₆	4⅜	4⁷⁄₁₆	4½	4⁹⁄₁₆	4⅝	4¹¹⁄₁₆
	Gallons per Minute							
¼	240	250	260	260	270	280	290	300
½	340	350	360	370	390	400	410	420
¾	420	430	450	460	470	490	500	510
1	490	500	520	530	550	560	570	590
¼	540	560	590	590	610	630	640	660
½	600	610	630	650	670	690	700	720
¾	640	660	680	700	720	740	760	780
2	690	710	730	750	770	790	810	840
¼	730	750	770	800	820	840	860	890
½	770	790	810	840	860	890	910	940
¾	810	830	850	880	900	930	950	980
3	840	870	890	920	940	970	1000	1020
¼	880	900	930	960	980	1010	1040	1060
½	910	940	970	990	1020	1050	1070	1100
¾	940	970	1000	1030	1050	1080	1110	1140
4	970	1000	1030	1060	1090	1120	1150	1180
¼	1000	1030	1060	1090	1120	1150	1180	1220
½	1030	1060	1090	1120	1160	1190	1220	1250
¾	1060	1090	1120	1150	1190	1220	1250	1290
5	1090	1120	1150	1180	1220	1250	1280	1320
¼	1110	1150	1180	1210	1250	1280	1320	1350
½	1140	1180	1210	1240	1280	1310	1350	1390
¾	1170	1200	1240	1270	1310	1340	1380	1420
6	1190	1230	1260	1300	1330	1370	1410	1450
¼	1220	1250	1290	1320	1360	1400	1440	1480
½	1240	1280	1310	1350	1390	1430	1470	1510
¾	1260	1300	1340	1380	1420	1450	1490	1540
7	1290	1330	1360	1400	1440	1480	1520	1560
¼	1310	1350	1390	1430	1470	1510	1550	1590
½	1330	1370	1410	1450	1490	1530	1570	1620
¾	1350	1390	1430	1480	1520	1560	1600	1640
8	1380	1420	1460	1500	1540	1580	1620	1670
¼	1400	1440	1480	1520	1570	1610	1650	1700
½	1420	1460	1500	1540	1590	1630	1680	1720
¾	1440	1480	1520	1570	1610	1650	1700	1750
9	1460	1500	1540	1590	1630	1680	1720	1770
¼	1480	1520	1570	1610	1660	1700	1750	1800
½	1500	1540	1590	1630	1680	1720	1770	1820
¾	1520	1560	1610	1650	1700	1750	1790	1840
10	1540	1580	1630	1670	1720	1770	1820	1870

WATER DISTRIBUTION SYSTEMS

DISCHARGE TABLE FOR CIRCULAR OUTLETS
Outlet pressure measured by pitot gage.

Outlet Pres. in lbs. per sq. in.	OUTLET DIAMETER IN INCHES							
	4¼	4⁵⁄₁₆	4⅜	4⁷⁄₁₆	4½	4⁹⁄₁₆	4⅝	4¹¹⁄₁₆
	Gallons per Minute							
10¼	1560	1600	1650	1700	1740	1790	1840	1890
½	1580	1620	1670	1720	1760	1810	1860	1910
¾	1590	1640	1690	1740	1790	1830	1880	1940
11	1610	1660	1710	1760	1810	1860	1910	1960
¼	1630	1680	1730	1780	1830	1880	1930	1980
½	1650	1700	1750	1800	1850	1900	1950	2000
¾	1670	1720	1760	1820	1870	1920	1970	2020
12	1690	1730	1780	1840	1890	1940	1990	2050
½	1720	1770	1820	1870	1930	1980	2030	2090
13	1750	1800	1850	1910	1970	2020	2070	2130
½	1790	1840	1890	1950	2000	2060	2110	2170
14	1820	1870	1930	1980	2040	2090	2150	2210
½	1850	1910	1960	2020	2080	2130	2190	2250
15	1880	1940	1990	2050	2110	2170	2230	2290
½	1910	1970	2030	2090	2150	2200	2260	2330
16	1940	2000	2060	2120	2180	2240	2300	2360
½	1970	2030	2090	2150	2210	2270	2330	2400
17	2000	2060	2120	2180	2250	2310	2370	2440
½	2030	2090	2150	2220	2280	2340	2400	2470
18	2060	2120	2180	2250	2310	2370	2440	2510
½	2090	2150	2210	2280	2350	2410	2470	2540
19	2120	2180	2240	2310	2380	2440	2510	2580
½	2140	2210	2270	2340	2410	2470	2540	2610
20	2170	2240	2300	2370	2440	2500	2570	2640
21	2220	2290	2360	2430	2500	2560	2630	2710
22	2280	2350	2420	2490	2560	2620	2700	2770
23	2330	2400	2470	2540	2610	2680	2760	2830
24	2380	2450	2520	2600	2670	2740	2820	2890
25	2430	2500	2580	2650	2720	2800	2870	2950
26	2480	2550	2630	2700	2780	2850	2930	3010
27	2530	2600	2680	2750	2830	2910	2990	3070
28	2580	2650	2730	2800	2880	2960	3040	3130
29	2620	2700	2770	2850	2940	3020	3090	3180
30	2670	2740	2820	2900	2990	3070	3150	3240
31	2710	2790	2870	2950	3030	3120	3200	3290
32	2750	2830	2929	3000	3080	3170	3250	3340
33	2790	2880	2960	3040	3130	3220	3300	3390
34	2830	2920	3000	3090	3170	3260	3350	3440
35	2870	2960	3040	3140	3220	3310	3400	3490
36	2910	3000	3080	3180	3270	3360	3450	3540

Coefficient .90

DISCHARGE TABLE FOR CIRCULAR OUTLETS
Outlet pressure measured by pitot gage.

Outlet Pres. in lbs. per sq. in.	OUTLET DIAMETER IN INCHES							
	4¾	4¹³⁄₁₆	4⅞	4¹⁵⁄₁₆	5	5¹⁄₁₆	5⅛	5³⁄₁₆
	Gallons per Minute							
¼	300	310	320	330	340	340	350	360
½	430	440	450	460	470	490	500	510
¾	520	540	550	570	580	590	610	630
1	610	620	640	650	670	690	700	720
¼	680	690	710	730	750	770	790	810
½	740	760	780	800	820	840	860	890
¾	800	820	840	870	890	910	930	960
2	860	880	900	930	950	970	1000	1020
¼	910	930	960	980	1010	1030	1060	1080
½	960	980	1010	1030	1060	1090	1110	1140
¾	1000	1030	1060	1080	1110	1140	1170	1200
3	1050	1080	1100	1130	1160	1190	1220	1250
¼	1090	1120	1150	1180	1210	1240	1270	1300
½	1130	1160	1190	1220	1250	1290	1320	1350
¾	1170	1200	1230	1270	1300	1330	1360	1400
4	1210	1240	1270	1310	1340	1380	1410	1450
¼	1250	1280	1310	1350	1380	1420	1450	1490
½	1290	1320	1350	1390	1420	1460	1500	1530
¾	1320	1360	1390	1430	1460	1500	1540	1580
5	1360	1390	1430	1460	1500	1540	1580	1620
¼	1390	1430	1460	1500	1540	1570	1610	1660
½	1420	1460	1500	1540	1570	1610	1650	1700
¾	1450	1490	1530	1570	1610	1650	1690	1730
6	1480	1520	1560	1600	1650	1680	1730	1770
¼	1510	1560	1590	1640	1680	1720	1760	1810
½	1540	1590	1620	1670	1710	1750	1800	1840
¾	1570	1620	1650	1700	1740	1790	1830	1880
7	1600	1650	1680	1730	1780	1820	1860	1910
¼	1630	1670	1710	1760	1810	1850	1900	1950
½	1660	1700	1740	1790	1840	1880	1930	1980
¾	1690	1730	1770	1820	1870	1910	1960	2010
8	1710	1760	1800	1850	1900	1940	1990	2040
¼	1740	1790	1830	1880	1930	1970	2020	2080
½	1770	1810	1860	1910	1960	2000	2050	2110
¾	1790	1840	1880	1940	1990	2030	2080	2140
9	1820	1870	1910	1960	2010	2060	2110	2170
¼	1840	1890	1940	1990	2040	2090	2140	2200
½	1870	1920	1960	2020	2070	2120	2170	2230
¾	1890	1940	1990	2040	2090	2150	2200	2260
10	1920	1970	2020	2070	2120	2170	2230	2290

DISCHARGE TABLE FOR CIRCULAR OUTLETS
Outlet pressure measured by pitot gage.

Outlet Pres. in lbs. per sq. in.	OUTLET DIAMETER IN INCHES							
	4¾	4¹³⁄₁₆	4⅞	4¹⁵⁄₁₆	5	5¹⁄₁₆	5⅛	5³⁄₁₆
	Gallons per Minute							
10¼	1940	1990	2040	2100	2150	2200	2260	2310
½	1960	2020	2060	2120	2170	2230	2280	2340
¾	1990	2040	2090	2150	2200	2250	2310	2370
11	2010	2060	2110	2170	2230	2280	2340	2400
¼	2030	2090	2140	2200	2250	2310	2360	2420
½	2060	2110	2160	2220	2270	2330	2390	2450
¾	2080	2130	2180	2250	2300	2360	2420	2480
12	2100	2160	2210	2270	2330	2380	2440	2510
½	2140	2200	2250	2320	2370	2430	2490	2560
13	2180	2240	2300	2360	2420	2480	2540	2610
½	2220	2290	2340	2410	2460	2530	2590	2660
14	2270	2330	2380	2450	2510	2570	2640	2710
½	2310	2370	2420	2500	2550	2620	2680	2760
15	2350	2410	2460	2540	2600	2660	2730	2800
½	2390	2450	2510	2580	2640	2710	2780	2850
16	2420	2490	2550	2620	2680	2750	2820	2890
½	2460	2530	2590	2660	2720	2790	2860	2940
17	2500	2560	2630	2700	2770	2840	2910	2980
½	2530	2600	2660	2740	2810	2880	2950	3030
18	2570	2640	2700	2780	2850	2920	2990	3070
½	2610	2680	2740	2820	2890	2960	3030	3110
19	2640	2710	2780	2860	2930	3000	3070	3150
½	2680	2750	2810	2890	2960	3040	3110	3190
20	2710	2790	2850	2930	3000	3080	3150	3230
21	2780	2850	2920	3000	3070	3150	3230	3310
22	2840	2920	2990	3070	3150	3230	3300	3390
23	2910	2980	3060	3140	3220	3300	3380	3470
24	2970	3050	3120	3210	3280	3370	3450	3540
25	3030	3110	3190	3270	3350	3440	3520	3620
26	3090	3170	3250	3340	3420	3510	3590	3690
27	3150	3230	3310	3400	3490	3570	3660	3760
28	3210	3290	3370	3460	3550	3640	3730	3830
29	3260	3350	3430	3530	3610	3710	3800	3900
30	3320	3410	3490	3590	3680	3770	3860	3960
31	3380	3460	3540	3650	3740	3830	3930	4030
32	3430	3520	3600	3700	3800	3890	3990	4090
33	3480	3570	3660	3760	3860	3950	4050	4150
34	3530	3630	3720	3820	3910	4010	4110	4220
35	3580	3680	3770	3880	3970	4070	4170	4280
36	3460	3730	3820	3930	4030	4130	4230	4340

Coefficient .90

CHAPTER FOUR

Fire service pumps

Pumps discharging water for the control and extinguishment of fire, as used by the fire service, can be grouped in seven classifications, the pumps in each class being designed for a specific use. The classes are: (1) fire pumps (mobile); (2) booster pumps; (3) high-pressure pumps; (4) portable pumps; (5) back-pack pumps; (6) tanker pumps; (7) municipal and industrial fire pumps (stationary).

Mobile fire pumps are mounted on a vehicle, self-propelled or on a trailer, to supply the volume and pressure of water required at the scene of a fire, or to act as a relay pumping station. Fire pumps are provided in several standard capacities, and are rated according to discharge capacity when operating at draft. The standard capacity ratings are as follows: 500 gpm; 750 gpm; 1,000 gpm; 1,250 gpm; 1,500 gpm; 1,750 and 2,000 gpm. These ratings are based on the ability of the pump to discharge its rated capacity, operating at draft with a vertical lift of not over 10 feet, as a net discharge pressure of 150 psi. Also, the pump must be capable of discharging 70 percent of the rated capacity at a net pressure of 200 psi, and 50 per cent of the rated capacity at a net pressure of 250 psi. In addition, a spurt test of not less than 10 minutes duration is required to prove the capability to discharge the rated capacity at 165 psi. This last requirement is to insure power reserve beyond that required to produce pump rated capacity.

FIRE SERVICE PUMPS

All new fire pumps undergo a 3-hour test, conducted at the factory of the fire apparatus manufacturer, prior to delivery to the fire department. Tests are conducted by engineers of Underwriters Laboratories, Inc. in the United States, and Canadian Underwriters' Association in Canada. The tests certify to the fire department that pump performance is in agreement with the requirements specified in NFPA (National Fire Protection Association) Specification No. 19, which may be supplemented by special pump performance requirements stated in the fire department purchase specification. The certification test does not include testing at a pump discharge pressure in excess of 250 psi.

The majority of fire pumps are mounted midship, that is, approximately midway between the front and rear axles. Two methods for driving the midship-mounted pump are employed. The most common method is mounting the pump drive transmission in the line of drive between the road transmission and the rear axle. The second method employs a flywheel-driven, full torque power take-off, that transmits engine power from ahead of the vehicle clutch and transmission, to the pump. This method of pump drive permits simultaneous pumping and vehicle movement, if desired. This design of drive serves well for grass and brush fire fighting, also for aircraft crash fire fighting, when the water supply is carried on the vehicle.

Front-mounting of the pump, particularly on commercial truck chassis is preferred in some areas to meet specific conditions. This mounting also permits vehicle movement simultaneous with operation of the pump. The fire pump is driven from the front end of the engine crankshaft, through a clutch and pump transmission. With this type of pump mounting, the maximum rated capacity available is 1,000 gpm, with the 750-gpm rating most popular.

Fire pumps, with only a few exceptions, are centrifugal type, with the single-stage and parallel-series 2-stage most common. There are other designs for specific uses: 2-stage; parallel-series 3-stage; and parallel-series 4-stage. Also a "Duplex" design which operates as a single-stage pump for rated capacity, a second smaller impeller operating as a single stage provides intermediate pressures; the two impellers operating in series provides for pressures of 250 psi, and higher.

The piston and rotary-type fire pumps once so popular, have been on restricted production since World War II due to the popularity of the centrifugal fire pump. On occasion a piston fire pump (Howe) or a rotary fire pump (Hale) will be purchased for use where all pumping operation is at draft, usually under high lift conditions such as deep wells.

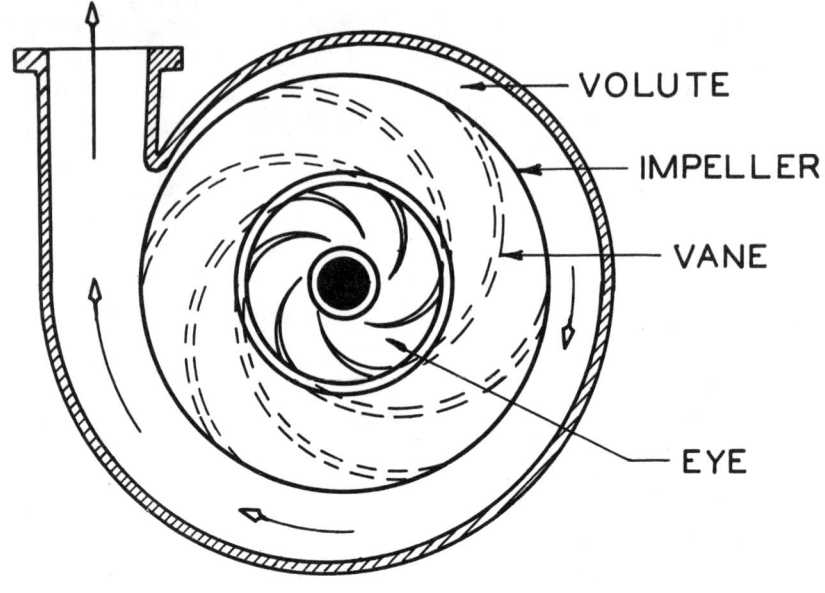

Figure 1

CENTRIFUGAL FIRE PUMP

The centrifugal fire pump consists essentially of one or more impellers mounted on a shaft (in a housing with one or more suction inlets), a volute-shaped discharge chamber for each impeller and two or more discharge outlets. A pump transmission is employed to transmit engine power to the pump and provide for increased impeller speed in relation to the engine crankshaft speed. The basic parts of the centrifugal fire pump are identified in Figure 1.

The centrifugal fire pump as designed for mounting on fire apparatus, is not self-priming. A priming device must be provided to exhaust air from the pump and suction line to permit operation at draft. (See section on priming devices for operating requirements.) When operating the fire pump from a source of water supply under positive pressure, a priming system is not used. The residual positive pressure at the pump suction inlet should never be less than 10 psi. If the pump inlet pressure drops below this value (10 psi), the discharge volume should be reduced to avoid damage to the water supply system. When operating the pump in relay, the residual pressure at the suction inlet of the second pump and others in succession, is usually maintained at 20 psi to prevent a possible collapse of the supply hose.

FIRE SERVICE PUMPS

The centrifugal pump received its name from its dependence on centrifugal action in the rotating impeller for the discharge of a liquid. In operation, water enters the eye of the impeller in an axial direction (parallel with the axis of impeller rotation), and is picked up by the vanes in the rapidly rotating impeller. Centrifugal force from the speed of impeller rotation, imparts energy and pressure to the water which is discharged into the volute housing surrounding the impeller. The water, discharged from the impeller at high velocity, enters the volute housing, which is designed to convert the kinetic energy of velocity to useful pressure energy with a minimum of shock and eddy losses. The sectional area of the volute increases, from a point adjacent to the outlet around the impeller to the discharge outlet, to provide a uniform or gradual reduction in flow velocity. Most centrifugal fire pumps convert more than 70 percent of the velocity energy to pressure energy, which added to the pressure energy as discharged by the impeller gives a high percentage of recovery.

As water enters the pump and the eye of the impeller, and is picked up by the vanes in the rotating impeller and discharged outward due to centrifugal force, a vacuum is created at the eye of the impeller. This vacuum causes water to continue to flow into the impeller eye replacing water being discharged. This action is continuous for the period of pump operation, provided:

1. No air enters the suction inlet of the pump.
2. The supply of water is adequate to supply required discharge volume (gpm).
3. The height of lift is not too great for the volume of water being discharged.
4. The suction or supply mains are not too small to supply the desired volume.
5. There is no obstruction at the entrance to the suction strainer or no collapse of the lining of the suction hose.

Pump horsepower

The rate of work done by a pump in a specified time is measured in terms of horsepower. The work equivalent of 1 horsepower is required to move a weight of 33,000 pounds, a distance of 1 foot, in a time of 1 minute. It is expressed by the equation:

$$\text{Horsepower} = \frac{\text{pounds} \times \text{feet}}{33,000}$$

The work done by a pump is usually designated as WHP (water

FIRE SERVICE HYDRAULICS

horsepower) to distinguish it from the more familiar BHP (brake horsepower) applied to the work done by an engine.

Since horsepower is expressed in terms of weight, distance and time, the values of pump discharge in gallons per minute and discharge pressure in pounds per square inch must be translated into values of weight and distance. A gallon of water (U.S.) has a weight of 8.34 pounds, and 1 psi of pressure will lift a column of water 2.31 feet. Substituting these values in the above equation, the work done by a pump expressed as water horsepower (WHP) can be determined by the following equation:

$$\begin{aligned} \text{WHP} &= \frac{(\text{gpm} \times 8.34) \times (\text{psi} \times 2.31)}{33{,}000} \\ &= \frac{8.34 \times 2.31}{33{,}000} \times \text{gpm} \times \text{psi} \\ &= \frac{\text{gpm} \times \text{psi}}{1716} \end{aligned}$$

No fire pump is 100 per cent efficient. Not all of the power delivered by the engine to the pump is converted into useful work. There are mechanical losses (gearing, bearings, packing and impeller disk friction), hydraulic losses (flow friction in inlet waterways, shock and eddy losses at impeller eye and roughness of passages in the impeller), and loss of discharge capacity by leakage from the discharge chamber to the suction chamber.

Brake horsepower

To determine the BHP (brake horsepower) required to produce a given volume (gpm) at a specific pressure (psi), the pump efficiency for the discharge and pressure must be known. Figures 4 and 17 are typical characteristic curves for pump efficiency at varying discharge and pressure. It will be noted the highest efficiency is at or near the rated capacity of the pump. As the volume of discharge is reduced, the efficiency also lowers. The lower efficiency at the reduced discharge rate is not usually a matter of concern to the pump operator as there is ample engine power available for delivering the reduced discharge. Loss in efficiency represents loss in power from the mechanical and hydraulic losses, which are converted to heat, about 25 percent being lost by radiation to the atmosphere around the pump (it will be noticed frequently by the darkening of paint on the pump housing) and the balance absorbed in the water being discharged. While the

FIRE SERVICE PUMPS

rise in water temperature is not usually noticed for normal operation at pressures below 300 psi, it does become a problem as the pressures increase beyond 300 psi. This condition only occurs when the regular major fire pump is used for high-pressure discharge (to 600 psi).

As an example, if a centrifugal pump with a rated capacity of 1,000 gpm is also to be used as a high-pressure pump, the discharge of 60 gpm at 600 psi represents 20.9 WHP. If a pump efficiency of 20 percent is assumed (and this is on the high side) for this discharge rate, 104.5 BHP will be required from the engine. The determination of heat loss and rise in water temperature is made as follows:

$$\begin{array}{r} 104.5 \text{ BHP required} \\ -\ 20.9 \text{ WHP useful work done by pump} \\ \hline 83.6 \text{ HP loss} \\ \times\quad .25 \\ \hline 20.9 \text{ HP loss by radiation} \end{array}$$

$$\begin{array}{r} 83.6 \text{ HP loss} \\ -\ 20.9 \text{ HP loss by radiation} \\ \hline 62.7 \text{ HP loss absorbed by water} \end{array}$$

One horsepower is the equivalent of 778 Btu (British Thermal Units). One Btu is the heat required to raise the temperature of 1 pound of water 1°F in 1 minute. Thus

$$\begin{array}{r} 778 \text{ Btu per horsepower} \\ \times 62.7 \text{ HP loss to water} \\ \hline 48{,}780 \text{ Btu loss absorbed by water} \end{array}$$

One gallon (U.S.) of water weighs 8.34 pounds. At a discharge rate of 60 gpm, the weight of water pumped is:

$$60 \times 8.34 = 500 \text{ pounds per minute.}$$

Dividing the total Btu loss by the weight of the water absorbing the heat, the temperature rise of the water is determined:

$$\frac{48{,}780}{500} = 97°\text{F temperature rise.}$$

With 60°F water entering the pump, the outlet temperature will be: 60 + 97 = 157°F.

A number of fire departments have resorted to the use of asbestos gloves to permit handling the discharge lines. Most departments simply reduce the discharge pressure to 400 psi with very satisfactory

FIRE SERVICE HYDRAULICS

results. This decrease in pressure, but with the same volume, reduces the water temperature rise to 64°F for a final discharge temperature of not over 124°F, which is handled without apparent difficulty.

The real problem with such a high-pressure operation arises when all discharge lines are shut off for 2 or 3 minutes with the pump operating, and with the bypass line from the pump discharge manifold to the water tank also closed. Superheated steam is quickly produced, which has caused some severe burns.

The pumps designed for high-pressure operation do not usually experience the heating problem because of higher pump efficiencies at small discharge rates at high pressure.

When the WHP is known, the required BHP is determined by the use of the pump efficiency at the specified discharge in the equation for determining WHP as follows:

The pump efficiency is assumed to be 70 percent at rated capacity of 1,000 gpm at 150 psi.

$$\begin{aligned} BHP &= \frac{\text{gpm} \times \text{psi}}{1716 \times .70} \\ &= \frac{1,000 \times 150}{1716 \times .70} \\ &= 124.8 \end{aligned}$$

Capabilities and limits

The general characteristics of a centrifugal pump are determined by the pump designer, and cannot be changed by the operator. Such built-in characteristics are: rated capacity; suction inlet diameter; diameter of the impeller eye; outside diameter, width between the sides, shape and number of vanes in the impeller; and design of the volute chamber. All affect and determine the pump capability. Within these design limits, the operator can control the volume of discharge and the pressure by throttle control of the impeller speed and control of the discharge gates.

There are three characteristics of a centrifugal pump which may be used for checking pump speeds and power requirements when a change is to be made in the volume discharged or in the discharge pressure. These are:

1. The volume (gpm) varies directly as the pump speed.
2. The pressure (psi) varies as the square of the pump speed.
3. The engine brake horsepower required to produce a given volume at a given pressure varies as the cube of the pump speed.

FIRE SERVICE PUMPS

These characteristics are based on the assumption that the pump efficiency is constant or uniform at all speeds, which would only be true over a very limited capacity range at or near the peak efficiency, as shown by the characteristic efficiency curves on Figures 4 and 17.

Any change in position of the engine throttle affects volume, pressure and power. A decrease in engine speed by throttle change, reduces volume (gpm), pressure (psi) and power output of the engine. An increase in engine speed will increase all three, unless the operator desires to maintain, for example, the pressure but increase the volume. To do this requires opening one or more discharge gates to accommodate the increased volume. If the pressure is to be increased, but not the volume, one or more discharge gates must be throttled or closed.

To actually demonstrate these characteristics with a pump is quite practical. However, to calculate the result of changes in advance, the engine speed must be substituted for the pump speed as no tachometer is provided that registers direct visual readings for pump speed. Only engine speed is visually indicated. An outlet is provided at the pump operator's panel to check engine speeds by a revolution counter, which is hand held. This is impractical for demonstration or training. If it is desired to know the pump speed, multiply the engine speed by the engine-to-pump gear ratio for that particular engine-pump combination. Many different gear ratios are used. One ratio is selected by the fire apparatus manufacturer (or the pump manufacturer) as giving the best overall performance for a specific combination of engine and pump.

To illustrate, assume the engine-to-pump ratio to be 1:1.4. Also, assume the pump to be discharging 750 gpm at 120 psi at an engine speed of 1640 rpm. Three 2½-inch lines are already stretched and a fourth line has been called for, requiring a pump discharge of 1,000 gpm. To determine the engine speed required to pump 1,000 gpm at 120 psi, the equation would be:

$$\frac{1000 \times 1640}{750} = 2186 = \text{new engine rpm.}$$

The pump speed at 750 gpm was 1640 × 1.4 = 2296 rpm.

The pump speed at 1,000 gpm will be 2186 × 1.4 = 3060 rpm.

To illustrate the second characteristic, the same substitution of engine speed for pump speed can be made. The pump designer pays little or no attention to engine speed when designing the pump. He works only with pump speeds. It is only when the pump is to be matched with a specific engine that ratios are selected to provide

FIRE SERVICE HYDRAULICS

required pump performance within the governed speed range of the engine, for best performance.

To demonstrate the second characteristic, use the same values as given above, but increase the pump discharge pressure from 120 psi to 150 psi, no change in volume. To find the engine and pump rpm at the increased pressure, the equation is:

$$\frac{150}{120} = \frac{(rpm)^2}{(1640)^2}$$

$$\frac{150}{120} = \frac{(rpm)^2}{1640 \times 1640}$$

$$(rpm)^2 = \frac{1640 \times 1640 \times 150}{120}$$

$$= \frac{403{,}440{,}000}{120}$$

$$= 3{,}362{,}000$$

$$rpm = \sqrt{3{,}362{,}000}$$

$$= 1834$$

The pump rpm at 150 psi pressure is $1834 \times 1.4 = 2568$ rpm.

As an example of the third characteristic, using the values given, to determine the WHP when the pressure is to be increased from 120 psi to 150 psi, the equation is:

$$\frac{WHP}{52.4} = \frac{(2568)^3}{(2296)^3}$$

The simpler method, mathematically is:

$$WHP = \frac{750 \times 150}{1716}$$

$$= 65.5$$

Positive pressure used

One of the reasons the centrifugal pump is preferred by the fire service is its ability to utilize to full advantage any positive pressure at the suction inlet to reduce the work required of the pump. For example, if the required pump discharge pressure is 120 psi and the residual or flow pressure at the suction inlet is 45 psi, the pump is only required to produce the difference in pressures, 75 psi, for a discharge pressure of 120 psi. This contributes to low engine and pump speeds

FIRE SERVICE PUMPS

with decreased maintenance. Another factor contributing to low maintenance costs is the few moving parts in the pump—only the impeller(s) and shaft.

Cavitation

The term *cavitation*, refers to an action within the pump when operating at draft under certain conditions. When the pressure in the pump suction drops below the vapor pressure of the water being pumped, vapor bubbles or vapor cavities form at the suction inlet to the impeller(s) and the vapor cavities are collapsed in a fraction of a second as they pass into the impeller and are subjected to its pressures.

This implosion or violent collapse of the vapor cavities is most pronounced near the leading edge of the impeller vanes where the vacuum is highest, which is usually the point of lowest pressure.

The result of these implosions is progressive damage to the impeller, as each implosion breaks away a small particle of the impeller. As the damage is cumulative, continued operation of the pump under conditions producing cavitation will cause eventual impeller failure.

The piston pump and the rotary pump are likewise damaged when operated under conditions producing cavitation. In the piston pump, the head of the piston is eroded by the loss of metal, and in a rotary pump the rotors will show loss of metal, and roughened surfaces produce increased rates of wear.

Cavitation is produced under several conditions:
1. When the lift (at draft) is too high for the volume and pressure being discharged.
2. Suction hose too small for the volume being discharged.
3. Restriction in suction line at strainer.
4. Partial collapse of lining in suction hose.
5. Temperature of water being pumped is too high.

The remedy for each of these potential causes of cavitation are:
1. Reduce the volume or pressure of discharge.
2. Use larger suction hose; shorten total length by removal of one length, if possible; or reduce volume of discharge.
3. Remove weeds or debris restricting entrance of water at the strainer.
4. Replace suspected section of suction hose for testing after return to quarters.
5. Reduce volume discharged unless another water source of cooler water is available.

Cavitation can usually be determined by the sound emanating from the pump while operating. If the pump sounds as if many small stones were passing through with increased vibration, cavitation usually exists.

A quick check can be made by watching the discharge pressure gage while slowly opening the engine throttle. If there is an increase in engine speed without a corresponding increase in discharge pressure, the pump is at the "run-away" point and is cavitating.

The throttle opening should be reduced to the original pressure setting and action initiated as outlined.

It is common practice in some fire departments to use 5-inch suction hose with pumps having a rated capacity of 1,000 gpm. There are two reasons for this practice: first, the hose is lighter and easier to handle than 6-inch suction hose; second, most pump operation is from a hydrant with positive pressure. When required to operate from draft, either for testing or fire fighting, the 5-inch suction limits the height of lift for capacity rating and this limitation must be observed to avoid cavitation.

An example of this limitation is provided by the use of a 1,000-gpm pump operating on a 10-foot lift using two lengths of 5-inch suction hose. Assuming sea level elevation and water temperature of 60°F for the most favorable conditions, the suction losses are as follows: Using the table for friction loss in suction lines (including strainer loss) as given in the table on page 67 in Chapter 2, the loss with 5-inch suction, two lengths, for 1,000-gpm flow is 9.5 feet. The lift is 10 feet. The pressure required to accelerate the water from zero velocity to the velocity in the suction hose (16 feet per second) is 4 feet. The sum of these losses is $9.5 + 10 + 4 = 23.5$ feet, the equivalent of 20.6 inches (mercury) of vacuum.

At sea level a pump in good condition would produce the required volume (1,000 gpm), but with an increase in altitude the height of lift would need to be reduced to discharge the rated capacity. An increase in water temperature could also produce cavitation.

By comparison, the total loss when using 6-inch suction under the same conditions of height of lift, altitude and water temperature will be: 4.5 feet, suction loss through hose; 10 feet lift; and 1.9 feet velocity head, for a total of 16.4 feet or 14.4 inches of mercury, vacuum. This is well within the limits of operation without cavitation.

Water temperature and pump performance

The temperature of water being pumped from a positive pressure source has negligible effect on fire pump performance. But, when operating at draft, the elevated water temperature does have a limiting effect. At water temperatures up to 85°F the capacity is not notice-

FIRE SERVICE PUMPS

ably affected. At water temperatures of 95 to 100°F a noticeable decrease in capacity occurs, which may be accompanied by cavitation.

This effect on pump performance is due to vapor pressure existing on the surface of the water in the pump. The vapor pressure increases with an increase in temperature and at 212°F equals atmospheric pressure, making it impossible to draft water at temperatures approximately 212°F. In the temperature operating range encountered by fire departments, we only need to consider water temperatures from 32 to 100°F. The vapor pressure in a pump varies from .204 feet at a water temperature of 32° to 2.19 feet at 100°F.

The values usually used by the fire service for hydraulic calculations are based on 50°F water temperature, at sea level with atmospheric pressure supporting a column of water 33.9 (33.88) feet with a barometer of 29.92 inches, mercury for a perfect vacuum.

As the water temperature increases, the vapor pressure in the pump also increases reducing the capability at draft. This is indicated in the following table of values for vapor pressure at various temperatures with corresponding height of water column at sea level and altitudes of 750, 1,000 and 1,500 feet above sea level.

Water Temperature	Vapor Pressure (feet)	Sea Level	Height of Water Column at Altitude		
			750 feet	1,000 feet	1,500 feet
50°F	.411	33.88	32.95	32.65	32.06
60°F	.591	33.70	32.77	32.47	31.88
70°F	.838	33.45	32.52	32.22	31.63
80°F	1.17	33.12	32.19	31.89	31.30
90°F	1.61	32.68	31.75	31.45	30.86
95°F	1.88	32.41	31.48	31.18	30.59
100°F	2.19	32.10	31.17	30.87	30.28

During the warm summer months a fire department may be required to use a swimming pool, stock watering tank or farm pond as a water supply source. Such sources are usually exposed directly to the sun. The water temperature, because the water is static and not flowing, often reaches the higher temperatures given in the above table.

Most of the constant values used in fire service hydraulics have been given, along with their basis and derivation in preceding chapters. There are, however, some terms related to pump operation at draft that should be clarified.

Static suction lift

When the free level source of water supply is below the center of the pump the static suction lift is the vertical distance measured in feet from the surface of the water to the center of the pump.

FIRE SERVICE HYDRAULICS

In practice, the center of the pump is assumed to be the center of the suction inlet. The pump center is not readily observed as it is concealed by panels carrying the control instruments. At most, it is usually not more than a few inches above or below the center of the pump suction side inlet.

Dynamic suction lift

Dynamic suction lift is the sum of (1) static suction lift, (2) entrance loss at strainer, (3) friction loss in suction hose and (4) velocity head.

In practice, the dynamic suction lift includes the height of static suction lift plus the combined entrance and suction hose friction loss. This total, as expressed in feet, is converted to the equivalent negative pressure in terms of pounds per square inch. The velocity head is ignored.

The value for the dynamic suction lift is added to the discharge pressure gage reading as it represents work done by the pump and is credited accordingly.

For example, a pump operating on a static suction lift of 6 feet, using 20 feet of 5-inch suction with a discharge of 1,000 gpm would have a suction allowance for the dynamic suction lift as follows:

$$\frac{6 + 9.5}{2.3} = 6.7 \text{ psi}$$

If the discharge pressure gage reading was 152 psi, it would be recorded as "pump pressure" and the 6.7 psi would be added to obtain the net pump pressure of 158.7 psi. It will be noted that reference to pump pressures in NFPA Specification No. 19 is to net pump pressure and not to gage pressure.

When a more accurate determination of suction allowance is required, the velocity head is included. The velocity head is the energy required to accelerate the water from zero velocity to the suction flow velocity.

Though the velocity head is normally ignored in fire service hydraulics, it does exist and is a factor in pump performance as indicated in the paragraphs on pump cavitation. The velocity head is recognized and credited in the performance of pumps in municipal water works.

SINGLE-STAGE CENTRIFUGAL FIRE PUMP

In a single-stage centrifugal pump the impeller may be a single-suction type or a double-suction type. In designing a pump for fire

FIRE SERVICE PUMPS

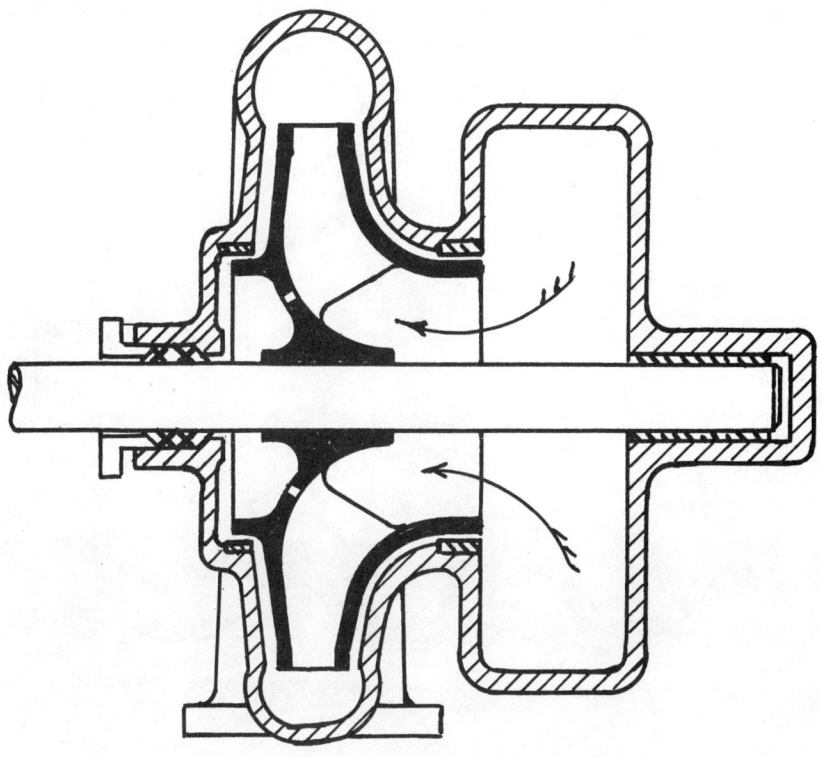

Figure 2

service use there is a practical limitation on physical dimensions to permit vehicle mounting within the space requirements. Considerable attention is given to design characteristics with reference to mounting location—front of vehicle, midship, or a direct mount on the driving engine (skid and trailer mounting). To efficiently satisfy capacity rating requirements and provide the high lift characteristics essential in a fire pump, the capacity rating for a single-stage centrifugal pump with a single-suction impeller (Figure 2) is usually limited to 750 gpm.

In operation, the single-stage centrifugal pump has a minimum of controls. The engine throttle control and discharge gate valves control the volume discharged at any desired pressure within the design capacity range. Water enters the eye of the impeller from the suction chamber, flowing in an axial direction (parallel with the impeller shaft), making approximately a 90 degree change in flow direction as it

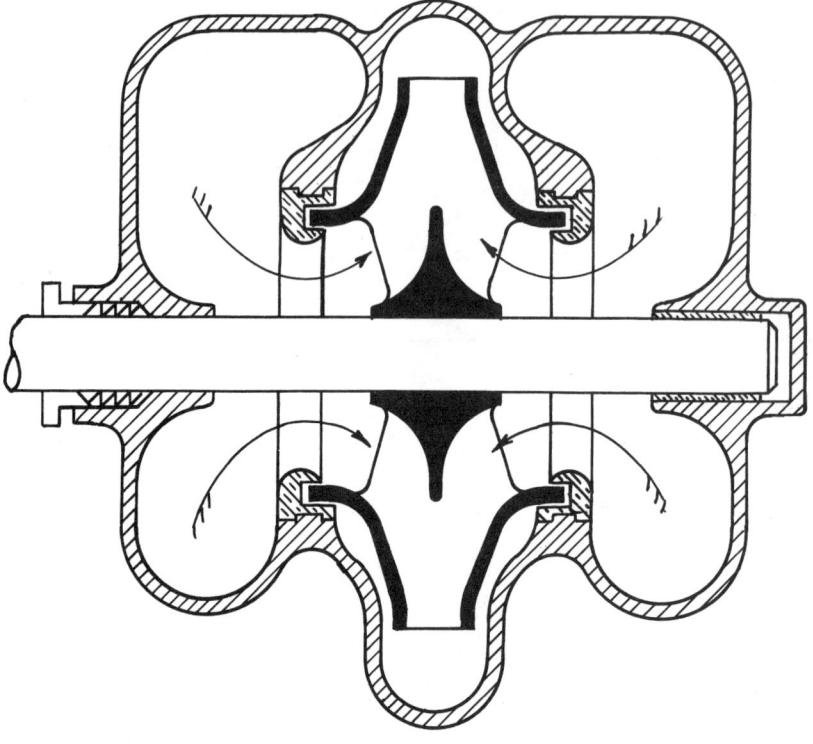

Figure 3

enters the impeller, which directs the flow in a radially outward direction to be picked up by the vanes in the impeller. An end thrust is produced as the flow direction is abruptly changed, for which some compensation must be made to keep the thrust load within the load limits of the bearings. To approximately balance this hydraulic thrust, clearance rings are provided in the housing at the back side of the impeller. The rings are about the same diameter as the clearance rings in the housing at the impeller eye. Holes are provided in the rear wall of the impeller, between the hub and the edge of the vanes to admit water to the back side of the impeller. This provision for partial hydraulic balance reduces end-thrust so that any difference can be absorbed by the impeller support bearings.

The double-suction impeller (Figure 3) is in hydraulic balance due to the symmetry of design of each suction inlet. Water enters each side of the impeller simultaneously, automatically balancing any end-thrust. Double-suction impellers are used in fire pumps with capacity

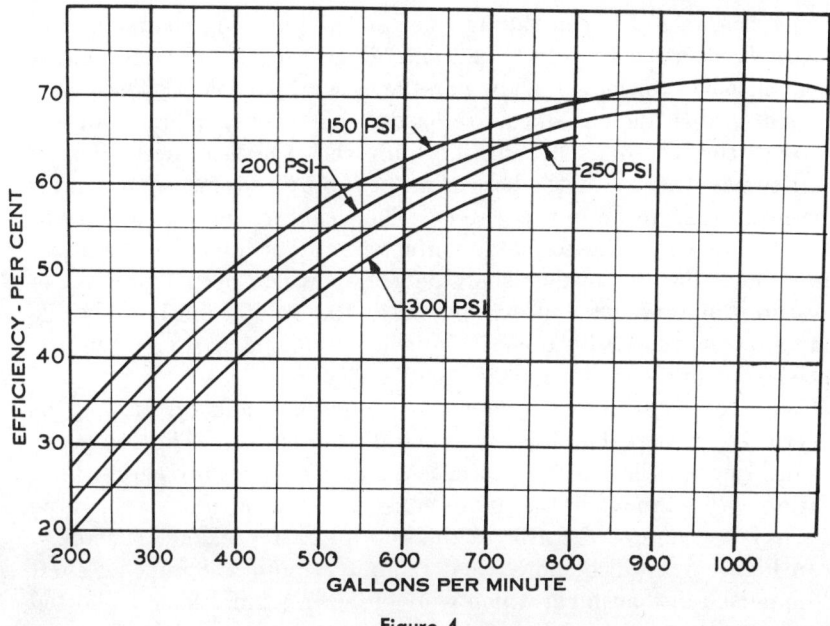

Figure 4

ratings of 1,000, 1,250 and 1,500 gpm. These impellers develop a high efficiency as shown by Figure 4. The double-suction impeller has also been used in the parallel-series 2-stage centrifugal pump.

Seal leakage

Clearance rings, or wear rings (wear ring is the term used in the nomenclature of the Hydraulic Institute) are provided in the pump housing to seal and control leakage from the discharge chamber to the suction chamber due to pressure difference. These rings are usually made of a high lead bronze alloy for toughness and freedom from scoring. A limited clearance must be provided between the rotating impeller and the stationary pump housing. Clearance rings provide the limited clearance and are usually of the replaceable type for economical renewal when excessive wear occurs (over 0.015 inches). Leakage is undesirable, but some leakage must be tolerated due to the operating clearances. The clearance between the impeller and the clearance ring in the housing is held to a close tolerance, 0.006 inch to 0.009 inch. While this clearance is relatively small, it is the equivalent area of a ½-inch-diameter hole. This size hole will vary according to the diameter of the clearance rings, which vary in size in different pumps.

FIRE SERVICE HYDRAULICS

The clearance between the impeller and the housing clearance ring cannot be decreased beyond the limits given because minimum clearance must be sufficient to allow passage of sand particles. These particles or grains, such as silica, are harder than bronze or iron and will cut and score the rings and impeller hub when the clearance is insufficient to pass them. In the best of water supply systems, some sand is present in suspension in the water. The clearance must also be sufficient to prevent metal-to-metal contact—impeller hub with the clearance ring—due to impeller shaft deflection which occurs because of pressure difference at opposite points in the volute housing. While manufacturers static-balance the impeller, dynamic or running unbalance also contributes to impeller shaft deflection.

Clearance rings may be plain type (Figures 2 and 3), or may be labyrinth type, replaceable or integral (Figure 5). The labyrinth sealing ring, in the design shown, is an integral part of the impeller, mating with a groove in the pump housing. The purpose of this type of seal is to increase the area of the seal for better leakage control.

In Figure 5 it will be noted that the outer end of the impeller shaft is supported in a bushing, which is mounted in a blind boss supported by three ribs. These three ribs are important to the high efficiency of the pump, as they control or eliminate any prerotation of the water entering the eye of the impeller. Such a rib or stationary vane is quite commonly used to control prerotation in both the single-stage and parallel-series 2-stage centrifugal fire pumps.

Packing glands or stuffing boxes are provided in several designs, most being of the adjustable type, but all are designed to seal against water leakage and entrance of air into the pump. Some seals are chevron type, some are lip-seal type using a spring to maintain proper contact with the shaft. Most manufacturers of fire pumps, however, provide an adjustable type of packing gland with square or rectangular packings. The square or rectangular packings are usually rings of metallic impregnated material, lead or graphite in a woven or molded composition with some asbestos. The base material may be a composition of synthetic compounds. Regardless of the actual material used, it is designed to perform the basic sealing function with ability to provide lubrication between the packing and shaft, and resist deterioration from heat generated by contact with the shaft. To provide a maximum of packing life, instructions are usually provided by pump manufacturers to avoid tightening the packing to prevent all leakage. The packing should be only tight enough to make a seal, with some droplets of water permitted when the pump is operating. Such slight water leakage acts to cool the packing and assists

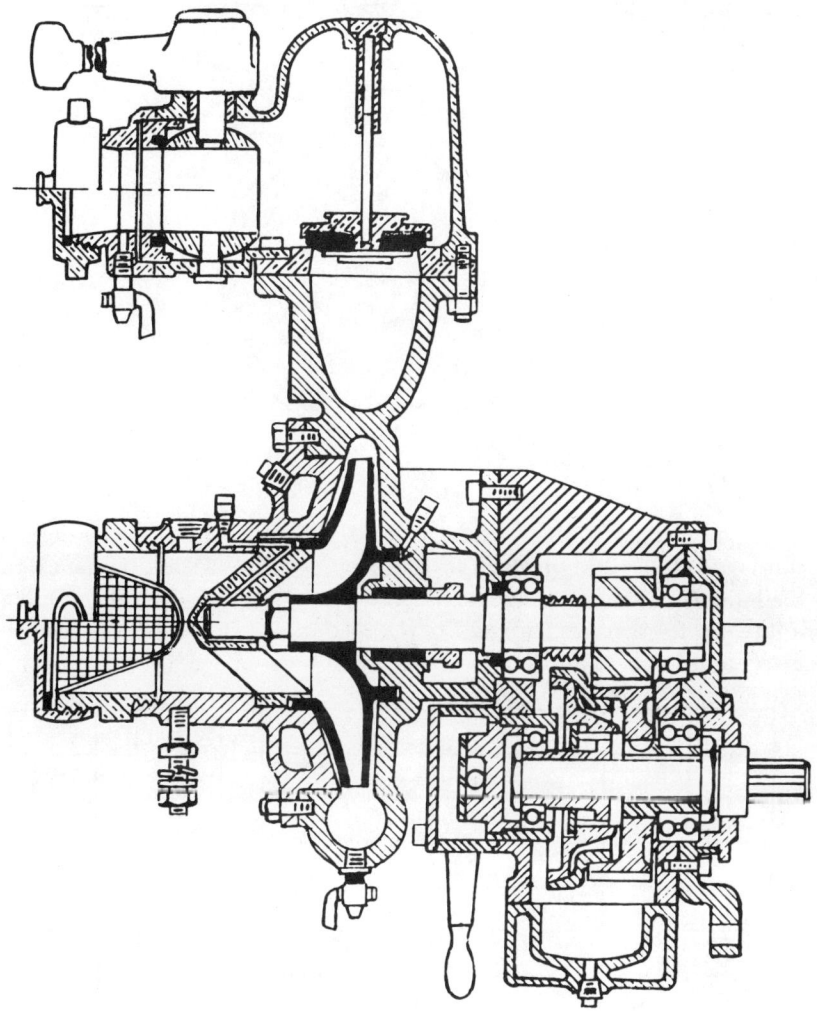

Figure 5

in lubrication. Adjusted too tight, the packing heats and usually causes a scored shaft, which then produces a rapid rate of packing wear. A silicone type of waterproof grease has been found that greatly extends packing life and reduces the frequency of adjustment.

Double-volute design

Impeller shaft deflection, from unbalanced pressures in the volute housing, produces vibration in the pump when operating and may

FIRE SERVICE HYDRAULICS

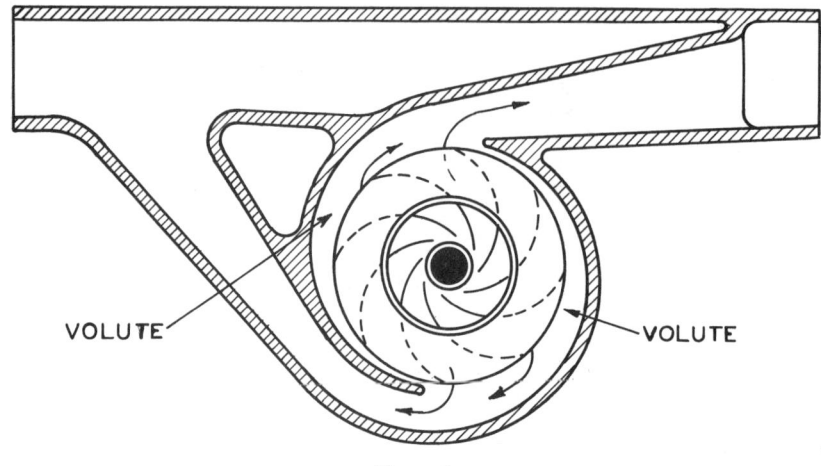

Figure 6

cause added wear at the clearance ring and impeller eye. Such deflection is also the cause of much of the packing wear, with a tendency to "bell mouth" the packing. To reduce these conditions, the double volute single-stage centrifugal fire pump was introduced for fire service use (Figure 6). The double-volute design has been used for many

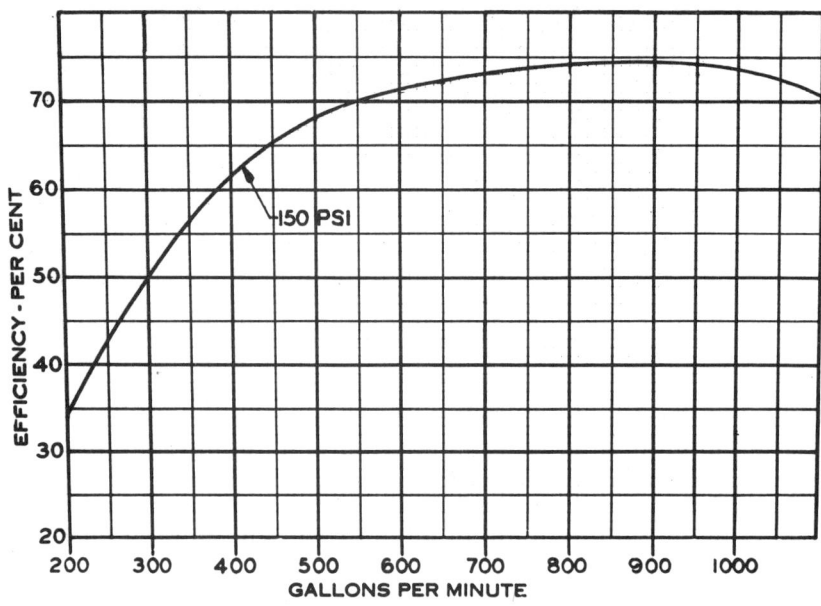

Figure 7

FIRE SERVICE PUMPS

Figure 8

years by pump manufacturers building pumps for municipal and industrial water supply systems and has been proven highly successful in reducing vibration and packing and clearance ring wear, with consequent less maintenance of the pump. The use of the double-volute design is confined to the single-stage pump since no satisfactory design has been developed to use the double-volute design in the parallel series multi-stage centrifugal pump.

One of the features of the single-stage pump is its high efficiency at or near rated capacity. The best efficiency is usually above 70 per-

FIRE SERVICE HYDRAULICS

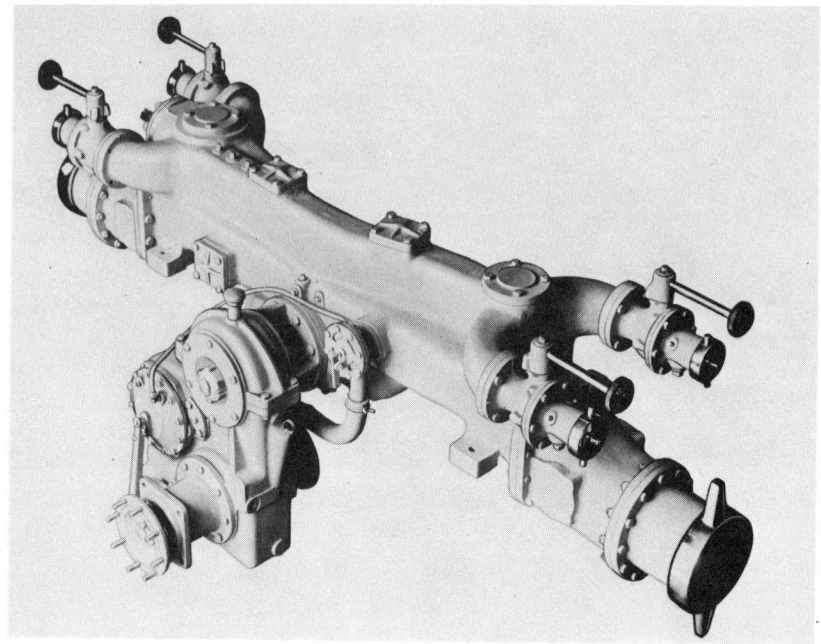

Figure 9

cent, a few percent higher in efficiency than any of the multistage centrifugal fire pumps. The efficiency at a given discharge rate (gpm) is subject to variation according to the discharge pressure (psi). And though the effect of pressure is small when operating at or near rated capacity, at lower discharge rates with higher pressures the efficiency drops quite rapidly (Figure 4). The maximum or cut-off pressure for most single-stage centrifugal fire pumps is in the range of 375 to 450 psi. At cut-off the discharge is zero gallons per minute.

The double-volute single-stage centrifugal pump has a very high efficiency at rated capacity, but more important, the high efficiency prevails over a considerable range of discharge (Figure 7).

Pump efficiency is important as it determines the amount of power required to be delivered to the pump for a given output (gallons per minute at a specified pressure). While over 90 percent of fire department pump operation is at reduced volume and pressure, the lower pump efficiency at these reduced flows is not too important at normal pressures; there is adequate engine power to produce the required volume and pressure.

FIRE SERVICE PUMPS

A single-stage centrifugal fire pump designed for vehicle front mounting is shown by Figure 8. This type of centrifugal is produced with capacity ratings of 500, 750 and 1,000 gpm.

A typical midship-mounted single-stage centrifugal pump is shown by Figure 9. This type of pump is produced with capacity ratings of 500, 750, 1,000, 1,250 and 1,500 gpm.

TWO-STAGE CENTRIFUGAL FIRE PUMP

The 2-stage centrifugal fire pump has two single-stage impellers, single-suction type, mounted on a single shaft, in a housing with a separate volute chamber for each impeller. In operation, water from the supply source enters only one impeller, which discharges to the inlet of the second impeller. The second impeller nearly doubles the pressure as received from the first impeller, and discharges to the pump outlets. Figure 10 illustrates the flow pattern for the straight 2-stage centrifugal pump.

The 2-stage centrifugal fire pump has had only limited use in fire service. The current 2-stage centrifugal fire pumps are front-mounted and have a capacity rating of 500 gpm. Their principal use is in areas requiring two lines (up to 200-gpm total discharge) at pressures of 300 to 400 psi, at moderate engine speed and with the vehicle in motion, carrying the water supply on the vehicle.

In a small size, the 2-stage pump is available as a high-pressure

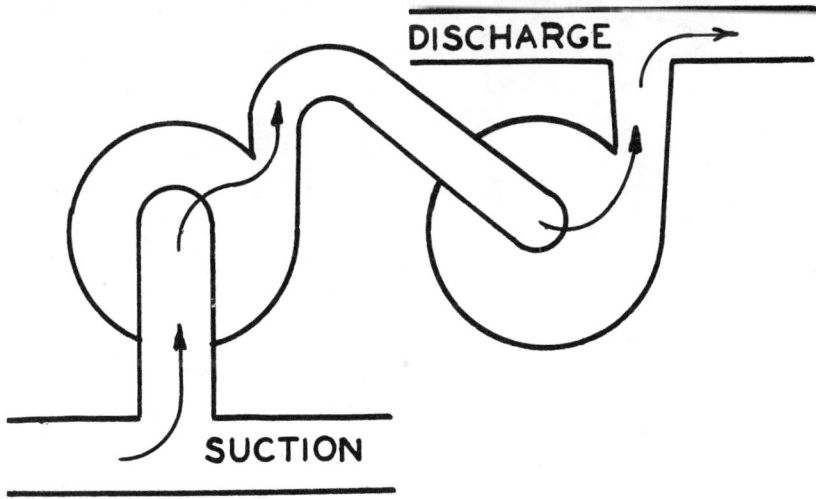

Figure 10

FIRE SERVICE HYDRAULICS

pump, with a power take-off that permits vehicle movement with the pump in operation. More detail is given in the section on high-pressure pumps.

PARALLEL-SERIES 2-STAGE CENTRIFUGAL PUMP

The parallel-series 2-stage centrifugal fire pump is built with two impellers of the same outside diameter that are mounted on a shaft in a housing, with a separate volute chamber for each impeller. The operation of a parallel-series 2-stage centrifugal varies from the straight 2-stage centrifugal pump in that it uses a transfer or changeover valve to direct water flow in the passages in the pump. In the parallel or "capacity" flow arrangement, each impeller receives water from a common source and discharges into a common outlet manifold. The volume discharged with this arrangement is the sum of the volume discharged by each impeller. The arrangement of flow is shown by Figure 11a.

A transfer or changeover valve is used to select the desired flow pattern. This can be either parallel for large volumes, or in series for volumes less than 70 percent of the rated capacity but at pressures

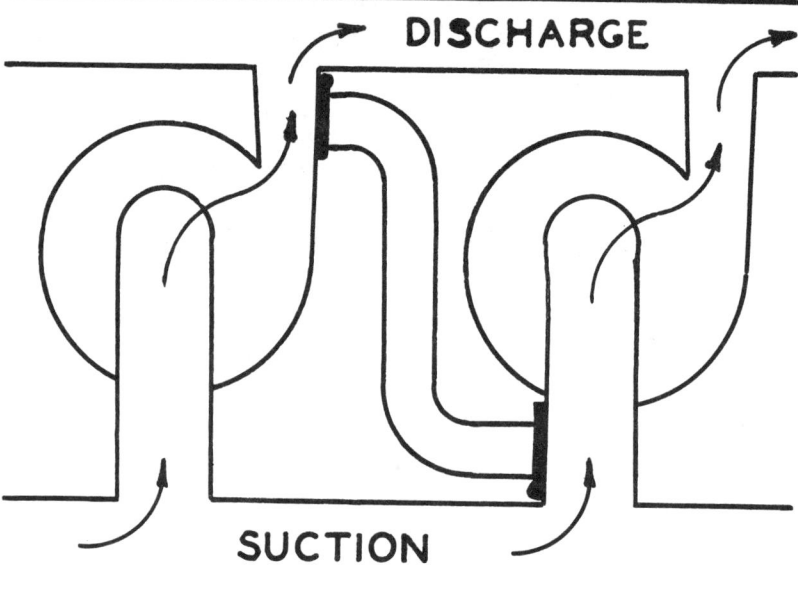

Figure 11a

FIRE SERVICE PUMPS

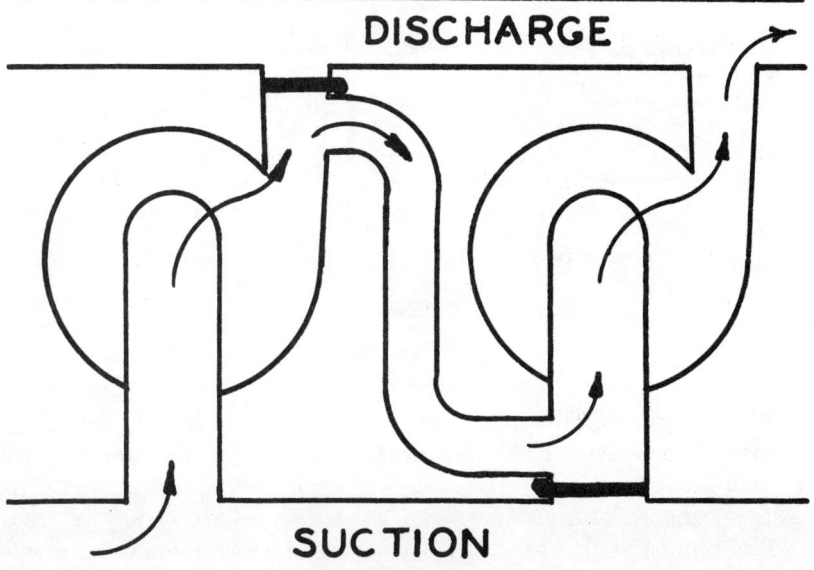

Figure 11b

over the entire range of requirements to the maximum of the pump capability. In the series setting of the transfer valve, one impeller receives water from the normal supply source but the normal suction supply passage to the second impeller is closed. The discharge from the first impeller is directed to the suction inlet of the second impeller. The second one nearly doubles the pressure before discharging into the pump outlet manifold. This arrangement is illustrated by Figure 11b.

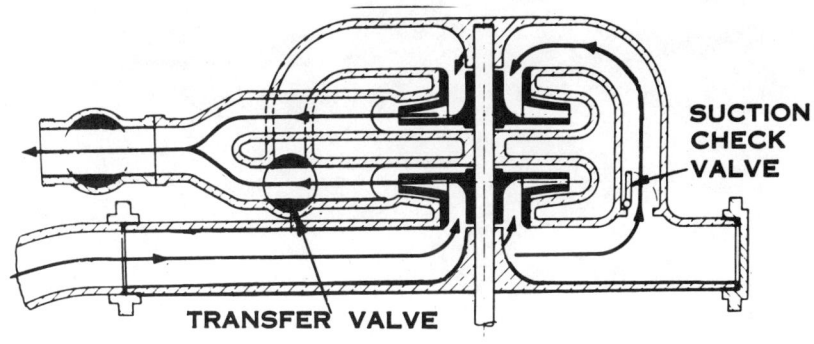

PARALLEL OPERATION

Figure 11c

FIRE SERVICE HYDRAULICS

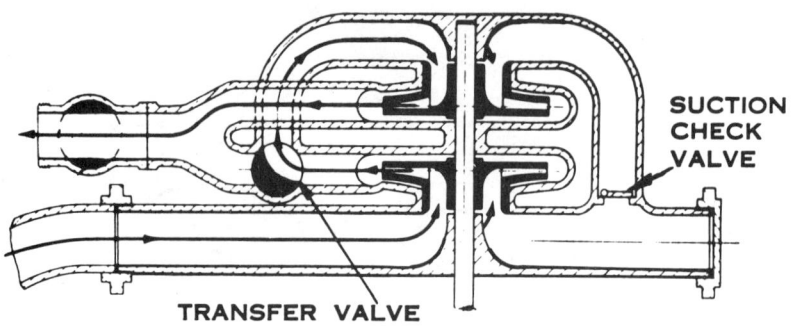

SERIES OPERATION

Figure 11d

The flow for parallel and series is clearly shown in Figure 11c and Figure 11d as a typical flow arrangement for most of the pumps of this type in fire service. Figure 12 is an external view of the parallel-series 2-stage centrifugal fire pump.

Fire pumps of this type have the transfer valve mounted either in the suction passage to limit flow to one impeller (for series operation) or in the discharge passage of one impeller to control the flow for parallel or series operation. The transfer valve may be either manual or power-operated at the option of the fire department when purchasing

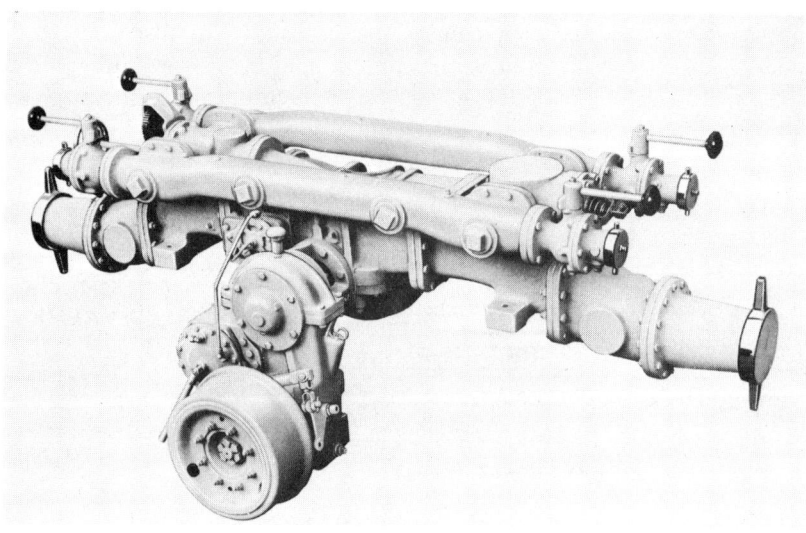

Figure 12

FIRE SERVICE PUMPS

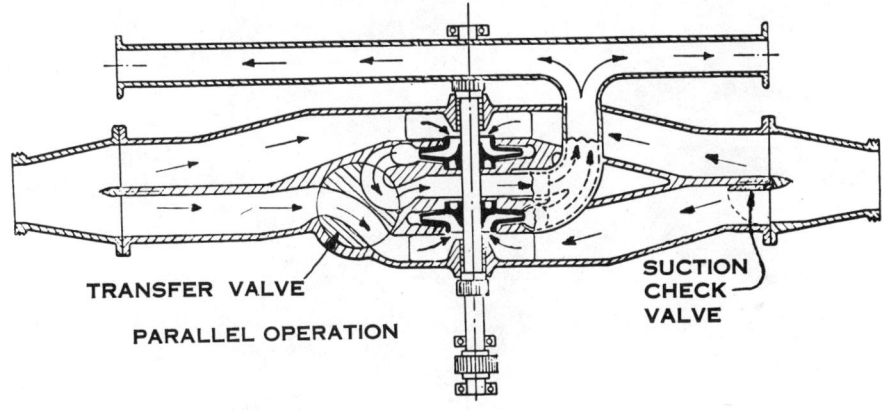

Figure 13a

the pump. One pump design with the transfer valve controlling flow in both the suction and discharge passage is shown in Figures 13a and 13b. In Figure 13a the transfer valve is set for parallel operation of the pump. One half turn of the transfer valve changes the operation to series as illustrated in Figure 13b. The transfer valve can be power-operated as shown in Figure 13c of the complete pump. Discharge gates and primer are also power-operated, electrically.

Transfer valves may be of the cylinder type as shown by the preceding illustrations. A disk-type transfer valve is also used for either manual or hydraulic operation. Figure 14a shows the disk-valve, manually operated for parallel flow. Figure 14b shows the position for series flow, the change being made by rotation of the handwheel. The hydraulic control for this valve is shown by Figure 14c which is

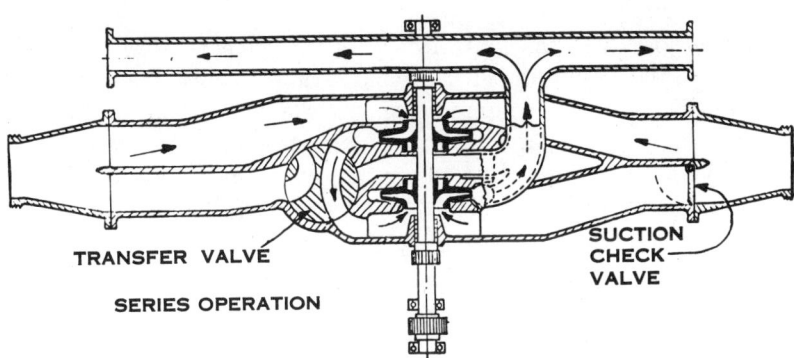

Figure 13b

FIRE SERVICE HYDRAULICS

Figure 13c

set for parallel operation. The controlling valve handle is on the pump operator's panel. The valve operates on a differential in pressures. The piston in the cylinder is larger in diameter than the transfer valve disk. When the control valve is set for parallel operation, the pressure from the second impeller acts on the large piston to hold the transfer valve disk on the seat, thus shutting off the passage from the first impeller discharge to the suction inlet of the second impeller. Both impellers discharge into a common header or manifold.

Series operation

For series operation, the control valve on the panel is moved to close the line from the second impeller discharge, and to open the line from the suction chamber of the pump housing to the cylinder of the transfer valve piston, Figure 14d. The discharge pressure from the first impeller which is greater than the pressure in the suction chamber, moves the piston to the end of the cylinder, sealing the passage to the discharge header, and the pressure on the transfer valve disk keeps the passage closed.

The spring in back of the piston has a rather unique function: it acts to return the piston and valve disk to the parallel operating position whenever the pump discharge pressure drops to zero. Thus, each time the pump stops operating, the transfer valve moves. When

FIRE SERVICE PUMPS

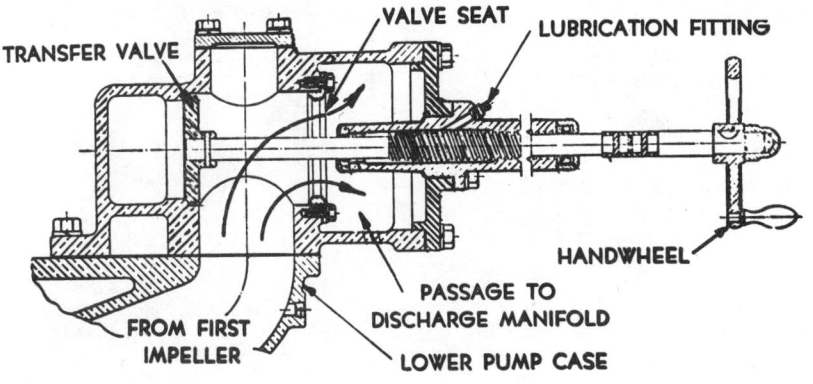

Figure 14a

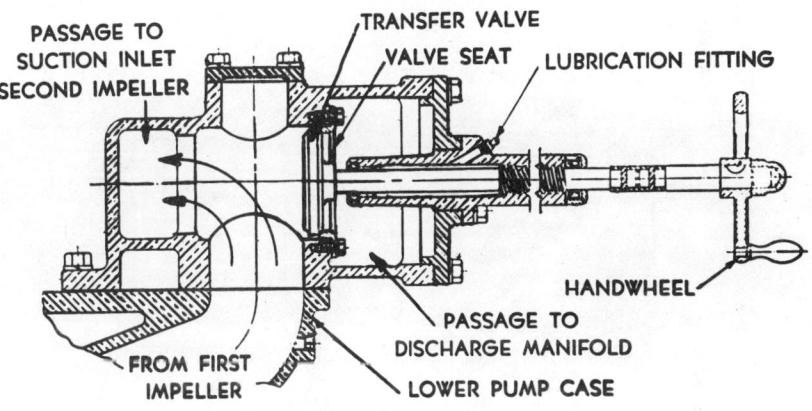

Figure 14b

pressure is built up in the pump again, the transfer valve automatically returns to the series position without any attention from the operator. If the panel control is in parallel setting, this action does not take place. Since most of the pump action is in series, the transfer valve should be moved frequently to prevent sticking.

For a given capacity rating, the impellers in the parallel-series 2-stage centrifugal pump are smaller than the impeller in a single-stage centrifugal pump. Since each impeller is of the same design, it is only required to deliver one half the capacity. For example, in a pump with a rated capacity of 1,000 gpm at 150 psi, each impeller is required to deliver only 500 gpm at 150 psi to produce the required total volume.

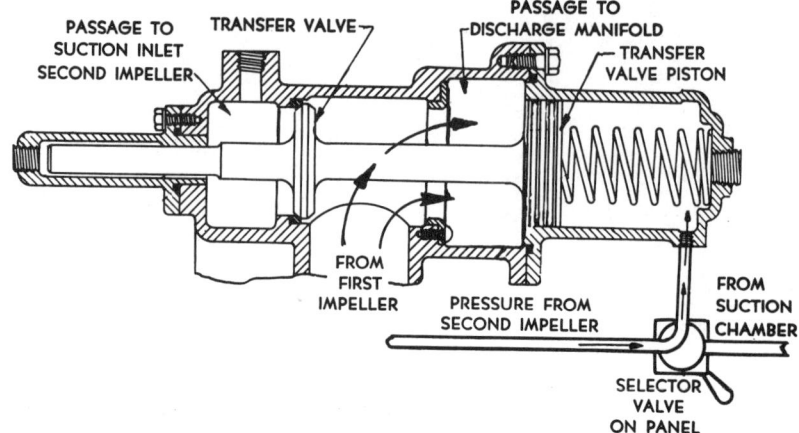

Figure 14c

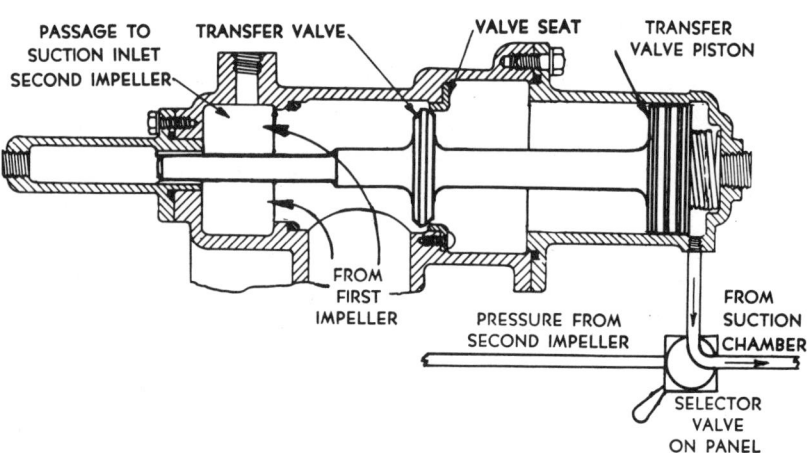

Figure 14d

It is customary practice for the current parallel-series 2-stage centrifugal pumps to operate in the parallel arrangement for the near rated capacities at 150 psi to 200 psi. The performance requirements, as specified in the NFPA Specification No. 19, include discharge of 70 percent of rated capacity at 200 psi. The majority of these pumps use impellers of the single-suction type, with the capacity in series operation limited to the capacity of one impeller. In the larger pumps, this requirement is best produced by parallel operation. Figure 15 is a sectional view of a typical parallel-series 2-stage centrifugal pump in

FIRE SERVICE PUMPS

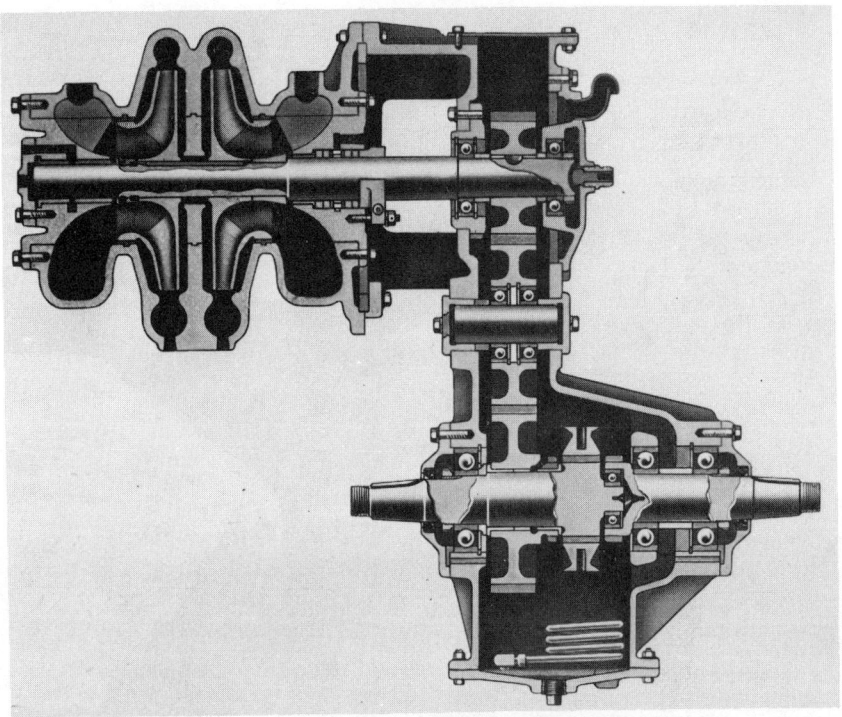

Figure 15

current use. Note the prerotation vanes, one at the entrance of each impeller to stop any whirling motion of the incoming water. While the piping is not shown in this illustration, a pipe is provided from the cover at the outer end of the shaft to the pump suction chamber. The purpose is to prevent a buildup of pressure by water leaking through the bearing and producing end-thrust on the impeller shaft.

The double-suction impeller which has the capacity to deliver more water than the single-suction impeller, has been used in the parallel series 2-stage centrifugal pump (Figure 16). With this arrangement the pump, when operating in the parallel setting, has 67 percent of the rated capacity delivered by the double-suction impeller and 33 percent delivered by the single-suction impeller. In the series setting, the double-suction impeller supplies 70 percent of the rated capacity and delivers it to the second impeller at a pressure of one half the final pressure. Under positive pressure at the inlet, the second impeller has only to increase the pressure for final discharge.

The parallel-series 2-stage centrifugal pump operates in hydraulic

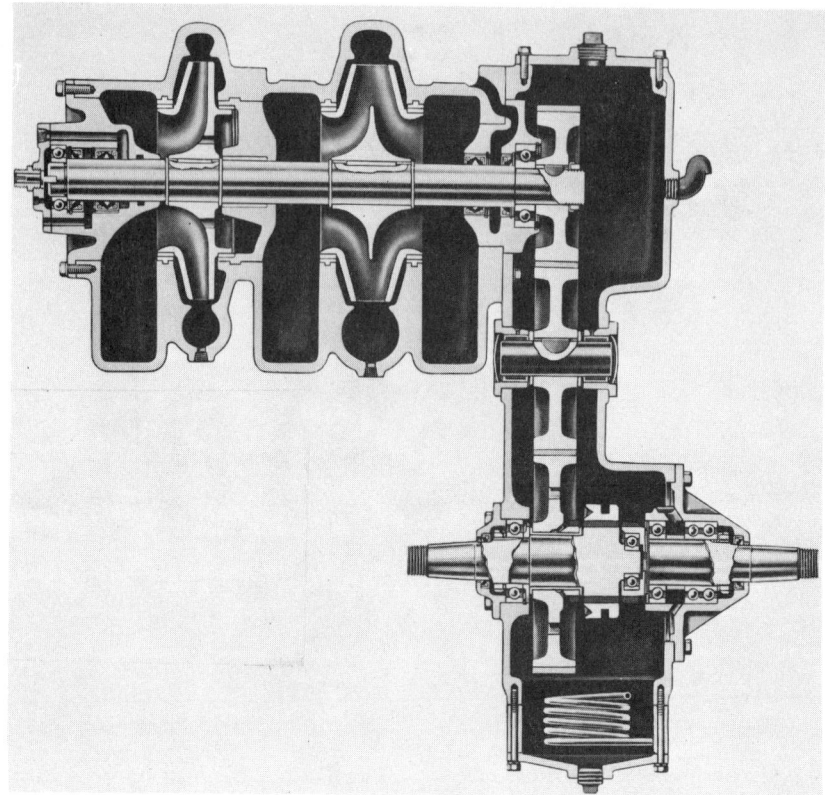

Figure 16

balance when in the parallel setting. In the series setting there is some end-thrust from hydraulic unbalance. But the use of clearance rings on each side of the impeller and the admittance of water at inlet pressure to the back side of the impeller limit the final end-thrust to small value. This is handled easily by the impeller shaft support bearings.

The efficiency of the parallel-series 2-stage centrifugal pump at rated capacity is 65 to 70 percent, as indicated on Figure 17 which shows typical efficiency characteristics. That the efficiency at rated capacity is slightly less than the single-stage centrifugal pump is due principally to the increased leakage loss from discharge to suction through the clearance rings. In the series operation, the parallel-series 2-stage pump has much higher efficiencies than the single-stage centrifugal pump. This is in the discharge range from 40 to 70 percent of rated

FIRE SERVICE PUMPS

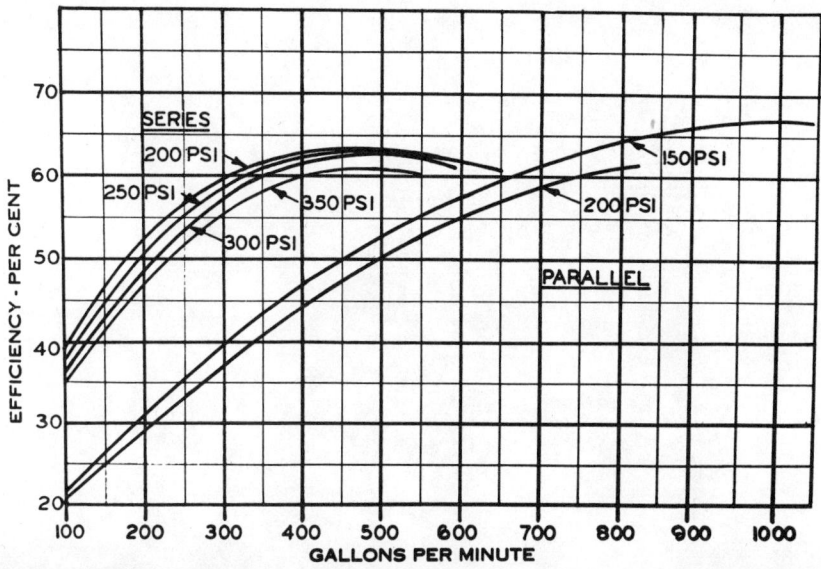

Figure 17

capacity and at pressures ranging from 100 to 350 psi. When discharge pressures higher than 350 psi are desired, special impellers are usually provided for higher pressure. When pressures of 400 to 600 psi are required, then some sacrifice must be made in maximum volume of pump capability because the impeller and volute design must be modified to provide the higher pressure. This usually leaves the reserve capacity above the rated capacity at a small margin, as little as 100 gpm. Over years of usage, wear of the clearance rings will further reduce the margin of capacity reserve, necessitating more frequent renewal of the clearance rings.

Operation at high pressures, above 400 psi, will also produce a problem of high temperature for the water discharged as previously discussed.

Pumps mounted on fire apparatus which have a capacity rating of 500 gpm to, and including 1,500 gpm are required by NFPA Specification No. 19 to be provided with at least one 2½-inch gated outlet for each 250 gpm of rated capacity. In practice the size of some of the gated outlets is increased for the 1,750 and 2,000-gpm pumps as the large-capacity pump is basically a portable pumping station to supply other fire fighting pumps. The discharge outlets, to handle the increased flow rates, are four 3½-inch and two 2½-inch discharge gates. Thus the pump (Figure 18), is equipped to handle

FIRE SERVICE HYDRAULICS

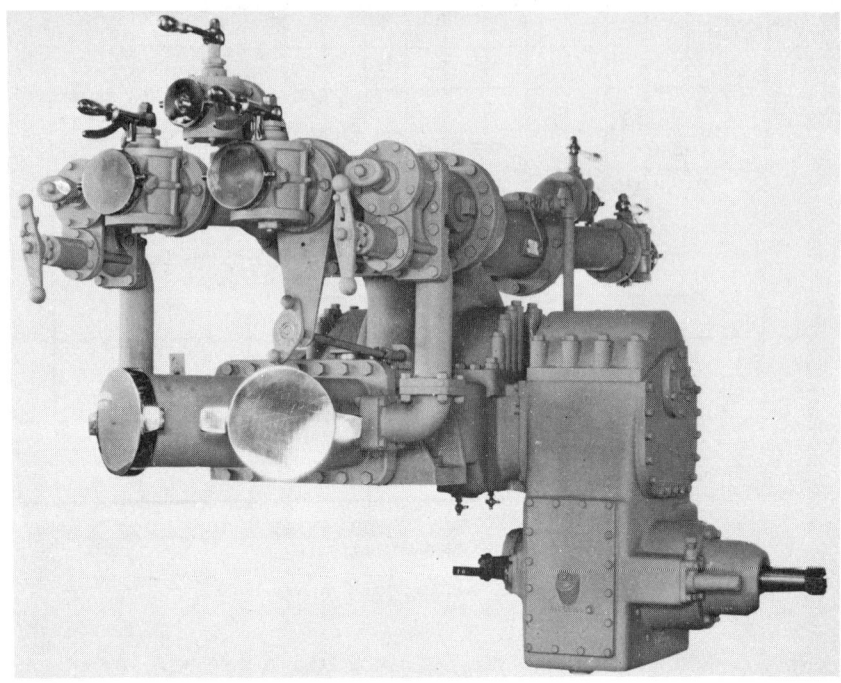

Figure 18

large-capacity discharge lines plus the regular 2½-inch lines without the use of a multitude of adaptors.

PARALLEL-SERIES 3-STAGE CENTRIFUGAL FIRE PUMP

The parallel-series 3-stage centrifugal pump is basically a parallel-series 2-stage centrifugal pump to which a third stage has been added for use only when discharge pressures above 300 psi are desired. This design was produced for fire service use to obtain higher efficiency for reduced flow at high pressure. The advantage of this design is reduced heat loss and lower engine speeds for a given pressure. The schematic flow arrangement is shown in Figures 19a, 19b and 19c.

One manufacturer provides a third stage with the third-stage impeller mounted on an extended 2-stage impeller shaft. While the third-stage impeller rotates whenever the pump is in operation, it does no work unless the control valve at the second-stage discharge outlet admits water for third-stage operation. The 3-stage pump is shown in Figure 20, with the third-stage housing and inlet piping shown in light color for readily identifying the third-stage mounting.

FIRE SERVICE PUMPS

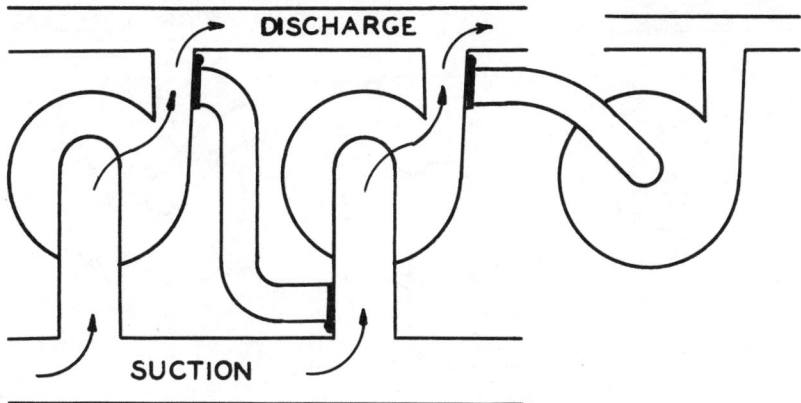

Figure 19a

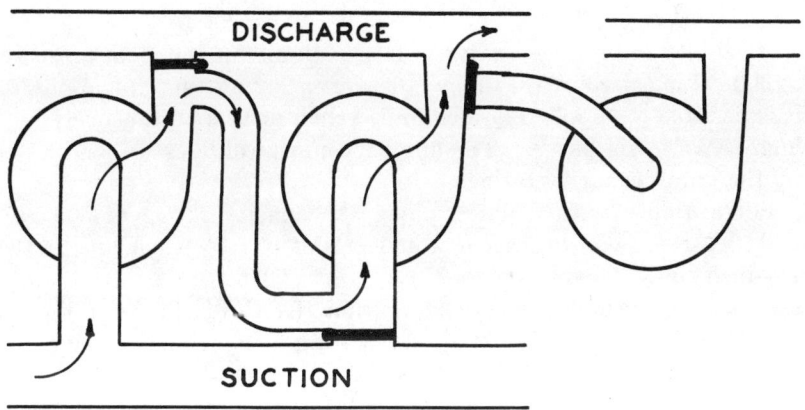

Figure 19b

A second manufacturer provides a similar mounting of a third stage, but with a difference. The third-stage impeller is operated only by engagement of a clutch mounted on the end of the 2-stage pump impeller shaft. Also, the passage from the second-stage discharge outlet to the third stage is open. No valve is used to close the line. Water enters the third stage whenever the regular pump is in operation, but as the impeller is not engaged and the third-stage outlet is closed, no power is consumed by the third stage since no work is done. This arrangement is shown in Figures 21a and 21b.

The parallel-series 3-stage centrifugal pump has some advantages for fire service use in addition to its high-pressure capability. Because

FIRE SERVICE HYDRAULICS

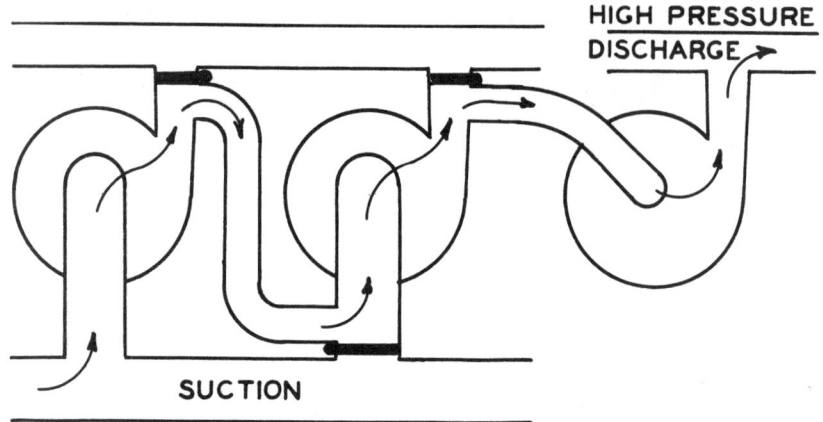

Figure 19c

of its ability to produce relatively large volumes of water, at pressures to 300 psi (a capability of the parallel-series 2-stage centrifugal pump), it is possible to supply 1½-inch or 2½-inch lines simultaneously with high-pressure discharge. The number and size of lines will be limited by the pump capacity rating.

For example, with a pump having a capacity rating of 500 gpm at 150 psi, parallel operation would permit one 2½-inch line or two 1½-inch lines (flowing up to 250 gpm at 150 psi) to be used simultaneously with high-pressure lines from the third stage for flow to 60 gpm at 300 psi. With the series setting of the transfer valve, one 1½-inch line, gated to reduce pressure, could be used simultaneously with a discharge of 30 gpm at 600 psi from the third stage.

The parallel-series 3-stage centrifugal has the capability of producing discharge pressures to 800 psi at the third stage.

Figure 20

FIRE SERVICE PUMPS

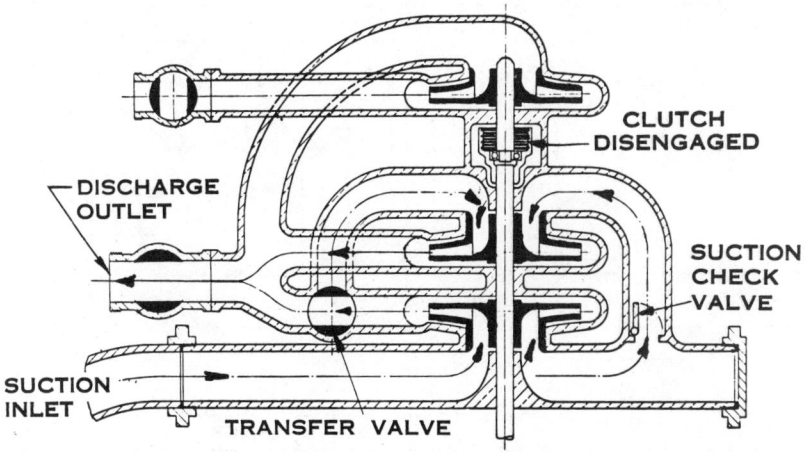

PARALLEL OPERATION

Figure 21a

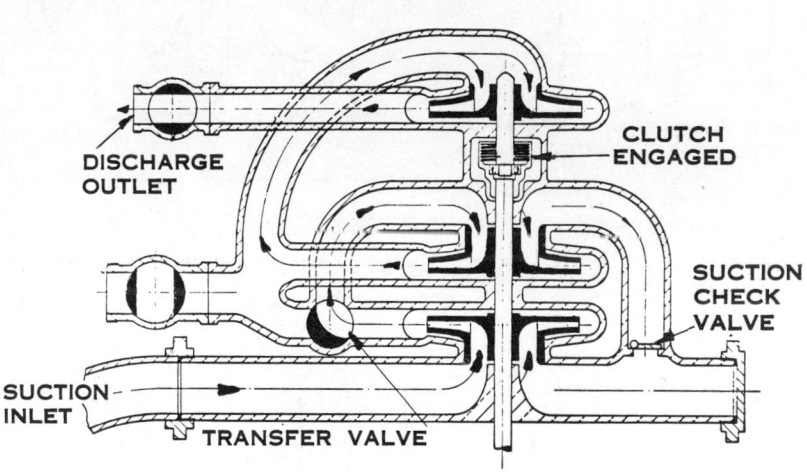

SERIES OPERATION

Figure 21b

PARALLEL-SERIES 4-STAGE CENTRIFUGAL PUMP

The parallel-series 4-stage centrifugal pump has limited use in the fire service as its primary use is for delivery of relatively large volume at high pressure (600 psi) for high-building standpipe supply. Build-

FIRE SERVICE HYDRAULICS

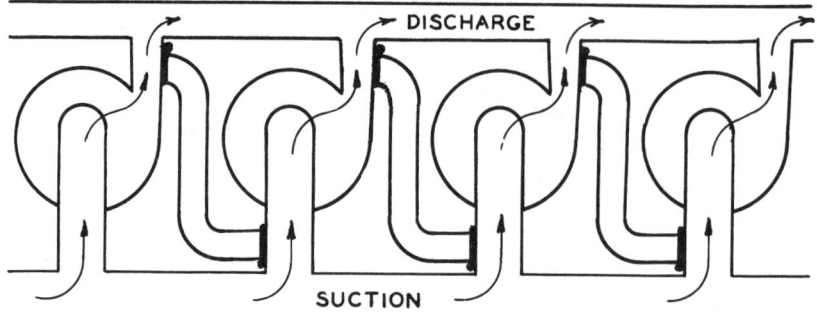

Figure 22a

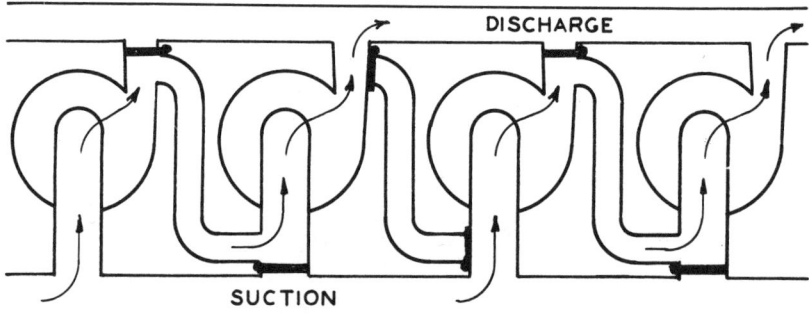

Figure 22b

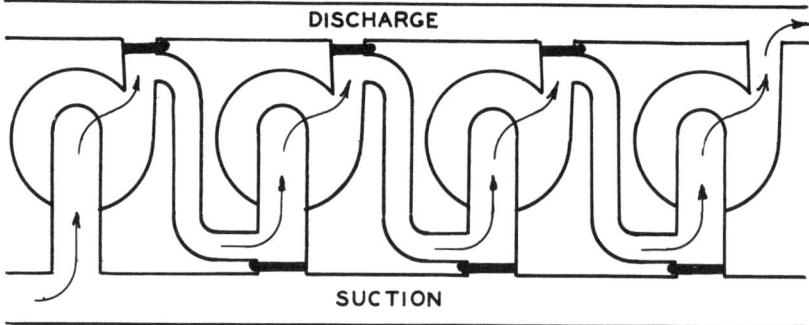

Figure 22c

ings of 50 or more stories normally depend on pumps electrically operated to supply the required volume for either general or fire needs. As fires frequently affect electric circuits, or the power is cut off for

FIRE SERVICE PUMPS

other reasons, dependence is placed on the fire pumps on apparatus to provide both the volume and pressure required to supply the standpipe system. Standard pumpers lack the volume of discharge required by these high pressures.

The parallel-series arrangement widens the useful range of operation. In parallel (Figure 22a) the pump provides its rated volume at pressures to 160 psi, so in this setting it finds its greatest usefulness. In the second setting the pump operates in series with two pairs of impellers discharging in parallel into a common outlet (Figure 22b) for discharge pressures of 300 to 350 psi. In the series setting (Figure 22c) the maximum pressures are provided with a discharge volume sufficient for at least two hand lines. Since the pump pressure is high, pressure loss, due to building height and friction loss, leaves the firemen with only a moderate working pressure on the uppermost floors.

An example would be a fire on the 60th floor of a building. With a pressure loss of ½ pound (approximate) per foot of elevation, and an average of 10 feet per story, a pressure loss from elevation would be about 300 psi. To this loss is added the friction loss in the lines from the pumper to the standpipe and friction loss in the standpipe. Thus

Figure 23

FIRE SERVICE HYDRAULICS

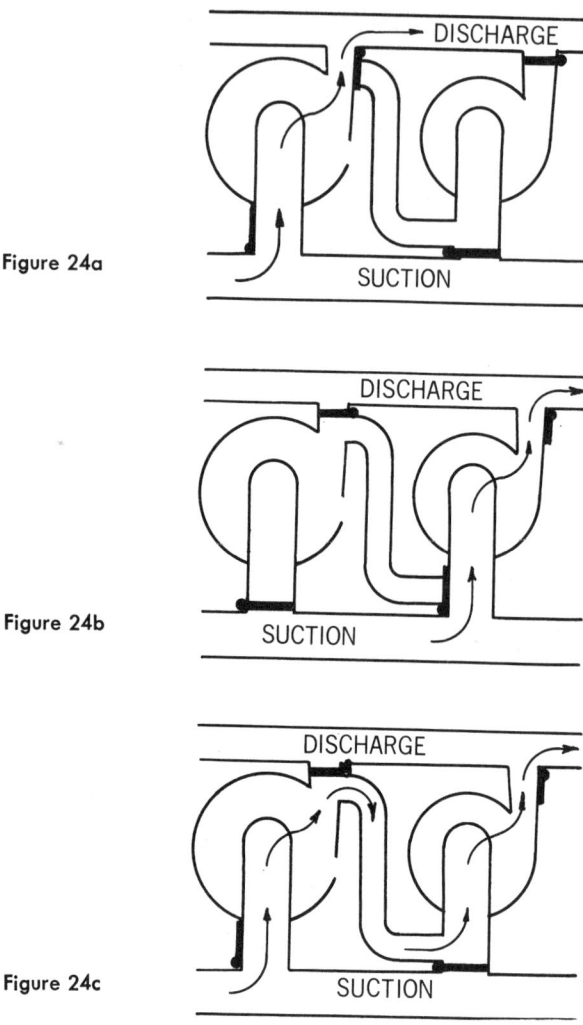

Figure 24a

Figure 24b

Figure 24c

even with short lines from the pumper to the standpipe, the pressure required at the pump to supply working pressure on the 60th floor will be from 400 to 450 psi.

"DUPLEX-MULTISTAGE" CENTRIFUGAL PUMP

The "Duplex-Multistage" centrifugal pump is an unusual combination of single-stage and series-type operation. The pump is built with two single-suction impellers, of different size and each mounted on a separate impeller shaft. Figure 23 illustrates the general construction.

When large volumes of water are required, the pump is operated with the single "capacity" impeller discharging the required quantity at pressures to approximately 160 psi. This arrangement is shown in Figure 24a. For pressures to 250 psi, the pump is operated with the single-stage intermediate pressure impeller engaged, and the "capacity" impeller disengaged (Figure 24b). For high-pressure operation, 250 to 350 psi, the pump is operated in series (Figure 24c).

Provision is made on the pump transmission case to mount a small two-stage pump, which is connected to the outlet of the "series" manifold passage to give a 4-stage, high-pressure (to 800 psi) discharge.

PISTON PUMP

The piston pump has a long record of dependable service to fire departments. It was the first pump developed for fire fighting. History records its use about 150 BC, and it continued to be the predominant type of fire pump for about 2,000 years. It was not until motorized fire apparatus came into general use, along with greatly improved water supply systems, that other types of fire pumps began to be favored. Where fire departments are required to operate mostly at draft, particularly under high lift conditions, the positive-displacement piston fire pump is still preferred by many.

The piston pump now in fire service use is a single-acting type (Figures 25a and 25b). In a single-acting pump, discharge occurs only with piston movement for one half the cycle of crankshaft rotation. A double-acting pump discharges twice for each cycle of crankshaft rotation. First discharge occurs as the piston moves on the half cycle away from the crankshaft with water entering behind the piston to fill the cylinder. Second discharge on the return stroke repeats the process. A set of intake and exhaust valves is required at each end of the cylinder in the double-acting pump. In the single-acting pump, only one set of valves to control inlet and exhaust is required.

Atmospheric pressure opens the inlet valve as the piston moves toward the crankshaft, creating a vacuum in the cylinder (Figure 25a). Water then enters to fill the cylinder. On the return stroke of the piston, the inlet valve closes and pressure created by the piston movement opens the discharge valve (Figure 25b).

Each rotation of the crankshaft causes a certain volume of water to be displaced or discharged, on the pressure stroke. To smooth out

FIRE SERVICE HYDRAULICS

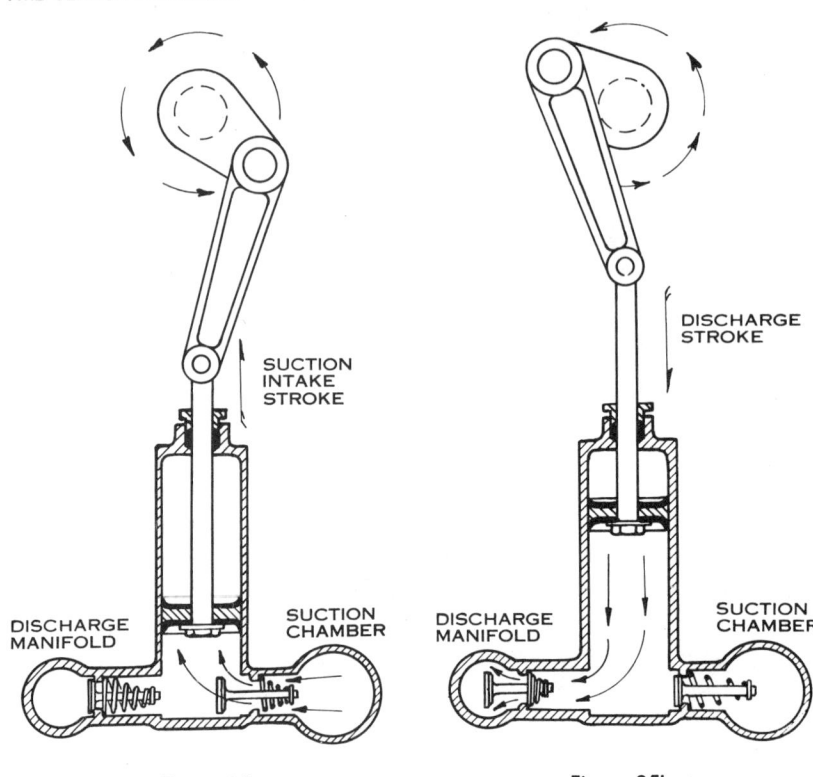

Figure 25a   Figure 25b

the flow pressure and obtain more continuity of flow with less pulsation and vibration, three cylinders are used. Each cylinder has a pressure stroke of 180 degrees of crankshaft travel. With three cylinders, discharge occurs at each 120 degrees of crankshaft rotation, an overlap that smooths and reduces any pulsation and vibration.

The theoretical displacement of the piston pump is determined by the formula:

$$D = \frac{A \times S \times \text{number of cylinders}}{231}$$

D = displacement in gallons per revolution
A = area of cylinder cross section
S = stroke or piston travel
231 = cubic inches in one gallon (US).

An example to determine the displacement of a pump, and both the theoretical and actual volume pumped, with the following assumed

FIRE SERVICE PUMPS

values would be:

$$6 \text{ inches} = \text{cylinder diameter (3 inches} = \text{radius)}$$
$$8 \text{ inches} = \text{stroke}$$
$$3 = \text{number of cylinders.}$$

$$D = \frac{3.1416 \times (3)^2 \times 8 \times 3}{231}$$

$$= \frac{28.274 \times 8 \times 3}{231}$$

$$= \frac{678.57}{231}$$

$$= 2.93 \text{ gallons per revolution of the pump crankshaft}$$

At an assumed pump speed of 190 rpm, the theoretical discharge would be $2.93 \times 190 = 556.7$ gallons per minute. To determine the actual discharge, the percent of slippage at 150 psi must be known. At a discharge pressure of 150 psi the slippage is approximately 5 percent. The actual discharge would then be: $556.7 \times .95 = 528.9$ gpm.

Gear ratios change

The piston fire pump now in current use (Figure 26), employs three different gear ratios, one for pressures to approximately 175 psi, one for pressures from 175 to 225 psi and a third ratio for pressures in excess of 225 psi.

If the different ratios were not used at the higher pressures (when the discharge volume is greatly reduced), the pump speed and corresponding engine speed would be so low that the engine would be unable to deliver adequate power for the required discharge and pressure.

The gear ratio for operation at pressures to 175 psi is approximately 11:1. All of the ratios will vary according to the engine characteristics, but for an example, using a ratio of 11:1 and with a pump speed of 190 rpm, the engine speed will be $190 \times 11 = 2,090$ rpm. At 200 psi pressure with a discharge of 350 gpm the pump speed will be approximately 125 rpm, allowing for a slippage of 2 percent. With the same gear ratio 11:1, the engine speed would drop to 1375 rpm but the brake horsepower requirement has *increased* from 64 BHP at 556 gpm to 68 at 350 gpm due to lower efficiency at the higher pressure.

Increasing the gear ratio for the pressures from 175 to 225 psi to approximately 19:1 for engine to pump, provides an engine speed for these increased pressures of 2375 rpm. For pressures above 225 psi,

FIRE SERVICE HYDRAULICS

Figure 26

the gear ratio is approximately 35:1, assuring ample engine power at all capacities and pressures.

One notable feature of the piston pump is the reduction in slippage as the pressure increases. This is due to the piston packing providing a more effective seal against leakage past the piston with increased pressure.

The piston pump has a relatively high efficiency at the capacity rating of 150 psi, 75 to 80 percent. The efficiency is well sustained over the operating range of discharge and pressures, the lowest at the one half capacity rating being in excess of 55 percent.

The piston pump is self-priming, eliminating any requirement for

FIRE SERVICE PUMPS

Figure 27

a supplemental priming system. As the pump is positive-displacement in operation, a relief valve is provided for discharge pressure control.

The piston pump performs best at draft although it can be used with a positive-pressure water supply with excellent performance. It does not, however, fully utilize positive pressure at the suction inlet, as does the centrifugal fire pump. The principal benefit to the piston pump from positive pressure at the suction inlet is a slight reduction in the brake horsepower requirement and reduced slippage.

ROTARY FIRE PUMP

The rotary pump was first used by the fire service on horse-drawn "tubs" beginning in the 17th century. During the 19th century the rotary pump was adapted to the steamer. In 1856 Silsby was the first to use it. Though it has an excellent record of fire service performance, its use has declined and it is now used only where nearly all pumping is at draft under conditions of high lift or long suction hose layouts. The rotary pump is self-priming and is positive-displacement. It is built in two designs, one with gear-type rotors and one with three lobe rotors. The gear type (Figure 27) uses one gear as the driver and the second gear as the driven rotor. A second design is shown by Figure 28. It uses 3-lobe rotors, both rotors being driven by externally mounted synchronizing gears. On some rotor designs a gib-type key or vane is inserted at the outer center of each lobe to seal against water leakage (slippage) between the outside diameter

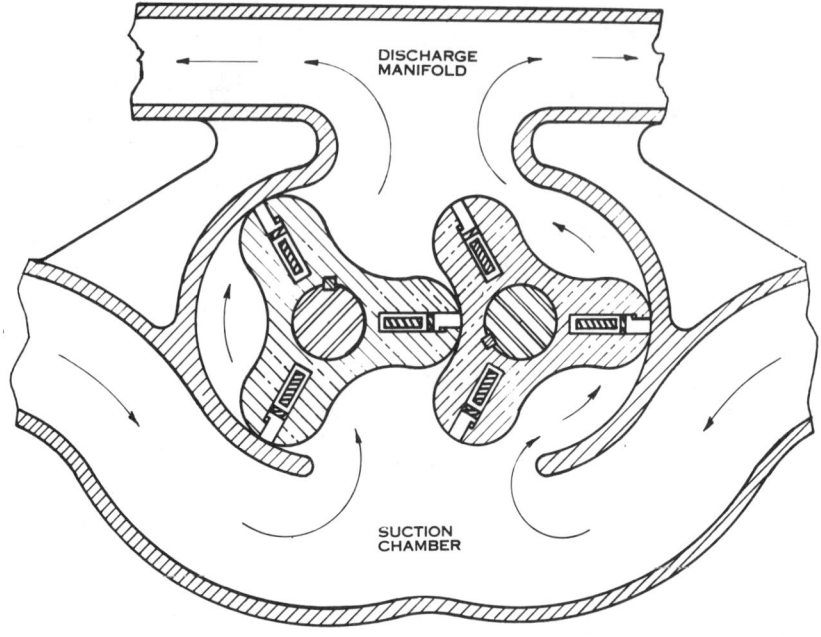

Figure 28

of the lobes and the housing, also between the lobes as they mesh at the center.

When operating, the rotors turn in opposite rotation, air or water in the suction chamber filling the space between the teeth or lobes. The turning rotors carry the air or water up to the discharge chamber where it is displaced as the rotor teeth or lobes roll into mesh, and is forced out through the discharge port. The number of displacements or pulsations at discharge is the sum of the number of teeth or lobes in both rotors for each revolution of the pump. The pulsation or pressure variation in the discharge from pumps with rated capacity to 750 gpm is not of sufficient magnitude to create any problem. For pumps with larger displacement and higher capacity rating, air domes are provided to cushion and partially absorb the pulsation for a smooth flow.

The displacement of a rotary pump varies with the design of the rotors and the discharge requirements. The pump manufacturer provides information on displacement, gear ratios and other basic information as a part of the test report submitted at the time of delivery of the fire apparatus. Such information should be carefully preserved for use in future testing of the pump.

FIRE SERVICE PUMPS

Rotary pump discharge

The discharge from a rotary pump can be determined by multiplying the pump displacement (gallons per revolution) by the pump speed (rpm). The slippage, which is given in terms of percent of the theoretical discharge, should be subtracted from the theoretical discharge to determine the actual output.

Engine speed is a product of pump speed multiplied by the engine to pump gear ratio. Two gear ratios are usually supplied, one for capacity operation to 175 psi and a second for use at pressures over 175 psi. If the second ratio were not supplied to increase engine speed in relation to pump speed, the engine speed would be too low at the lower discharge rates to provide adequate power for pressures above 175 psi.

An example of an operating pump, with the following values is:

- .71 = displacement, gallons per revolution of the pump.
- 755 = pump speed (rpm) observed at capacity discharge at 150 psi pressure.
- 440 = pump speed (rpm) observed at capacity discharge at 250 psi pressure.
- 2.3:1 = engine to pump gear ratio, capacity to 175 psi pressure
- 4.5:1 = engine to pump gear ratio, pressures above 175 psi.
- 5% = slippage at 150 psi pressure.
- 18% = slippage at 250 psi pressure.

.71 × 755 = 536 gpm theoretical discharge
536 × .95 = 509 gpm actual discharge.
755 × 2.3 = 1,736 engine rpm at 509-gpm discharge at 150 psi.

$$\frac{509 \times 150}{1716} = 44.4 \text{ WHP}$$

With a pump efficiency of 73 percent as shown on the characteristic curve (Figure 29), the pump requires 60 BHP at rated capacity. When operating at 250 psi, discharging 250 gpm the following values are found:

.71 × 440 = 312 gpm theoretical discharge
312 × .82 = 255 gpm actual discharge.
440 × 4.5 = 1,980 engine rpm at 255 gpm discharge at 250 psi

$$\frac{255 \times 250}{1716} = 37.1 \text{ WHP}.$$

With a pump efficiency of 54 percent (Figure 29), the pump requires 68.7 BHP to produce the lower volume at higher pressure.

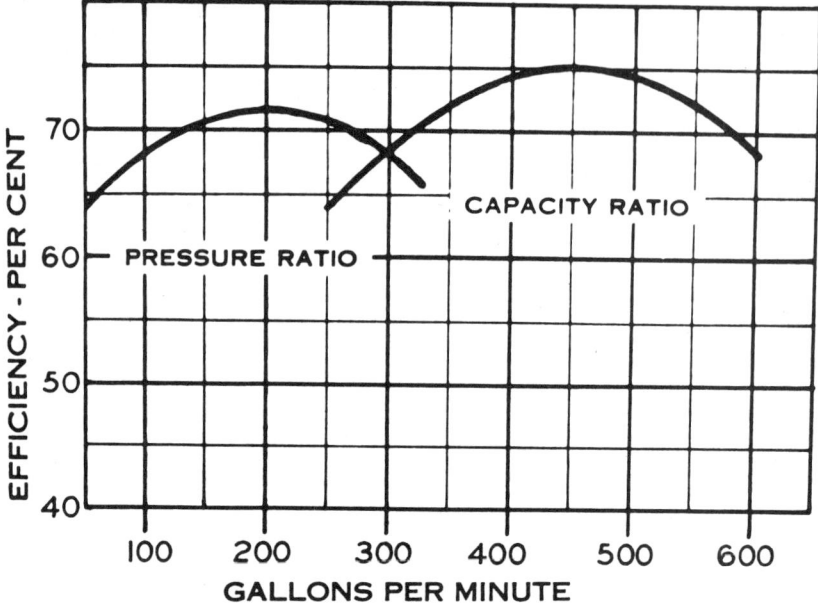

Figure 29

BOOSTER PUMP

A booster pump, by definition in the NFPA Specification No. 19, is a pump permanently mounted on the fire apparatus with a discharge capacity rating of less than 500 gpm at 150 psi. There are two types in current use in this classification. These are: (1) rotary gear and (2) single-stage centrifugal.

The rotary pump is a positive-displacement type, self-priming, with a capacity rating of 200 gpm at 150 psi. The pump is mounted on the chassis frame, usually back of the cab, and is driven by a power take-off mounted on the side of the road transmission. This drive arrangement permits apparatus movement, when desired for fighting grass and brush fires.

The suction side of the pump is piped to the water tank on the apparatus for the primary source of water supply. Two suction inlets, one on each side of the apparatus, may also be provided for water supply from a hydrant, or used with hard suction hose—usually 3-inch inside diameter—to supply water from draft. As the pump is self-priming, no auxiliary primer is required for operation at draft.

The capacity rating for all booster pumps, like the major fire pumps,

FIRE SERVICE PUMPS

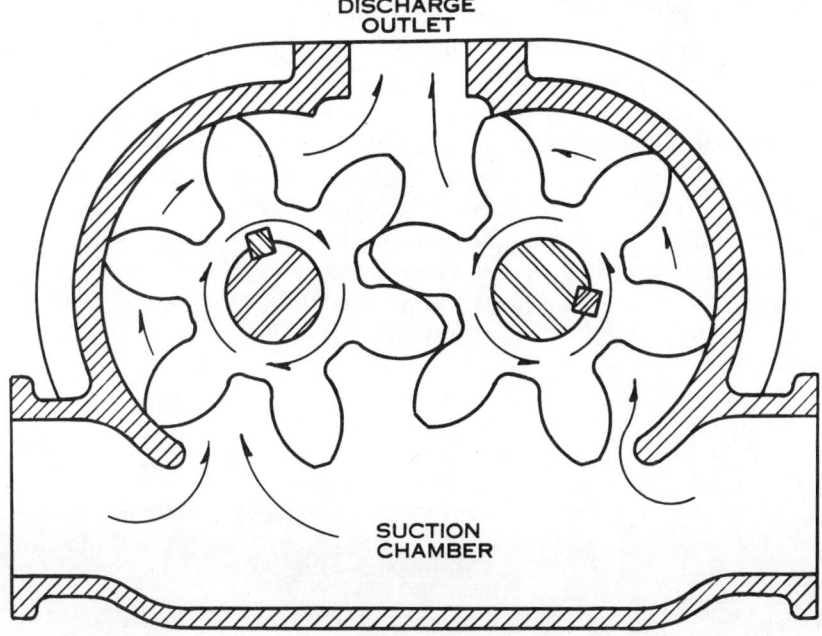

Figure 30

is established by test at draft on a lift up to 10 feet. The capacity rating at 200 psi pressure is one half the rated capacity at 150 psi. The discharge volume and pressures are adequate to supply two booster lines (1-inch) or one 1½-inch line for small fires which can be controlled and extinguished with the volume discharged from such lines.

The rotary pump (Figure 30) consists of two meshing gears in a housing, a relief valve to control discharge pressure at a pre-set value, valves to control water flow from the tank, and discharge gate valves. One rotary gear is the driver. When the driving gear rotates, the rotation of the driven gear is opposite. The rotor gears pick up air or water from the suction chamber, filling the space between the gear teeth. In about 180 degrees of rotation (½ revolution), the air or water carried between the teeth is displaced by the meshing of the teeth, forcing air or water out the discharge port. Figure 30 illustrates the action in the pump.

A definite volume of air or water is displaced as each pump gear tooth rolls into engagement and this displaced volume multiplied by the total number of teeth in the two rotors or gears is the total displacement per revolution of the driving shaft. This displacement value for any specific pump is provided by the manufacturer, on request. It is also usually available from brochures describing the pump.

The displacement value given by the manufacturer is usually the theoretical displacement. That is, no allowance has been made for "slippage" which is a term describing the water lost from the discharge side of the pump back to the suction side. This leakage, which occurs at the ends of the rotors, plus a small amount trapped between the teeth and the end of the engaging tooth, varies with the discharge pressure, usually 5 percent at 150 psi and 10 to 15 percent at 200 psi.

An example of the determination of the discharge with a rotary booster pump having a displacement of 0.166 gallons per revolution, and operating at a speed of 1270 rpm, would be as follows:

$$1270 \times 0.166 = 210.8 \text{ gpm}$$

Allowing for a slippage of 5 percent, the net output at 150 psi would be:

$$210.8 \times .95 = 200.3 \text{ gpm, net discharge.}$$

The engine speed will vary according to the gear ratio through the power take-off. A pump with larger displacement would usually operate at a lower pump speed and the gear ratio between the engine and pump would be somewhat higher, so that the engine speed would be in the range of 1200 to 1500 rpm for most rotary booster pumps.

The rotary booster pump is quite efficient, the efficiency being 70 to 75 percent. The brake horsepower requirements are low enough so that only a partial throttle will provide ample power. For a discharge of 200 gpm at 150 psi which is an output of 17.3 WHP, and with a pump efficiency of 70 percent, only 24.7 BHP is required of the engine, well within the limitation of power transmission of the power take-off.

Centrifugal booster

The centrifugal type of booster pump is provided in either of two designs: front-mounting on the vehicle with drive from the front-end of the engine crankshaft, or midship frame-mounting with drive by a power take-off on the side of the road transmission.

The front-mounted booster pump usually has a capacity rating of 300 to 400 gpm at 150 psi, with the 350-gpm rating the most common. It is a single-stage centrifugal pump, with the same design features, including clutch, transmission, primer and pressure control (governor) as the larger standard capacity rated fire pump described in the section on single-stage centrifugal fire pumps.

The front-mount centrifugal booster pump can be operated for

FIRE SERVICE PUMPS

Figure 31

fire fighting purposes simultaneously with vehicle movement, if desired by the operator.

The pump is capable of supplying two booster lines at pressures to 350 psi, or one 2½-inch line or three 1½-inch lines at moderate pressures.

The midship-mounted centrifugal booster pump (Figure 31) is single

FIRE SERVICE HYDRAULICS

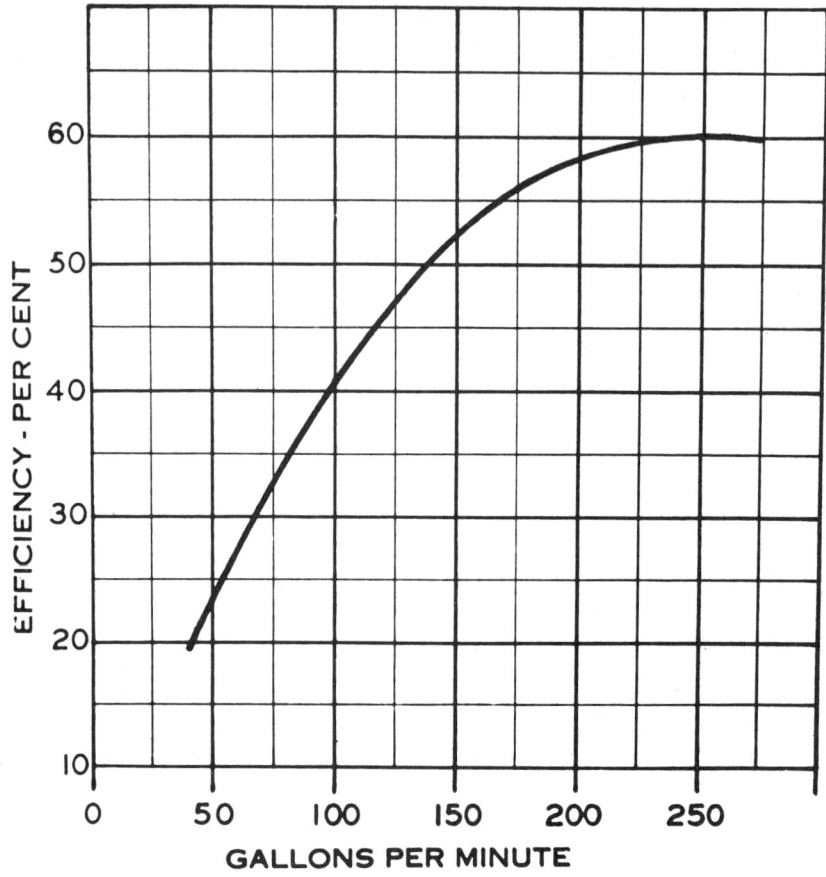

Figure 32

stage and is driven by a power take-off mounted on the side of the road transmission. This drive arrangement permits vehicle movement while pumping, if the type of fire fighting makes this feature desirable.

Most of the centrifugal booster pumps have a dual rating, 250 gpm at 150 psi using 3-inch suction hose (20 feet) on a 10-foot lift, or 200 gpm at 150 psi using 20 feet of 2½-inch suction hose on a 10-foot lift. The capacity is greatly increased when the pump is operated with the water tank on the apparatus as the source of supply. This increased capacity, which may be limited by an excess in the number of elbows and tees, is due to a small but positive flow pressure because of elevation difference between the pump and water tank which is mounted above the chassis frame.

FIRE SERVICE PUMPS

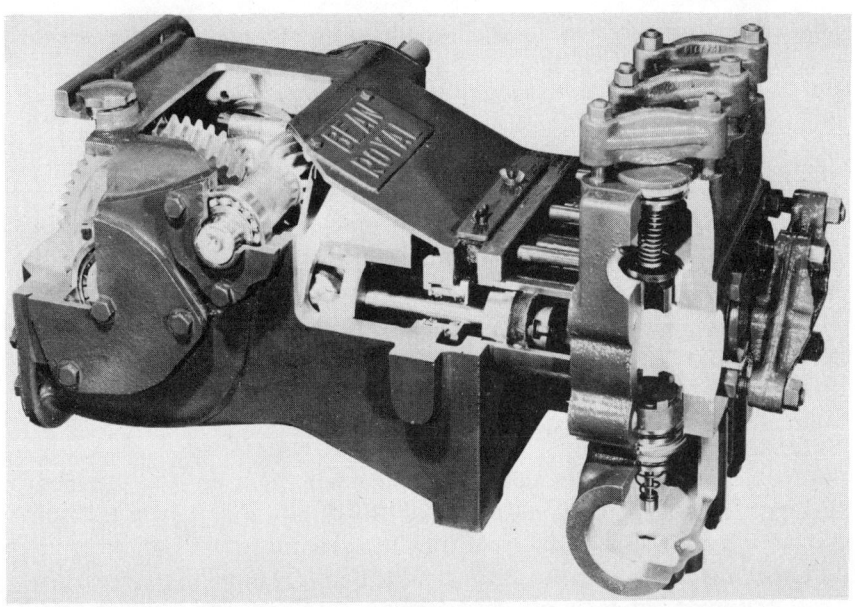

Figure 33

The limitation of performance may not be entirely in the design of the pump but in the very definite limitation of power that can be transmitted through the power take-off. The desirable maximum is 25 BHP. The absolute maximum is 40 BHP. With booster pump efficiency as shown on the characteristic curve (Figure 32) the pump can be expected to provide the following performance without exceeding the 40 BHP limit through the power take-off:

gpm	Pressure (psi)
250	150
200	175
175	200
125	250
60	300

HIGH-PRESSURE PUMPS

High-pressure pumps for fire service use are of two types, (1) piston and (2) centrifugal. They are capable of operating in the pressure range of 350 to 850 psi, with most in the 600 to 850-psi pressure range.

The piston pump is positive-displacement, self-priming, and those

FIRE SERVICE HYDRAULICS

used in the fire service are single acting. The pumps have three cylinders (Figure 33) and are rated at 70 gpm at 850 psi. The pump is provided with 2½-inch suction inlets for drafting water or connecting to a hydrant for water supply. An automatic operating relief valve protects against pressure rise should one or more lines be shut off while the pump is in operation.

The pump is driven from a power take-off mounted on the side of the road transmission or through a power divider at the rear of the road transmission. With either type of drive, the pump can be in operation simultaneously with vehicle movement, if such operation is desired.

High-pressure centrifugal

Most high-pressure centrifugal pumps are built in two designs; 2-stage and 4-stage, midship-mounted and driven by a power take-off mounted on the side of the road transmission.

There are design variations to produce high-pressure discharge using the basic major fire pump with the addition of one or more auxiliary stages. The addition of a third stage as an auxiliary to a parallel-series 2-stage major fire pump is described and illustrated in a preceding section of this chapter.

Some manufacturers offer an auxiliary 2-stage pump mounted on the major fire pump transmission case to convert a single-stage major fire pump to a 3-stage high-pressure pump. Also a similar mounting of a 2-stage pump on a Duplex-Multistage pump provides 4-stage operation for high pressure.

All of these auxiliary stages are optional for operation. As the name implies, they are auxiliary and only used for high-pressure discharge. They do not affect the normal operation of the major fire pump.

The 2-stage high-pressure centrifugal pump (Figure 34) has two impellers designed specifically for high-pressure operation in a series arrangement (Figure 10). The pump is designed to operate at pressures to 1,000 psi, although the discharge at this pressure is too small to be of a fire fighting value. At 850 psi the discharge is sufficient to supply one high-pressure line. These performance figures apply only when the drive system power transmission is limited to 40 BHP. Actually, the pump has the capability of discharging up to 80 gpm at 1,000 psi. Such performance is available only when the drive line is not a limiting factor, such as a skid mount with the pump directly connected to the drive engine.

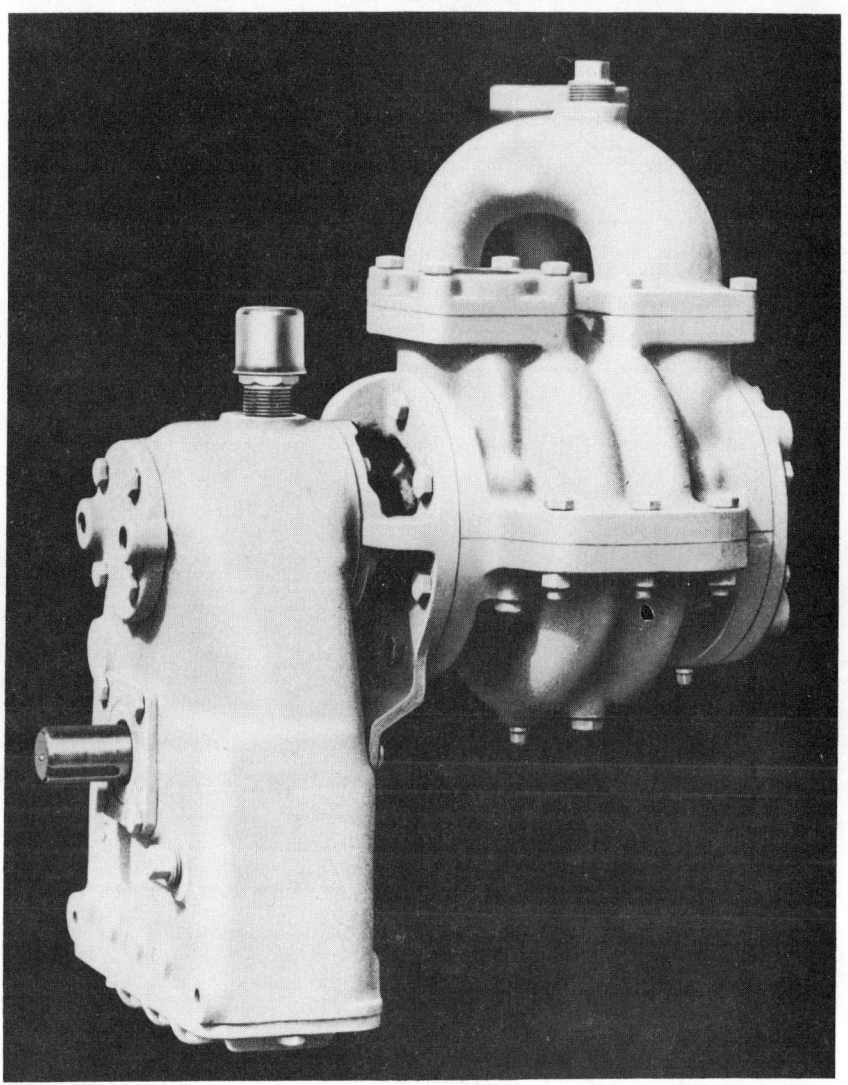

Figure 34

The efficiency of the 2-stage centrifugal booster pump is shown for various discharge rates (Figure 35). These efficiencies cover discharge rates and pressures within the normal fire service operating requirement, and within the power transmission maximum of 40 BHP.

The range of discharge rates at various pressures when operating

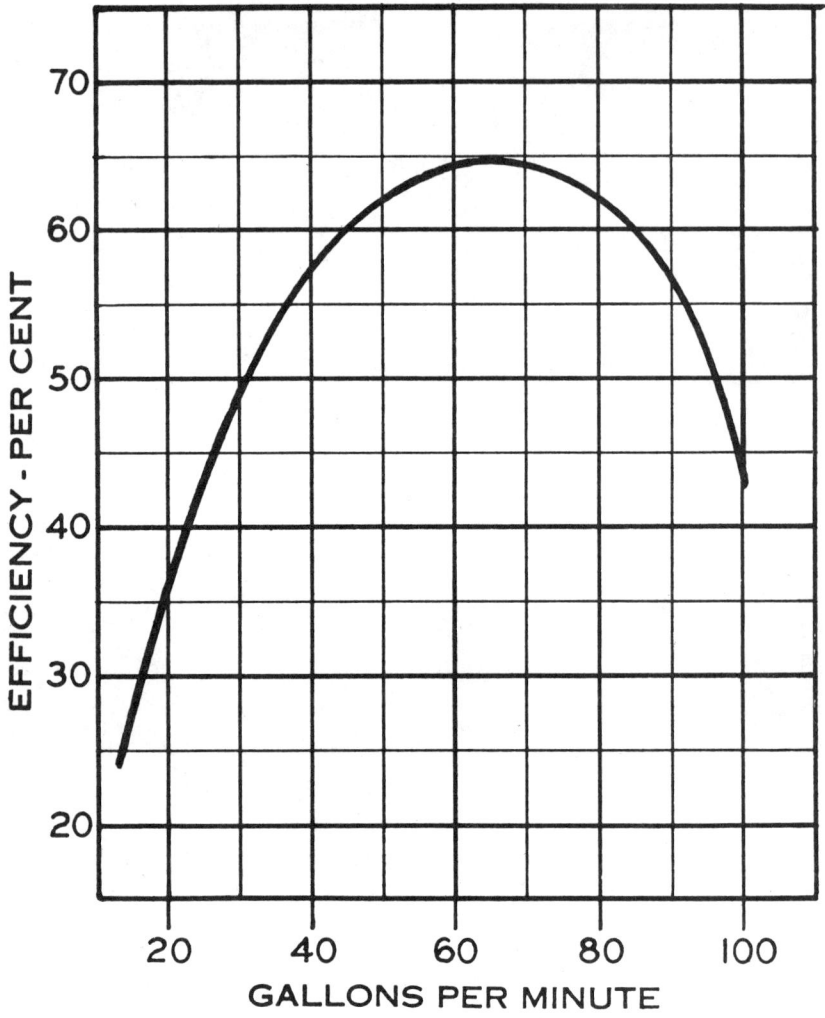

Figure 35

at draft on a 10 foot lift, are as follows:

gpm	Pressure (psi)
100	300
80	400
70	500
50	600
30	700
20	800
18	850

FIRE SERVICE PUMPS

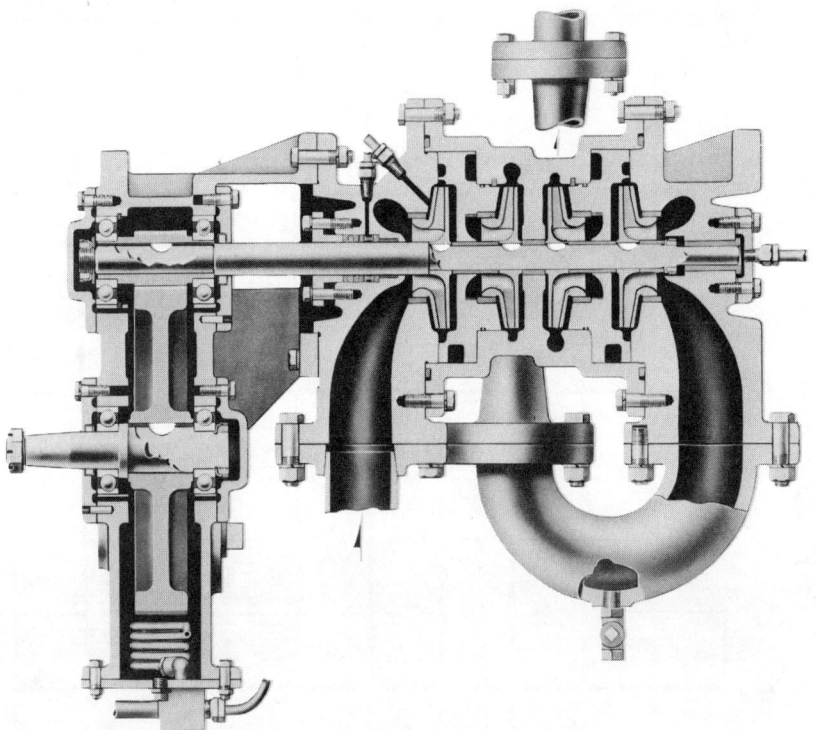

Figure 36

When operating with the water tank on the apparatus as the supply source some increased performance will be obtained over the values given above.

The 4-stage high-pressure centrifugal pump has four impellers operating in series (Figures 22c and 36). The pump is designed to operate at pressures to 1,000 psi, but any pressure in excess of 850 psi produces such finely divided droplets that the stream or fog range becomes too short to be useful. Tests have shown that at pressures of 1,300 psi and higher, using current fog guns, the discharge approaches a true fog. The particles move only a few feet from the gun and stay in suspension in the atmosphere for a considerable time. Tests conducted by the Underwriters' Laboratories, Inc. (Research Bulletin No. 10) indicate the droplet size should be between 150 microns and 300 microns to be effective in fire extinguishment. The droplet must be large enough to travel from the nozzle to the heated area surrounding the fire. The primary function of pressure in a fire pump is to supply sufficient

FIRE SERVICE HYDRAULICS

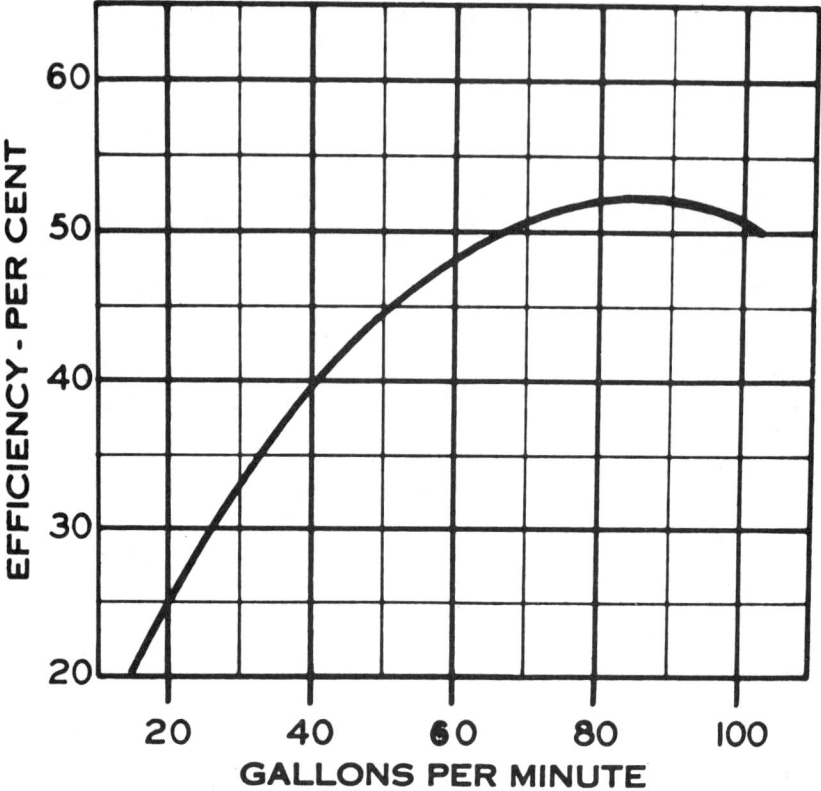

Figure 37

pressure energy to the water stream to overcome friction loss in the discharge hose, the loss in energy at the nozzle as the stream is broken into small droplets and to carry the droplets to the fire area. Any pressure in excess of these requirements does not serve a useful purpose.

The efficiency of the 4-stage centrifugal high-pressure pump is shown by the typical characteristic curve (Figure 37).

The capacity ratings at various pressures, with the pump operating at draft, on a 10 foot lift, are as follows:

gpm	Pressure (psi)
100	200
80	400
70	600
60	800
50	900

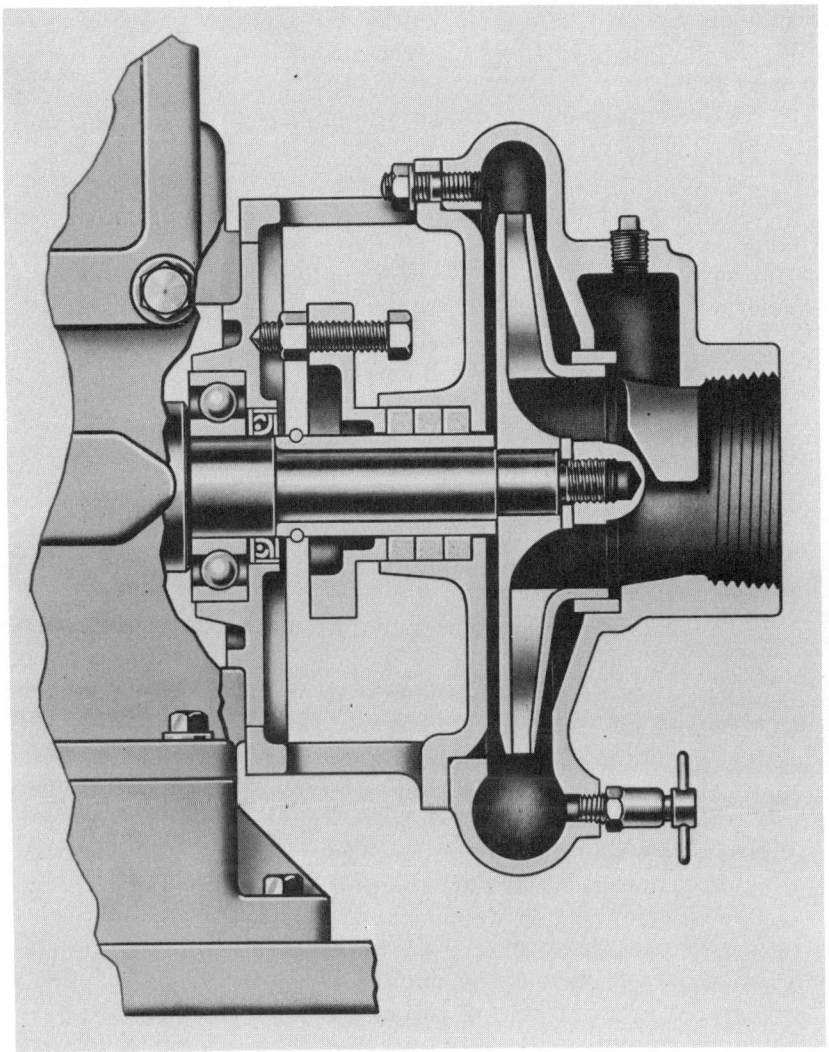

Figure 38

PORTABLE PUMPS

The portable pump is a valuable auxiliary pump in fire service because of its many important uses. The portable pump is a single-stage centrifugal pump directly connected to an air-cooled gasoline engine. Figure 38 illustrates the engine mounting.

Three styles of mounting the pump-engine unit for portability are

available:
1. Tubular frames with swing-out handles to permit two to four men to carry it.
2. Unit mounting on a small platform provided with two long handles and a pair of truck-type wheels.
3. Unit mounting on two steel channels for a skid-type assembly.

The first type is most favored by the fire service although all have a definite use.

To insure the portability of the pump unit, weight limits have been established for the five classifications. These are given in NFPA No. 191, "Specifications for Portable Pumps,"[1] and are as follows:
1. Small volume, relatively high pressure.
    15–25 gpm capacity at 200 psi pressure.
    Maximum weight, 175 pounds.
2. Medium volume, medium pressure.
    100 gpm capacity at 60 psi pressure to
    50 gpm capacity at 90 psi pressure.
    Maximum weight, 150 pounds.
3. Large volume, relatively low pressure.
    250 gpm capacity at 20 psi pressure to
    100 gpm capacity at 50 psi pressure.
    Maximum weight, 150 pounds.
4. Medium volume, relatively high pressure.
    160 gpm capacity at 90 psi pressure to
    90 gpm capacity at 250 psi pressure.
    Maximum weight, 200 pounds.
5. Extra large volume, medium pressure.
    500 gpm capacity at 100 psi pressure.
    Maximum weight, 200 pounds.

In practice, a large variation exists between the capacity and pressure listings for any classification and the actual performance as stated by the several manufacturers of portable pumps.

All pump capacities are rated with the pump operating at draft on a 10-foot lift, using 20 feet of suction hose of the size listed for that pump.

The pumps are equipped with a priming device and are required to be capable of priming the pump when on a 10-foot lift with 20 feet of suction hose, in a period not to exceed 30 seconds. A speed-limiting governor is supplied on the engine for protection in event the pump runs out of water.

Figure 39 illustrates the construction of the self-priming pump.

[1] By permission of copyright owner, National Fire Protection Association.

FIRE SERVICE PUMPS

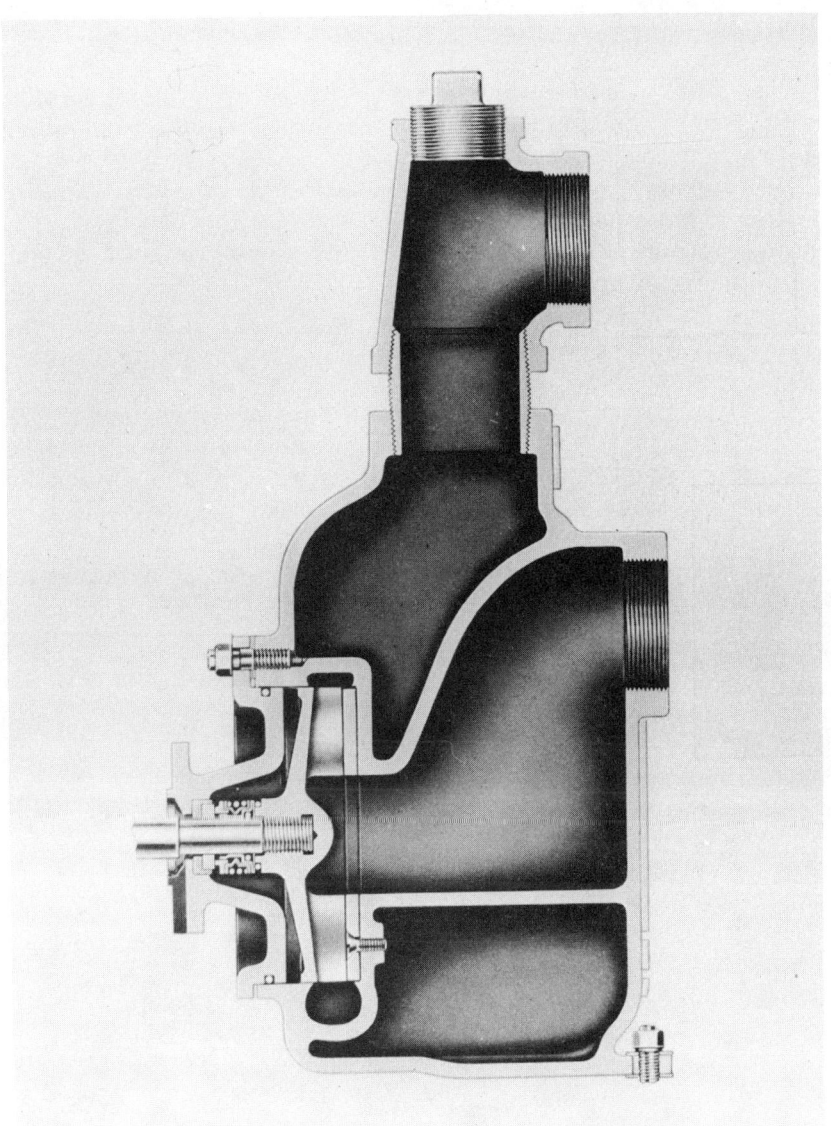

Figure 39

FIRE SERVICE HYDRAULICS

The suction chamber holds sufficient water to prime on lifts in excess of 10 feet.

The range of performance available in the self-priming portable pump is not as great as the performance range in the pumps requiring a priming device for operation at draft.

Self-priming portable centrifugal pumps come in several capacity ratings. The capacity range is from 100 to 300 gpm at 15 psi. The highest pressure offered is in a model with a capacity rating of 130 gpm at 15 psi and 30 gpm at 45 psi.

BACK-PACK PUMPS

The back-pack pump is a lightweight centrifugal pump and engine unit that can be carried on the back of a fireman to an area where motorized equipment cannot reach. It is basically a pump unit for forestry fire fighting, and as such usually meets the requirements of the U.S. Forest Service Specification No. 5100-274.

The pump (Figure 40) is single stage and has capacity ratings for

Figure 40

continuous duty as follows:

gpm	Pressure (psi)
200	10
175	20
150	27
100	42
50	50

The weight of the complete unit is approximately 30 pounds. The engine is 2-cycle and equipped with an automatic speed governor.

The back-pack pump is usually primed by a detachable hand-type vacuum pump, which has a quick-release connection and check valves to prevent loss of prime when the priming pump is detached from the centrifugal pump.

Figure 41

FIRE SERVICE HYDRAULICS

Some back-pack pumps are multi-stage (up to 4 stages) to provide increased discharge pressure at a flow of 50 gpm. Such pumping units will necessarily increase in weight up to 60 pounds, which is about the limit for this type of a pump.

TANKER PUMPS

The tanker is growing in importance as a unit of the fire department in areas of water scarcity or inaccessibility. Its pump has two functions:

1. To fill the tank as rapidly as possible. When the tank is unloaded, the elapsed time to travel back to the water source, fill the tank, return to the fire, and transfer the water to the fire fighting vehicle, will determine the effectiveness of the whole fire fighting operation.

2. To empty the tank on the tanker in a minimum of time, and if necessary, furnish sufficient pressure to supply one or two hose lines for fire fighting.

The tanker pump is a single-stage centrifugal driven by a power take-off mounted on the side of the road transmission, or more usually driven by a split shaft or full torque power divider. The skid mount is also used with an engine directly coupled to the pump.

The pump may be of the self-priming type or with a separate primer. Capacities vary over a wide range, from 500 gpm at 40 psi to low-performance pumps discharging 50 gpm at 125 psi.

One of the larger capacity skid-mounted tanker pumps is illustrated in Figure 41. A lower capacity power take-off driven self-priming tanker pump is illustrated by Figure 42.

MUNICIPAL AND INDUSTRIAL FIRE PUMPS (STATIONARY)

Pumps for municipal or industrial fire service are centrifugal type and are produced in a wide range of capacity ratings. The designs include horizontal impeller shaft, single-stage and multi-stage, and vertical shaft turbine pumps.

Fire pumps, and the complete installation, should be in agreement with the requirements in NFPA No. 20, "Standard for the Installation of Centrifugal Fire Pumps." Fire pumps are required to be approved for this class of service. Such approval and certification usually are given after satisfactory tests are conducted by engineers of Underwriters' Laboratories, Inc. (U.S.)

The standard capacity ratings are 500, 750, 1,000, 1,500, 2,000 and

FIRE SERVICE PUMPS

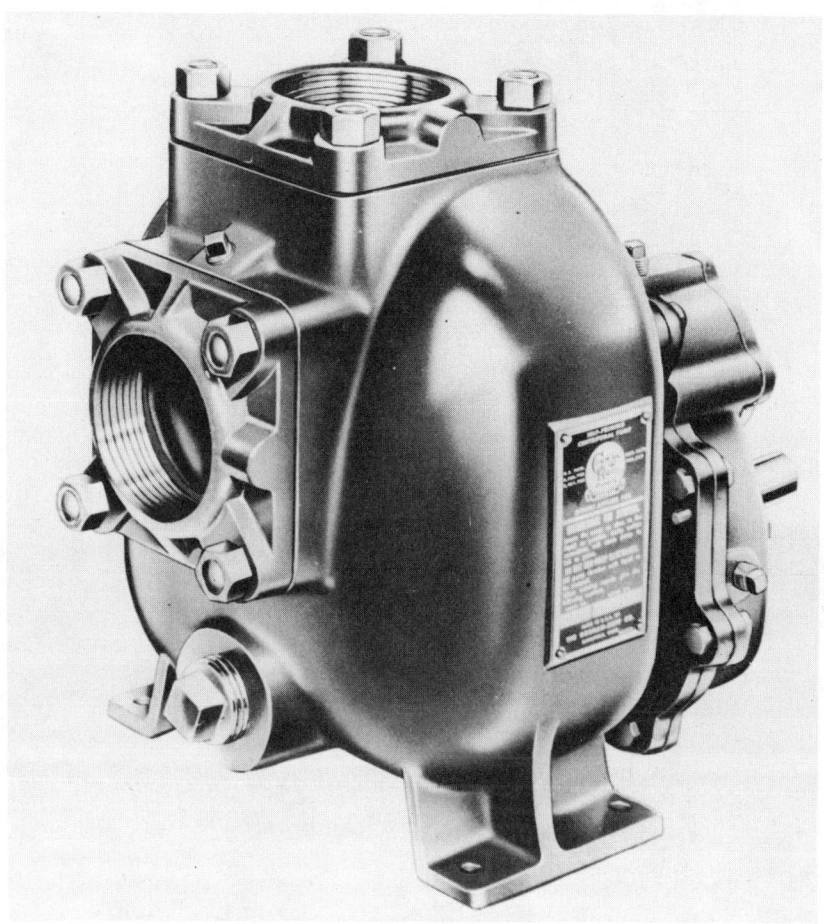

Figure 42

2,500 gpm. Larger capacities may be used in specially engineered installations. The standard capacity ratings are at a discharge pressure of 100 psi. Higher pressures may be required for specific installations to meet service needs.

Discharge pipes are sized according to pump capacity rating. Provision for testing the pump includes 2½-inch hose valves in the discharge header or manifold. The number of 2½-inch hose valves required depends on the rated capacity of the fire pump, as given in

FIRE SERVICE HYDRAULICS

the table below:

Pump Capacity gpm	Size Discharge Pipe	Number Hose Valves
500	6 inches	2
750	8 inches	3
1,000	8 inches	4
1,500	10 inches	6
2,000	10 inches	6
2,500	12 inches	8

Above data by courtesy of copyright owner, National Fire Protection Association.

For the larger capacity pumps, a metering device may be provided to measure flow, or a fixed nozzle or pipe outlet with discharge at an appropriate location can be used, reducing the number of 2½-inch outlets required.

A priming system and a relief valve are required as part of the installation.

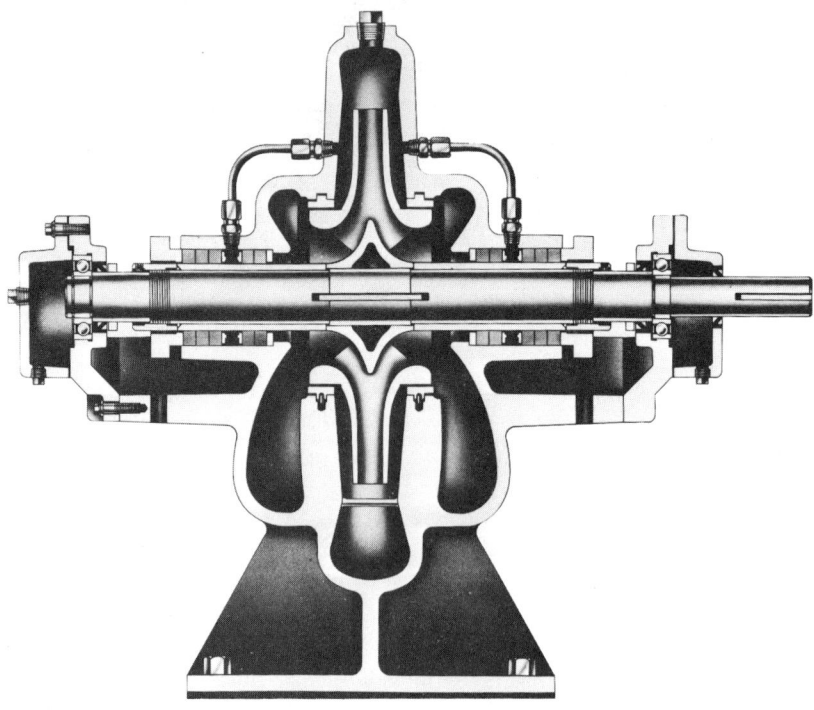

Figure 43

FIRE SERVICE PUMPS

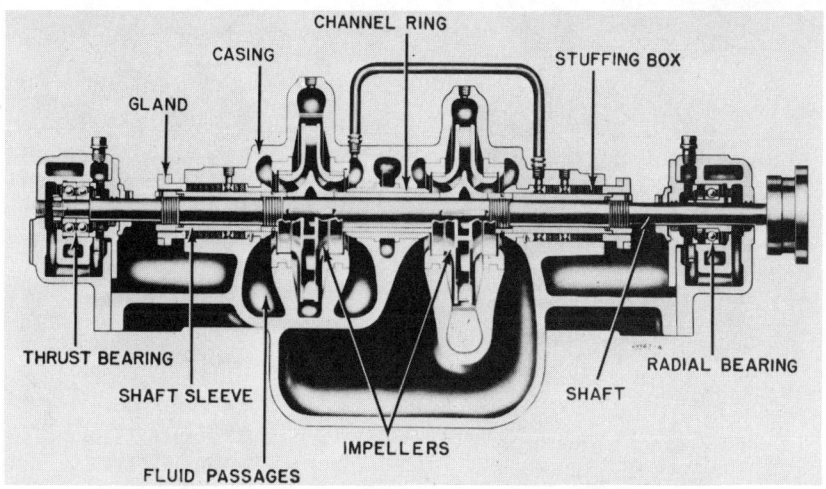

Figure 44

A noticeable difference exists between hydraulic terminology in fire departments and in municipal water supply departments. The almost complete absence of the word "pressure" is very noticeable in literature and in conversation with water works people. Where the fire service refers to "pressure" the water works personnel discuss pump performance in terms of "head." Actually, the end result is the same, but some mental agility is required to quickly reduce "feet of head" to the more familiar "pressure." This only means dividing the feet of head by 2.31 to know the equivalent pressure.

The horizontal impeller shaft fire pumps are usually single-stage, double-suction (Figure 43) or larger capacity pumps which may be 2-stage double-suction (Figure 44).

The vertical-shaft turbine-type fire pump is gaining favor as it is available for both wet pit use where the height of lift is relatively low and for deep wells where the lift requirements exceed lift capability with the pump above the ground.

The wet-pit pump is shown by Figure 45. This type of pump does not require suction piping as the pump impellers are submerged in water. No priming is required and the pump drive requires a minimum of floor space in the pumping station.

The deep-well pump is quite similar to the wet-pit pump, but the impellers are located much farther below the ground level and a casing or barrel encloses the column discharge pipe, line drive shaft and bowl assemblies. The bowl assembly includes the impeller and the

FIRE SERVICE HYDRAULICS

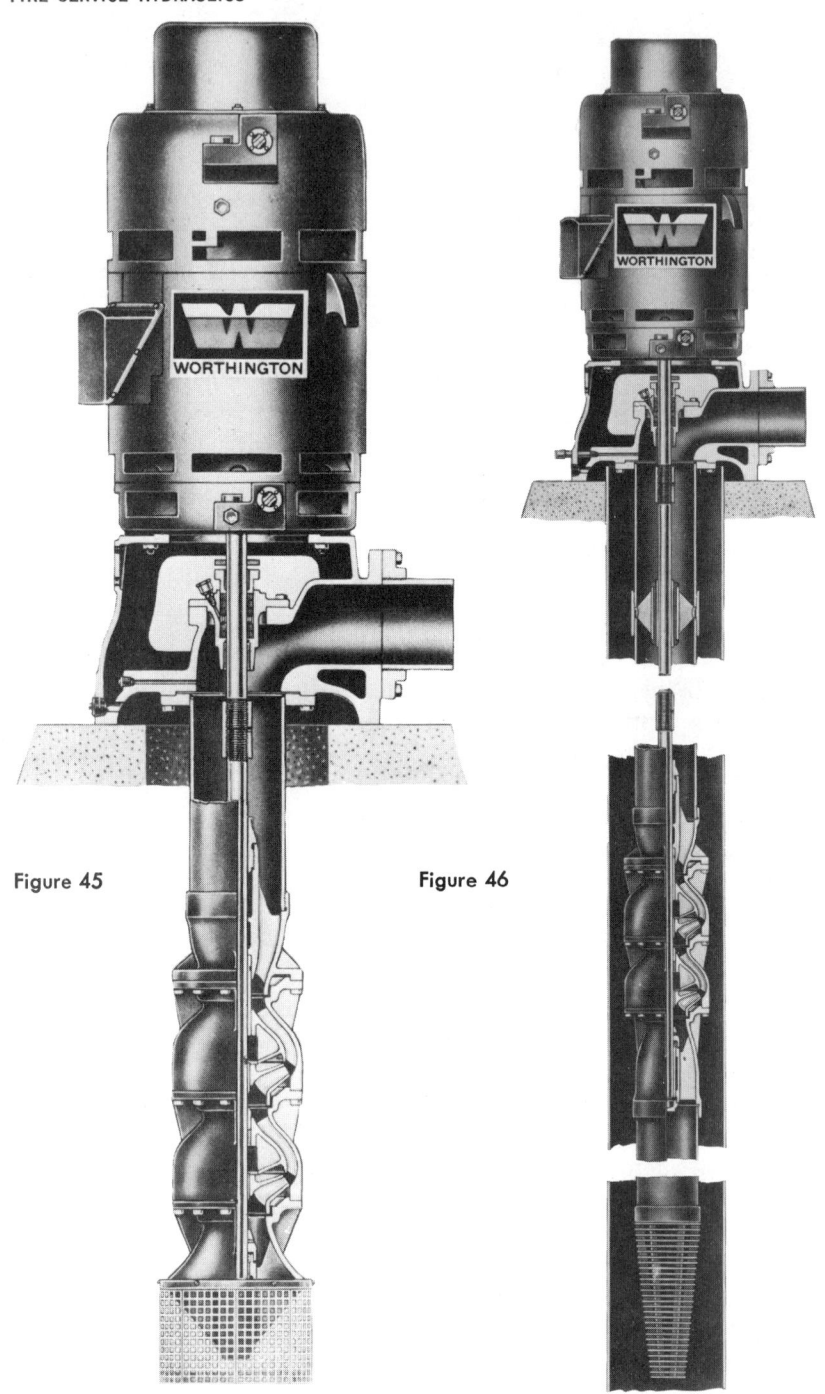

Figure 45

Figure 46

FIRE SERVICE PUMPS

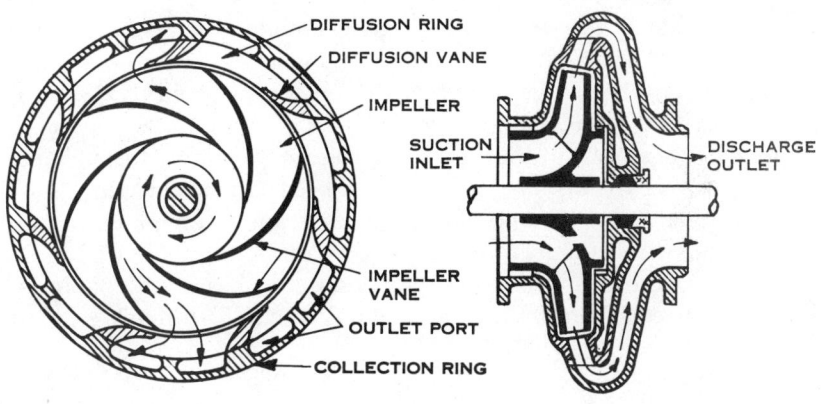

Figure 47

outlet housing. The design has great flexibility. If more pressure is required for either service or due to increasing the depth, it is only necessary to increase the number of bowl assemblies. Figure 46 illustrates a deep-well turbine-type fire pump.

In the vertical-mount turbine-type fire pump, the discharge from the impeller is guided by a number of diffusion vanes to the outlet instead of discharging into a volute chamber as in horizontal-shaft centrifugal pumps. The reason for using the turbine-type pump is to have radial balance in the discharge and obtain the high efficiency possible with the diffusion vanes outside the impeller.

From 1912 to about 1930 most centrifugal fire pumps were turbine or diffusion vane. This type of pump is ideally suited for use when the discharge and pressure are practically constant as they are for municipal and industrial fire pumps. On fire apparatus, with the wide range in both capacities and discharge pressures that are required under changing fire fighting requirements, the volute type has proven more satisfactory.

The diffusion-vane pump is shown in Figure 47. It has been given the name "turbine" because the elements are similar to a steam or water turbine producing usable power. In the turbine pump the flow direction is reversed to the direction of flow in a steam or water turbine.

PUMP GAGES

Two instruments are provided with all fire pumps which are necessary for proper pump operation. One is a compound gage registering either positive pressure or negative pressure (vacuum) in the pump

FIRE SERVICE HYDRAULICS

suction chamber. The second gage registers pump discharge pressure. When the pump is a centrifugal, the second gage may also be a compound or it may be a pressure type without the stop pin at the zero pressure position. Omission of the stop pin is necessary when using the pressure gage due to the negative pressure (vacuum) throughout the pump during the priming operation. Specifications covering these gages are given in NFPA Specification No. 19, and briefly, require the gages to be not less than 3½ inches in diameter.

The compound gage piped to the suction chamber of the pump is calibrated to read from 30 inches vacuum (hg) to 0 to not less than 300 and not more than 600 psi. The pressure gage is usually calibrated for pressures from 0 to 300 psi, but not over 600 psi, except for pumps designed to operate at higher pressures.

In the fire service the assumption is made, for convenience, that reference to pressure means pressure above atmospheric pressure. When positive pressure is not being transmitted to the gage, the pressure reading will be zero since atmospheric pressure inside and outside the gage is in balance. Under conditions of a perfect vacuum, the pressure is defined as zero absolute, and the word "absolute" is used to identify the scale. (See Chapter 1.) The absolute pressure will be the gage pressure plus 14.7 as zero pressure on the gage is 14.7 psi absolute pressure. The term "negative pressure" refers to a pressure less than atmospheric, more commonly referred to as a vacuum.

The relationship of atmospheric pressure to its capability of supporting a column of water 33.9 feet (at sea level) when all air has been exhausted from the pipe has been discussed in foregoing chapters. As an accurate medium of calibrating vacuum, the water tube is impractical due to its physical size. Mercury, which has a specific gravity of 13.5 permits the use of a tube with a vertical height of slightly over 34 inches. With the tube or glass bent in the form of a U, with one end open to atmospheric pressure and the other end sealed after all air above the mercury is exhausted, the atmospheric pressure will maintain a difference in the height of the columns of 29.92 inches at sea level. As atmospheric pressure changes daily, the average height of 29.92 inches has been standardized for sea level. This instrument is called a barometer.

Such a barometer, while accurate and useful in a laboratory or weather station, is too bulky for fire service use. A more practical means of registering both vacuum and positive pressure is the compound gage used on pumpers. A hollow curved tube of rectangular section is employed to activate the gage to register pressure. This tube is known as a Bourdon tube (Figure 48). When a vacuum is

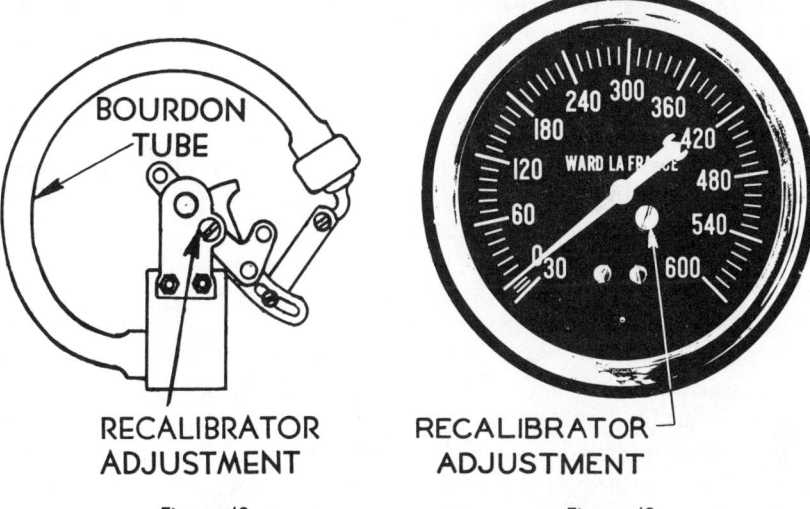

Figure 48

Figure 49

being created in the pump, as during a priming operation, the curve of the Bourdon tube decreases and this movement is transmitted through the linkage to register the vacuum, which is shown on the gage in inches of mercury rather than as a negative pressure in pounds per square inch. When a positive pressure (above atmospheric) exists in the pump, the Bourdon tube tends to straighten, reducing the curvature, and this movement will be proportional to the positive pressure, and will be registered by the indicator on the dial, as pounds per square inch pressure. Figure 49 illustrates the gage graduation for a compound gage.

This gage is standard for fire service use. It is reasonably rugged, but it can be thrown out of calibration and damaged by improper control of the fire pump, and by freezing. Fast shutoff of nozzles while the pump is discharging induces shock loading in the pump, which affects the pressure gages due to water hammer. This is particularly frequent when operating from water mains where the residual or inlet flow pressure is in the 10 to 20-psi range.

Freezing of the gage usually results in permanent deformation of the Bourdon tube. The tube curvature is reduced, resulting in a higher pressure being indicated, and the indicator hand does not return to zero pressure when the pump is not operating. A slight deformation can be tolerated, that is +5 psi in excess of the correct reading. This error in reading will not be constant over the complete calibrated

FIRE SERVICE HYDRAULICS

range and any error greater than 5 psi at the zero setting should be corrected by re-calibration of the gage. Some gages can only be recalibrated by resetting the indicator hand and using a gage tester to check. A new compound pressure gage should read within a tolerance of ±3 psi at all pressures.

Some gages are of the nonrepairable type, and damage to the Bourdon tube requires a gage replacement. Other manufacturers supply a gage which is equipped with a recalibrator adjustment screw. This is shown on Figure 49. Some adjustments can be made internally for the linkage (Figure 48). The best practice is return the gage to the manufacturer for repair when the recalibrator adjustment will not fully correct the reading at all pressures. In the gage connection on the back side, the liquid passage is restricted by a plug with a very small hole, about $1/64$ inch, which acts to dampen pressure changes and results in a smoother gage action. If the gage does not respond to pressure changes, check this small restriction, as sand or scale can plug it making the gage slow to respond.

From basic values already established, some equivalent values can be determined that will clarify the readings and better serve as a guide for the operator. For example, 1 inch of mercury equals:

$$\frac{33.9}{29.92} = 1.13 \text{ feet}$$

This is the height of a water column, or lift, when operating at draft. Another and more useful value expresses the lift in terms of inches of mercury. One foot of height equals:

$$\frac{29.92}{33.9} = .88 \text{ inches of mercury}$$

The figure .9 is usually used for convenience. For example, if a pump is to operate at draft on a 10-foot lift, the operator will know the primer must develop $10 \times .9 = 9$ inches of vacuum on the gage to prime the pump.

After the pump is primed and in operation, as discharge gate valves are opened, the vacuum reading will rapidly increase. The pump must create additional vacuum to sustain the height of lift, overcome the friction loss in the suction hose, entrance loss at the strainer, and provide the energy as velocity head to accelerate the water from no motion to the flow velocity in the suction hose. This vacuum reading with the pump in operation will vary according to the height of lift, volume of water being discharged, the size and total length of suction hose.

If, during the time the pump is operating, the vacuum reading begins to decrease without having reduced the discharge rate, an air leak in the suction line is indicated. This may be due to a loosened suction hose coupling, defective suction hose gasket, or a drop in the surface level of the water supply causing a small whirlpool developing above the suction strainer, admitting air with the water. Other causes may be an air leak through the pump packing, a drain valve partly open, or a partly open valve in the line from the water tank to the pump suction chamber. This last would only be a source of trouble when the water tank is empty. In any event, the operator must quickly determine the cause of the air leak as the pump will soon lose its prime.

A sudden increase in the vacuum reading, without a change in discharge volume, indicates that a restriction has developed in the suction line at some point. The most logical point to check is the outer suction strainer for partial plugging by leaves, grass, stones, mud, old newspapers or clothing. If the strainer at the pump suction inlet is also clear of obstruction, the lining in the suction hose can be suspect. Partial collapse of the lining will restrict the flow and in some cases has completely shut off the flow.

Where the water supply is limited, continued operation will lower the surface level, increasing the lift with a corresponding increase in vacuum reading. Such increase will be gradual and not in a short period of time (2 to 5 minutes) as occurs from other causes.

Fire apparatus manufacturers usually furnish a barrel or bell-mouth type of suction strainer, which is satisfactory for most draft operation where the strainer can be supported by ropes or a ladder underneath, to keep it off the bottom by at least 12 inches and under the surface by at least 18 inches. When drafting from a stock tank, a shallow creek or pond, the pan-type strainer is preferred. It allows suction to be taken at the lowest point without danger of plugging with bottom debris, loose rocks or mud. It also provides the maximum depth of water above the strainer. In a shallow creek the depth can be increased by the use of a short ladder and a salvage cover as a temporary dam.

In places where the water depth is greater, but the outer end of the suction hose cannot be supported by ropes or other means, the floating type of suction strainer is preferred. This type also permits the maximum use of available water as the level can be lowered to within a few inches of the bottom without drawing air into the suction line. The floating strainer also works well when a portable reservoir is used, the reservoir being supplied by a portable pump or transfer from a tanker.

The discharge gage is also an indicator of developing trouble and the competent operator quickly recognizes the warning and can usually determine the cause and take corrective measures before a shutdown becomes necessary.

A surge or fluctuation of the pressure gage is usually accompanied by an alternating increase and decrease in engine speed as will be noted on the tachometer. Such action usually indicates the pump is running away from water or is losing its prime. The cause may be one of the following conditions:

1. Water mains too small to supply required volume.
2. Suction lift too high for the volume being discharged.
3. Water temperature too high for height of lift or volume being discharged. Cavitation may also be occurring.
4. Pump only partially primed before discharge gates were opened.
5. Suction hose too small for volume being discharged.
6. Outer suction strainer not properly submerged, admitting air.
7. Suction strainer partially plugged, either outer strainer or strainer at pump suction inlet.
8. Loose suction hose couplings or defective gaskets admitting air.
9. Pump drain valve not fully closed.
10. Water tank to pump line valve partly open.
11. Worn pump packings admitting air.

PRIMING SYSTEMS

Since rotary and piston pumps are self-priming, no separate priming system is required. This requirement, however, is made for centrifugal fire pumps, which are not self-priming, as all fire pumps are tested for rated capacities at draft with lift to 10 feet. All subsequent, in-service testing by the rating organization is also at draft. While many fire departments operate almost entirely from positive-pressure water sources, the majority of fire departments are required to operate at draft on either part or all responses.

Certain performance requirements for the priming system are given in NFPA Specification No. 19. Briefly, some of these requirements are:

1. The priming system must be capable of developing 22 inches vacuum (Hg) at elevation to 1000 feet above sea level. For each 1,000 feet additional elevation, or fraction of 1,000 feet, the requirement is reduced 1 inch.
2. The system must be capable of developing a vacuum and holding it for 10 minutes with a loss during this test period of not over

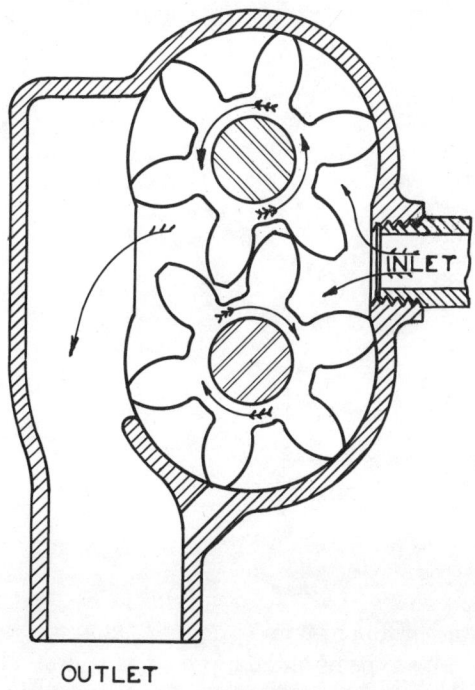

Figure 50

10 inches vacuum (Hg). The test is conducted with 20 feet of hard suction hose attached to the pump, with the outer end capped.

3. The system must have a priming capacity capable of priming the pump when operating on a lift to 10 feet and using 20 feet of suction hose, in 30 seconds for pumps of rated capacity to and including 1,250 gmp. For pumps of 1,500-gpm rated capacity and larger, the priming time limit is 45 seconds. The reason for the increased allowance for priming time on the larger capacity pumps is that dual suctions are usually used when operating at rated capacity at draft.

Many types of priming systems have been used by the fire service, but only two types are in current use: (1) rotary and (2) engine manifold vacuum.

The rotary primer pump is positive-displacement and is furnished in two designs: (a) rotary gear and (b) rotary vane. The rotary gear primer (Figure 50) may be either gear-driven at the pump transmission, or may be coupled directly to an electric drive motor. It varies in size and displacement according to the type of drive which governs the speed of rotation.

FIRE SERVICE HYDRAULICS

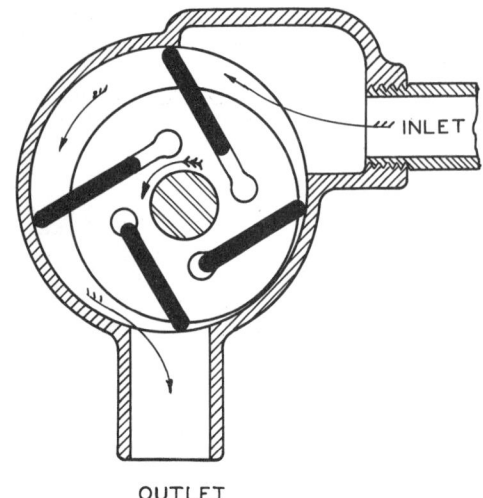

Figure 51

The rotary vane priming pump (Figures 51 and 52) is driven by an electric motor. This type of pump has been in continual use for priming centrifugal fire pumps since 1912, which is an indication of its high efficiency and excellent service record.

Both designs of rotary primers are capable of developing 26 inches

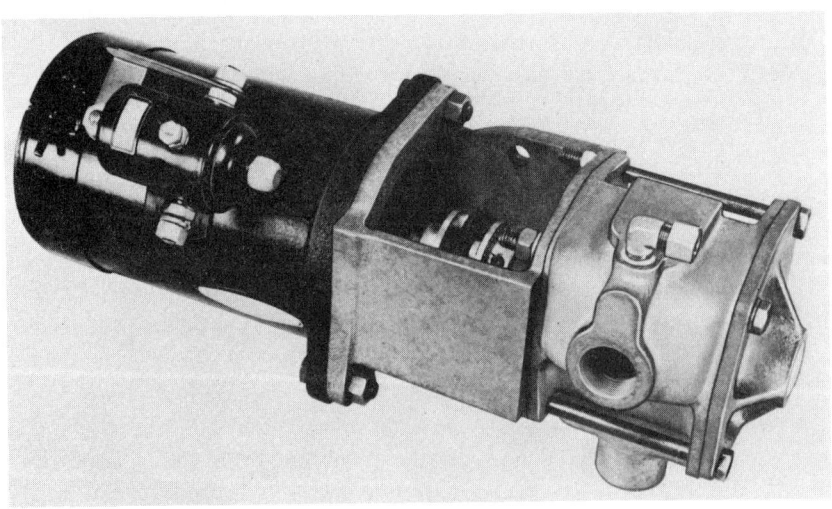

Figure 52

FIRE SERVICE PUMPS

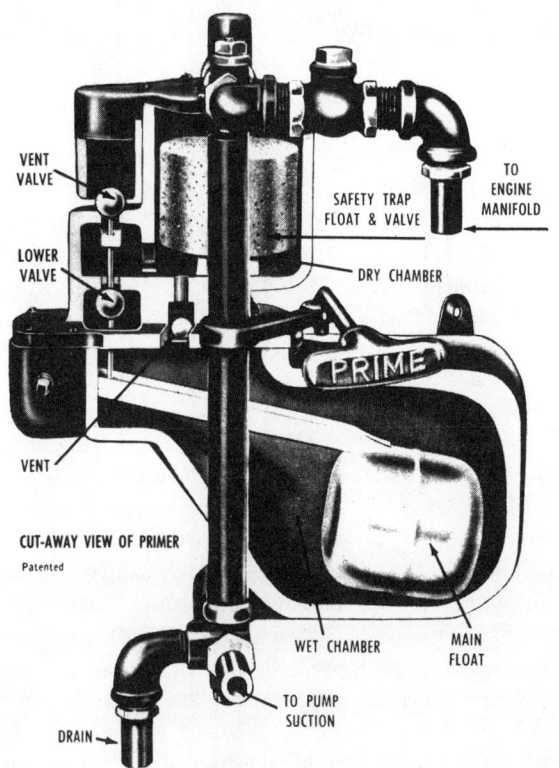

Figure 53

vacuum (Hg) at elevations to 1,000 feet above sea level. This is the equivalent of $26 \times 1.13 = 29.38$ feet of static lift. This is a lift beyond the capability of any pump to actually discharge a fire stream. The 22-inch vacuum (Hg) required is the equivalent of a static lift of $22 \times 1.13 = 24.86$ feet which is the limit for drafting water and discharging approximately one fire stream (250 gpm).

The engine manifold vacuum priming system takes advantage of the vacuum in the engine (gasoline engine) manifold with control as shown by the priming device in Figure 53. A float operates valves that prevent water being drawn into the engine. This system has been used continuously by the fire service for over 40 years.

This system can be used only with the gasoline engine. The diesel engine, even with natural aspiration, does not have a vacuum in the air intake manifold sufficient to serve as a source for air removal for priming a fire pump.

PRESSURE-CONTROL DEVICES

A pressure-control device is an essential part of fire pump equipment. Its basic purpose is to protect men handling hose lines against an undue pressure rise in the line. This would occur when two or more lines are in use and one or more is shut off at the nozzle. The pump operator may be unaware of the closing of one or more of the lines, and without automatic pressure control, with reduced discharge and no change in throttle setting, the discharge pressure in the remaining lines can increase to make them unmanageable. This condition can result in serious injury to the nozzlemen, and cause burst hose lines and mechanical damage.

Limits of allowable pressure rise have been established and the test procedure is outlined in NFPA Specification No. 19. The recommended procedure for testing the control device is to operate the pump at draft, lift to 10 feet, with pump discharging rated capacity at 150 psi. Set the control device to operate according to the manufacturer's instructions, which in nearly all manuals is at only a few psi above the operating pressure. After closing all pump discharge outlets, the pressure rise shall not exceed 30 psi above the initial operating pressure.

In practice, a pressure-control device in proper working order will limit the pressure rise to approximately 10 psi with one or more lines still discharging.

The pressure-control device shall operate over a range of 90 psi to 300 psi pressure, when the pump is operating at draft.

Pressure-control devices are of two types: relief valves and pressure-operated engine governors.

Relief valves

The relief valve consists of an adjustable pilot valve that actuates its opening and closing. It opens to bypass sufficient water from the discharge manifold back to the suction chamber to maintain a stable operating pressure when one or more discharge lines are closed without a change in the engine throttle setting.

All relief valve operation is based on a pressure differential between the pressure (negative or positive) in the pump suction chamber and the pump discharge manifold. The maximum pressure differential will occur when the pump is operating at draft, and the relief valve will be most responsive.

When a pump is operated from a positive-pressure supply source, such as a municipal water system, the pump operating pressure differential is greatly reduced. Unless the pressure differential between

FIRE SERVICE PUMPS

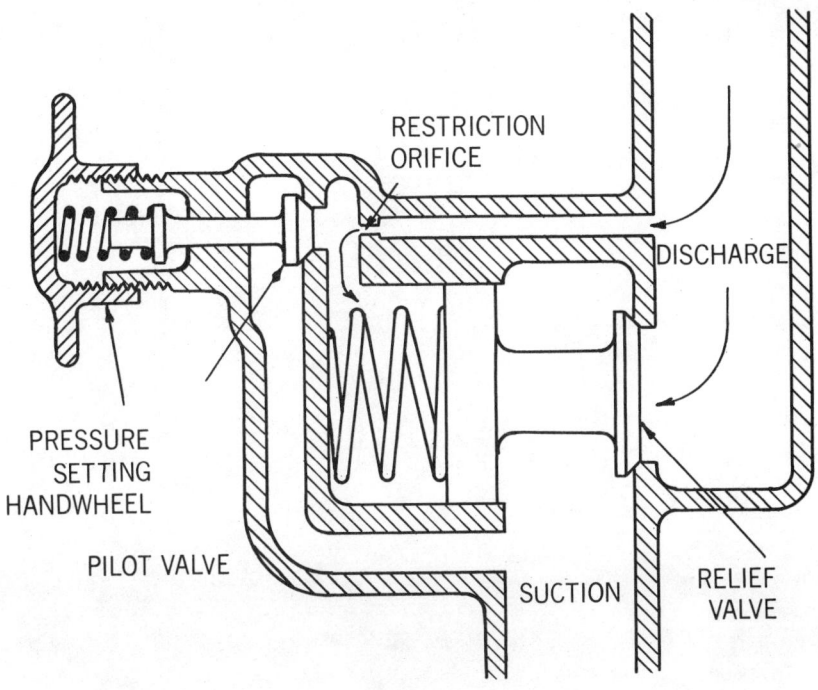

Figure 54

the residual pressure at the pump suction inlet and the discharge pressure exceeds 90 psi, the relief valve will not operate. The operation of the relief valve is affected most when the mains are small—4 and 6-inch—and are long without suitable grids or loops. For example, assuming values found in fire service operation, a pump is discharging 750 gpm through three 2½-inch lines, at a discharge pressure of 120 psi, and the residual flow pressure is 20 psi. Two lines are

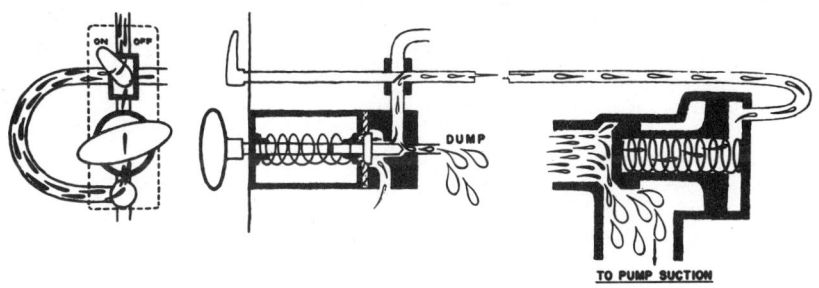

Figure 55

215

FIRE SERVICE HYDRAULICS

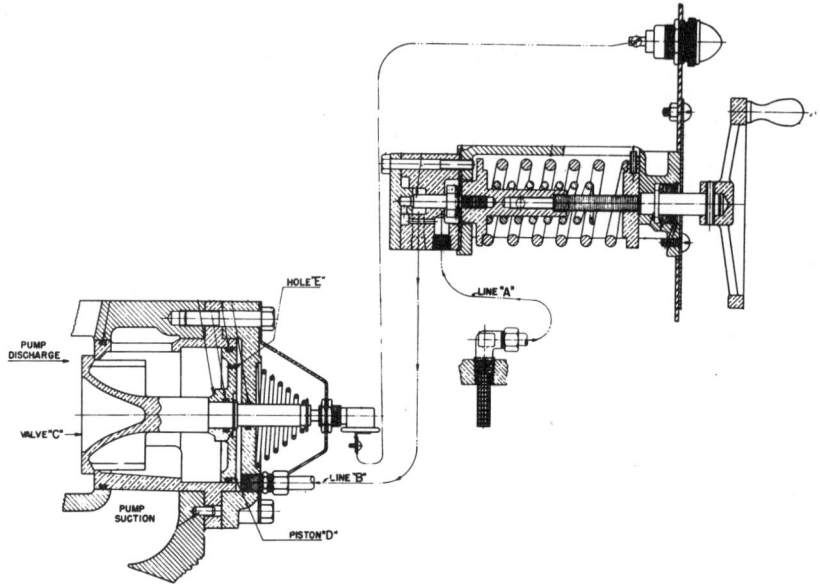

Figure 56

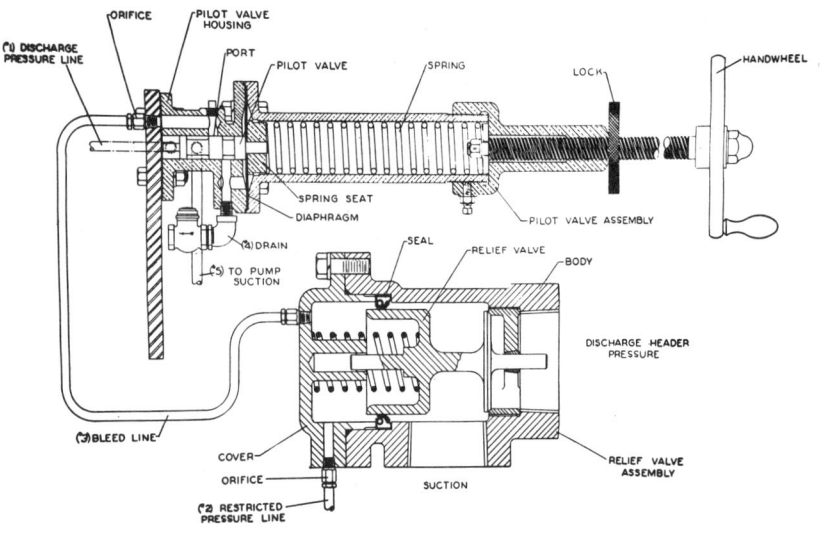

Figure 57

FIRE SERVICE PUMPS

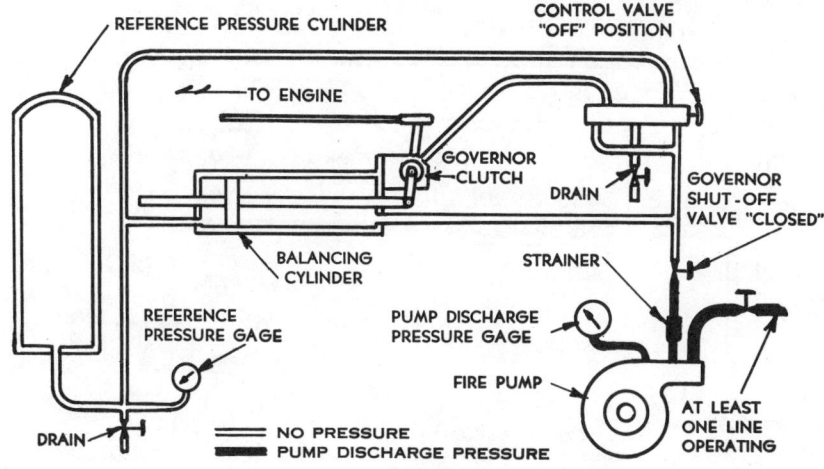

Figure 58a

shut off, reducing the total flow to 250 gpm. There is an immediate rise in the residual pressure to 60 psi due to the decrease in flow. The discharge pressure has increased to 140 psi, an increase of 20 psi. The differential in pressure is now $140 - 60 = 80$ psi. This is insufficient to activate the governor, but the increase in discharge pressure is within the 30-psi limit so the lack of relief valve operation is not important in this example.

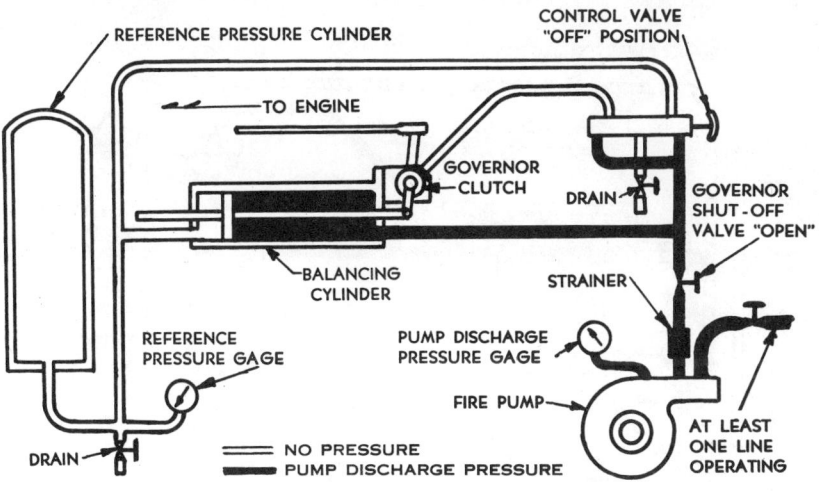

Figure 58b

FIRE SERVICE HYDRAULICS

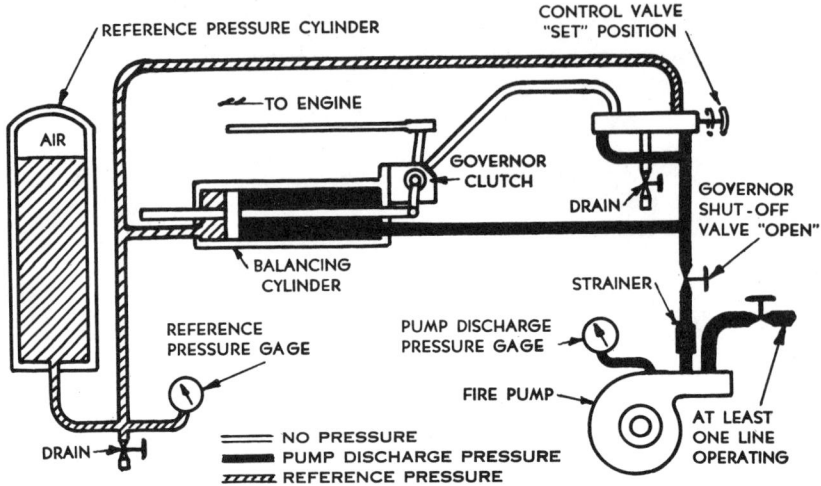

Figure 58c

The actual increase in residual pressure will be affected by such factors as the size of mains, length of mains, volume of water flowing before shut-off of any lines and after shut-off of one or more lines. In small mains of considerable length having static pressures of 80 to 100 psi, care should be taken when shutting down discharge lines to avoid water hammer with its potential damage to mains, hydrant connections, hose and household equipment.

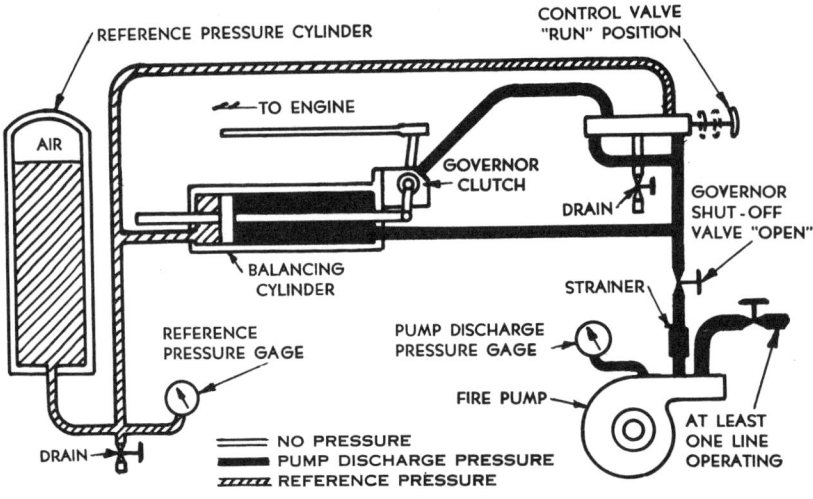

Figure 58d

FIRE SERVICE PUMPS

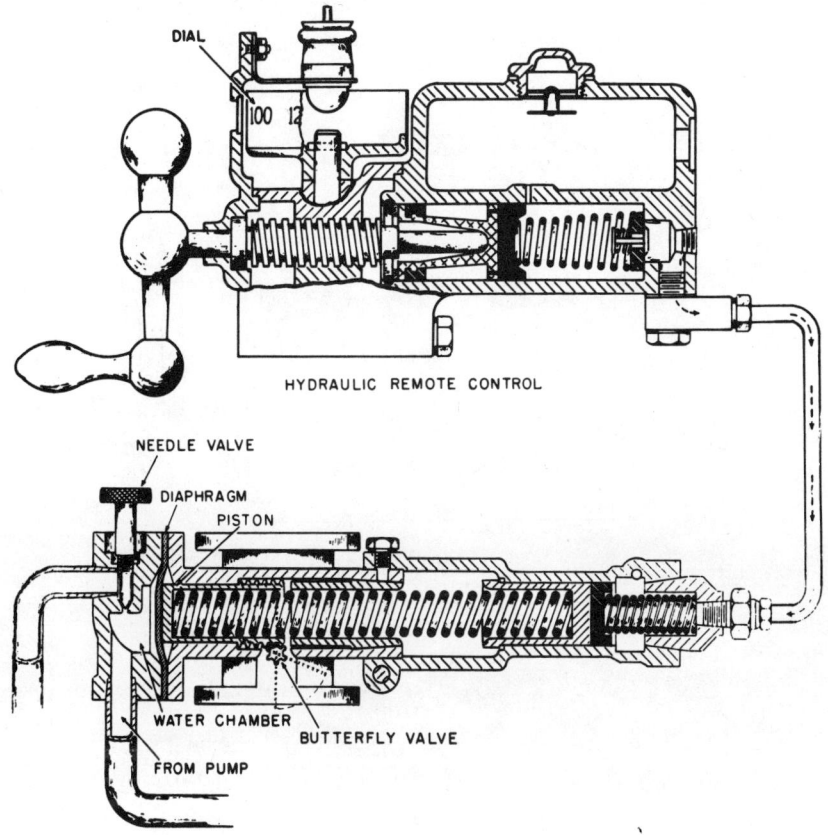

Figure 59

The relief valve, shown in schematic section by Figure 54, is representative of the basic principle of operation of most relief valves used by fire departments. Water at discharge pressure acts directly against the relief valve to open the passage to suction. Water at the same pressure also enters the chamber in back of the relief valve piston through a passage and restricted orifice. The pressure in back of the relief valve piston acts on a larger diameter piston to hold the valve in the closed position. The pilot valve is spring-loaded to hold it on the seat, and the pressure holding the valve on the seat is varied by a screw thread.

When the water pressure exceeds the pre-set spring pressure, the valve lifts from its seat, permitting water to flow from the chamber back of the relief valve piston to the suction outlet. This produces a

FIRE SERVICE HYDRAULICS

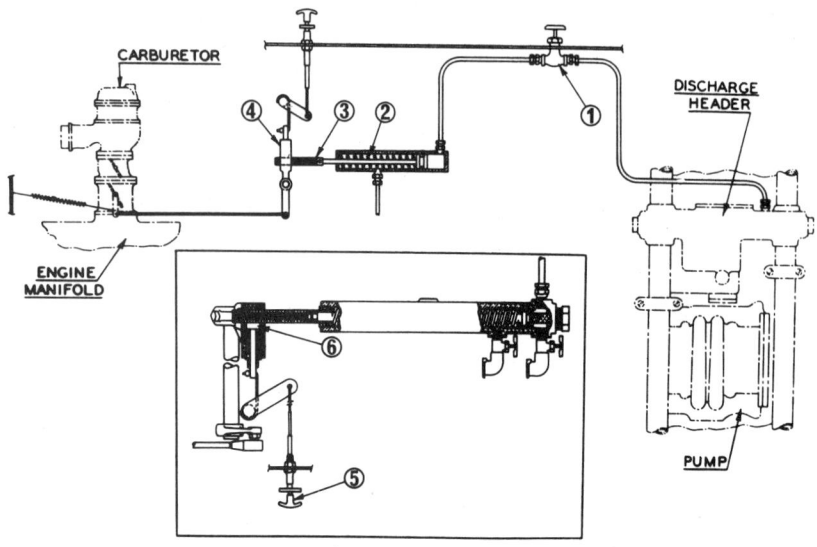

Figure 60

drop in pressure in the chamber and the discharge pressure can now move the relief valve off its seat to bypass water directly from discharge to suction.

The key to successful operation is a restriction in the flow passage from the discharge manifold to the chamber in back of the relief valve. The water cannot flow fast enough to balance the loss in pressure because the pilot valve is open. When the discharge pressure drops below the pre-set pilot valve setting, as when a line is opened again to increase the flow, the spring pressure closes the pilot valve opening and the pressure builds up fast to close the relief valve again. If the orifice becomes clogged with dirt, the relief valve will not function properly. A screen is usually provided for this orifice, but this requires periodic attention.

Three commercial relief valve operating diagrams are shown in Figures 55, 56 and 57.

Pump discharge pressure control by the engine throttle has been used by the fire service for many years, but only with gasoline engines. Diesel engines have a different arrangement for controlling the engine speed and power output.

A pressure-operated governor for use with either diesel or gasoline engines has recently been developed. It is automatic in controlling the engine throttle to match the power output with the pump dis-

FIRE SERVICE PUMPS

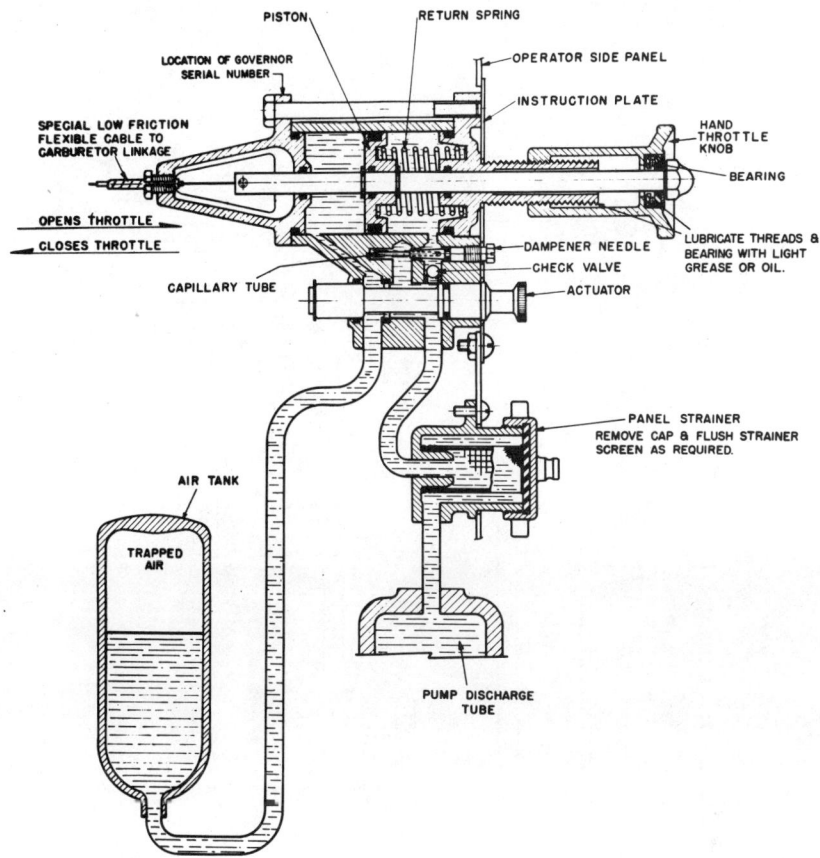

Figure 61a

charge requirements above the minimum flow setting, and protects the engine if there is a loss of water supply to the pump.

The governor can be set at the desired discharge pressure at zero discharge or set at the minimum rate of discharge, at the option of the operator. If the operation requires total shutoff of all lines, the setting is made with zero discharge. If at least one line is operating, the setting is made with one line open. The governor automatically adjusts the power output for any increase in number of lines for discharge.

Figure 58a is a schematic diagram of the governor system and is shown with one line discharging at the desired pressure. The governor system is completely shut off. Figure 58b shows the pump pressure admitted to the governor system, but the governor control valve

FIRE SERVICE HYDRAULICS

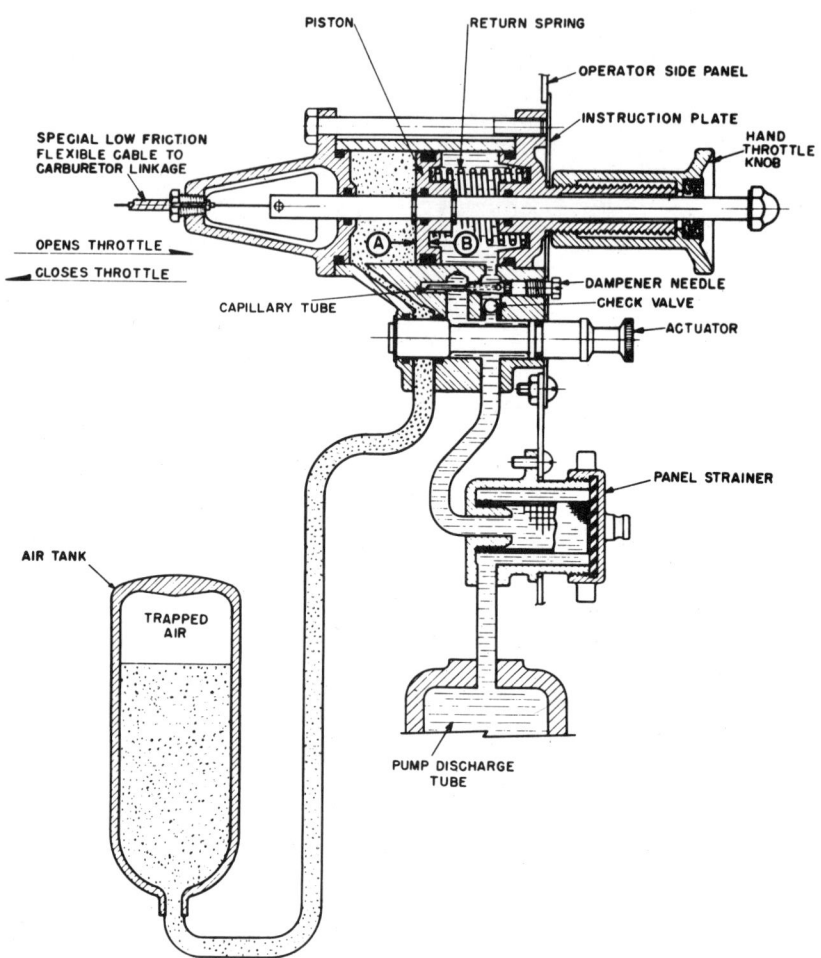

Figure 61b

has not been activated. Figure 58c shows the system water flow to balance the governor control piston and pressurize the reference pressure reservoir. This is the SET position for the governor control handle on the operator's pump panel. Figure 58d shows the complete system in operation with the control valve now in the RUN position. The governor clutch now locks the piston movement with the engine throttle control. In event of pressure loss, the clutch disengages automatically to return the engine throttle to idle position.

The pump pressure control governor has the same basic objective as the relief valve: protection of men on operating discharge lines against

FIRE SERVICE PUMPS

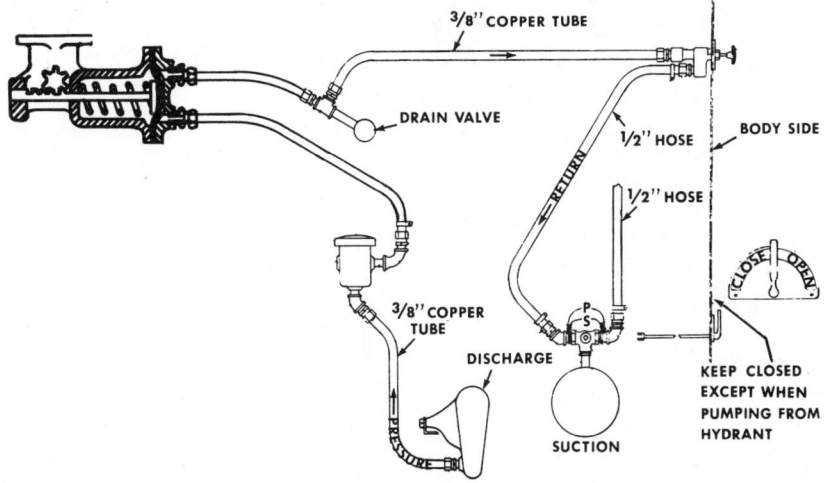

Figure 62

undue pressure rise when one or more lines are shut off. The governor uses a different method to accomplish this purpose. Instead of bypassing water to maintain the discharge pressure, the pressure-operated engine governor is activated by any increase in pump discharge pressure, above the pre-set pressure. In action it reduces the engine throttle setting, which reduces engine speed and power to match the power requirements for the supply of lines still in operation.

The basic principle of operation is the same for all the various pressure governors in current use in fire departments. Water pressure acts against a diaphragm or piston which is mechanically connected to the engine throttle. The diaphragm or piston acts to close the throttle to maintain a pre-set pressure whenever the discharge pressure exceeds it.

Four of the pump pressure control governors in current use by fire departments are shown in Figures 60, 61a, 61b and 62.

ILLUSTRATION CREDITS

American Fire Pump Company—Figures 5, 8, 23, 53 and 62
American LaFrance—Figures 14 (a, b, c, d), 15, 16, 27, 36, 57, 58 (a, b, c, d), and 60
Bilgram Gear & Machine Works—Figure 41
Food Machinery Corporation, John Bean Division—Figure 33
Gorman-Rupp Company—Figures 40 and 42
Hale Fire Pump Company—Figures 6, 7, 9, 11c, 11d, 12, 21 (a, b), 28, 30, 31, 34, 38, 39, 50, 51, 52, 56, 61 (a, b)

FIRE SERVICE HYDRAULICS

Howe Fire Apparatus Company—Figure 26
Ingersoll-Rand Corporation—Figures 43 and 44
Marsh Instrument Company—Figure 48
Seagrave Fire Apparatus, Inc.—Figure 59
Ward LaFrance Truck Corporation—Figure 49
Waterous Company—Figures 13 (a, b, c), 18, 20, 55
Worthington Corporation—Figures 45 and 46

PART THREE: practice

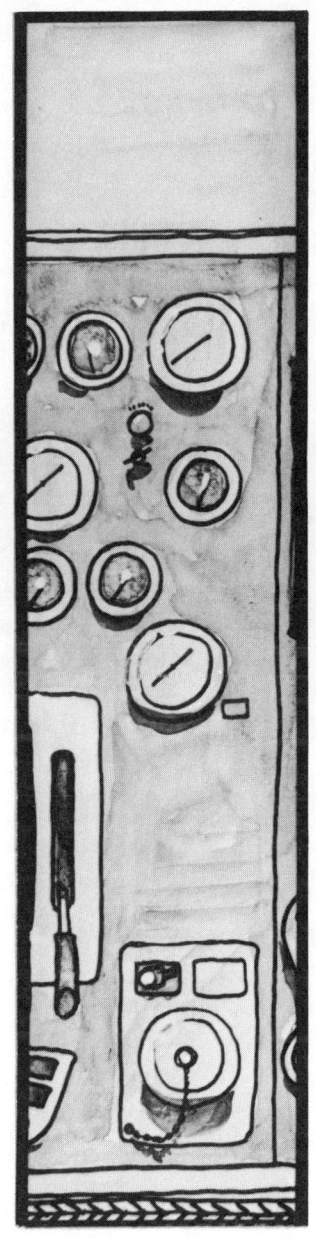

CHAPTER FIVE

Friction loss calculations

Friction can be described as the rubbing of one body against another. Loosely, this is what happens when water flows through a hose or a pipe and the water next to the inside surface of the hose or pipe becomes roiled. At the same time, this turbulent skin of the moving column of water tends to "grease" the passage of the inner volume of water. (There is a more scientific explanation in Chapter 2.)

As a result of the friction created when two substances rub together, there is a dissipation of energy. When this occurs in fire hose, standpipes or sprinkler system piping, we refer to it as friction loss.

In operating fire department pumpers, the direct result of friction loss is a nozzle pressure that is less than the engine (pump) pressure. Therefore, we have to calculate the expected friction loss in a hose layout to determine the engine pressure necessary to produce the desired nozzle pressure.

Friction can also be defined as disagreement. And that definition is applicable to the present situation in which we have friction loss tables that were developed in the latter half of the 19th century and hose that has been developed in the middle of the 20th century. The condition of the rubber lining in fabric fire hose and the degree of smoothness of the weave of the fabric are variables in the production of friction loss that cannot be evaluated on the fireground. Because the rubber lining is relatively thin, the higher the pressure, the more

FRICTION LOSS CALCULATIONS

closely the lining will conform to the weave of the hose fabric. The degree to which corrugations of the fabric will be mirrored on the water surface of the lining depends on the smoothness, or coarseness, of the fabric weave and the water pressure.

The friction loss for a specific gallons per minute flow rate in a hose line will vary slightly according to the different degrees of smoothness of the individual lengths making up the line. However, the friction differences in lengths of serviceable fire hose average out to a degree that makes the use of friction loss tables practicable.

The increasing use of 100 percent polyester jacket, rubber-lined hose and the introduction of polymer hose, however, have raised some questions about friction loss figures. Remember, the friction loss tables were developed from John R. Freeman's experiments with cotton jacket, rubber-lined hose. There have been no extensive, scientific tests of a representative variety of the new types of hose to determine beyond doubt the accuracy of the standard friction loss tables when applied to polyester hose. However, it is the belief of some underwriters that any difference in friction losses that may exist are too insignificant to cause us to throw the old tables out of the window.

The manufacturers of polymer hose report lower friction losses for this hose as compared with the same sizes of fabric, rubber-lined hose. Again, no extensive tests have been conducted to develop friction loss tables.

However, whatever the friction losses for the new types of hose are finally determined to be, the physical laws for friction losses will continue to reign unchallenged.

The four hydraulic laws governing friction in hose and pipes are as follows:

1. Friction loss in hose varies directly as the length of the line, provided all other conditions are equal.

If the same gallonage per minute is flowing, the friction loss in 500 feet of hose will be five times the friction loss in 100 feet of the same size and quality hose. To put it another way, if you double the length of the line, you double the friction loss.

2. In the same size hose, friction loss varies approximately as the square of the flow velocity.

This means that the resultant friction loss increases more rapidly than an increase in the flow velocity. For example, if the flow velocity is doubled, the friction loss becomes 2×2, or 4 times as much as it was originally. If the flow velocity is tripled, then the friction loss becomes 3×3, or 9 times as much; if the flow is quadrupled, then the friction loss becomes 4×4, or 16 times as much as it was originally.

3. For the same discharge, friction loss varies inversely as the fifth power of the diameter of the hose.

This cites the fact that if the discharge remains the same, increasing the size of the hose can drastically reduce the friction loss. Or looking at it from a different angle, with the same volume flowing, the larger the hose, the smaller the friction loss. For example, if the discharge remained the same but the size of the hose was tripled, the friction loss would be a mere

$$\frac{1}{3 \times 3 \times 3 \times 3 \times 3} = \frac{1}{243}$$

of the original friction loss. This underscores the fact that the most effective way to reduce friction loss is to increase the size of the hose. If you wish to compare two sizes of hose, such as $2\frac{1}{2}$ and 3-inch (in regard to their friction losses) for the same flow, the friction loss of $2\frac{1}{2}$-inch hose is to the friction loss of the 3-inch hose inversely as the fifth powers of the diameters:

$$\frac{\text{Friction loss in } 2\frac{1}{2}\text{-inch}}{\text{Friction loss in 3-inch}} = \frac{3 \times 3 \times 3 \times 3 \times 3}{2\frac{1}{2} \times 2\frac{1}{2} \times 2\frac{1}{2} \times 2\frac{1}{2} \times 2\frac{1}{2}} = 2.49$$

So we learn that the friction loss in $2\frac{1}{2}$-inch hose is 2.49 times that in 3-inch hose.

4. For a given flow velocity, the friction loss in hose is approximately the same no matter what the water pressure may be.

This means that when water is flowing through hose at a certain number of linear feet per minute, the friction loss is the same whether it is, for example, 240 gpm through $2\frac{1}{2}$-inch hose or 400 gpm through 3-inch hose, or nearly 625 gpm through $3\frac{1}{2}$-inch hose, although the pressure will differ. In other words, it is the flow speed (velocity) of the water and not the gpm that governs the friction loss. If the water velocity in a hose is 10 linear feet per second, the friction loss will be the same whether the pressure is 100 psi or 150 psi. The velocity of the water along the hose lining and the condition of the lining govern the friction loss.

Other friction loss sources

The hose itself is the major but not the only source of friction loss. Couplings, bends in the hose, valves, siameses, wyes and other accessories are responsible for pressure losses. Washers that protrude into

FRICTION LOSS CALCULATIONS

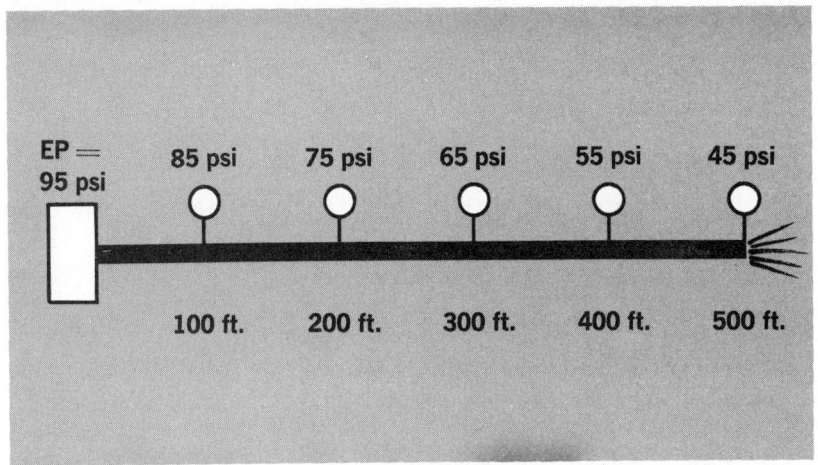

Figure 1

the waterway can create unnecessary friction losses. In a report to the American Society of Civil Engineers, Freeman stated that in 2½-inch hose with 200 gpm flowing, a washer with a 2⅜-inch hole was responsible for a loss of 0.10 psi; a washer with a 2¼-inch hole, 0.52 psi; and a washer with a 2⅛-inch hole, 1.16 psi.

Fire hose laid in the common snakelike pattern, but not kinked, will cause a friction loss of about 6 percent more than when it is laid perfectly straight. With 200 gpm flowing in a 2½-inch hose, a sinuous line will have about ½ psi additional friction loss per 100 feet. With 300 gpm flowing, the extra friction loss in a snakelike line will be about 1¼ psi.

How friction loss works

We have mentioned that friction loss reduces the amount of pressure available at the nozzle. Let's look at the way friction loss works.

If a 2½-inch line with a 1-inch tip is attached to a pumper and stretched straight out on a level surface, we can determine the action of friction loss by connecting line gages between couplings at 100-foot intervals (Figure 1) and using a pitot gage at the nozzle.

If we have a nozzle pressure of 45 psi, 198 gpm will be flowing with a friction loss of 10 psi. The engine pressure will be 95 psi. At 100 feet from the pump, the gage will read 85 psi; at 200 feet, 75 psi; at 300 feet, 65 psi; at 400 feet, 55 psi; and at the nozzle, 45 psi. This experiment can be done with any size and length of hose line and with various volumes of water flowing. Gages inserted in a line at 100-foot intervals will show an equal drop in pressure for each 100 feet of hose. The amount of drop in pressure will depend on the friction loss for the specific hose and water flow. In the field, minor variations may be experienced because of differences in the conditions of the hose lengths and protruding washers.

In the fire service, friction loss tables show the amount of friction loss in 100 feet of hose under a multitude of conditions. And the friction loss figures are for pounds per square inch. Thus, if we say that the friction loss for 250 gpm in a 2½-inch hose is 15, we mean that it is 15 psi for each 100 feet of this size hose.

Example: If the friction loss for a 250-gpm flow in 2½-inch hose is 15 psi, what is the friction loss in 800 feet of hose?
Answer: Friction loss varies directly as the length of the line. Inasmuch as the friction loss of 15 psi is quoted for 100 feet of hose and we have 800 feet, the friction loss for the entire line will be 8 × 15 = 120 psi.

Example: A 3-inch hose has a friction loss of 10.5 psi with 340 gpm flowing. What is the friction loss for 1,200 feet of 3-inch hose at this flow rate?
Answer: Again, the friction loss of 10.5 is for 100 feet of hose, so 12 × 10.5 = 126 psi.

Friction loss calculations

The simplest method of determining friction loss is to ascertain the engine and nozzle pressures and subtract the nozzle pressure from the engine pressure. The difference is the friction loss. This gives us the formula:

$$FL = EP - NP$$

There are only two hitches in this. First, this formula is true only when the hose layout is on level ground. There is no provision for back pressure, which is positive when the nozzle is higher than the pump and negative when the nozzle is lower than the pump. The second hitch in this formula is that it isn't useful on the fireground. It is hardly practical to measure nozzle pressure on the fireground.

FRICTION LOSS CALCULATIONS

In operating a pump at a fire, we base friction loss calculations on a predetermination of how much water (in gallons per minute) we intend to flow through a hose. Then we can use this formula, which is only for 2½-inch hose:

$$FL = 2Q^2 + Q$$

in which Q is the gallons per minute flowing divided by 100.

This is the one hydraulics formula that, above all others, should be memorized by every pump operator because it is the only one he can easily use while standing in the street at the operator's position of a pumper. This is the formula that will get him out of hot water when he forgets the friction losses for seldom used flows.

Example: What is the friction loss for a flow of 400 gpm in 2½-inch hose?
Answer: The flow of 400 gpm divided by 100 is 4. Therefore Q is 4.
Substituting:

$$FL = 2(4)^2 + 4$$
$$FL = 2(16) + 4 = 32 + 4 = 36$$

Example: What is the friction loss for a flow of 200 gpm in 2½-inch hose?
Answer: The 200-gpm flow divided by 100 is 2. Therefore, Q is 2.
Substituting:

$$FL = 2(2)^2 + 2 = 2(4) + 2 = 8 + 2 = 10$$

For flows under 100 gpm, the formula is modified as follows:

$$FL = 2Q^2 + \tfrac{1}{2}Q$$

The results of the friction loss formulas are sufficiently accurate for fireground use. Getting into decimal points would only complicate the solution beyond the fireground necessity of having to do the computations in your head.

So far, we have a quick way of figuring friction losses for 2½-inch hose, but other sizes are used in fire fighting. Our basic formula remains useful because we can determine the friction loss for any gpm flow in 2½-inch hose and convert the result to provide the friction loss for other hose sizes. We can do this in one of two ways, depending on whether we are sitting at a desk or standing beside a pumper.

An approximation of the friction loss in hose sizes other than 2½-inch can be obtained by dividing the 2½-inch hose friction loss for a specific flow by the factor for the desired hose size. The factors for common hose sizes are given in the following table:

2½-Inch Hose Friction Loss Conversion Table

Hose Diameter	Factor
Single Lines	
¾-Inch booster	0.0029
1-Inch booster	0.011
1½-Inch rubber-lined	0.074
1½-Inch unlined linen	0.039
2½-Inch unlined linen	0.47
3-Inch rubber-lined, 3-inch couplings	2.6
3-Inch rubber-lined, 2½-inch couplings	2.5
3½-Inch rubber-lined	5.8
4-Inch rubber-lined	11.0
4½-Inch rubber-lined	19.5
5-Inch rubber-lined	32.0
Siamesed Lines of Equal Length (Rubber Lined)	
2-2½-Inch	3.6
3-2½-Inch	7.75
4-2½-Inch	12.4
6-2½-Inch	27.0
1-2½-Inch and 1-3-inch	6.1
2-2½-Inch and 1-3-inch	11.5
2-3-Inch	9.35
3-3-Inch	20.4
4-3-Inch	32.0
2-3-Inch and 1-2½-Inch	15.0

From the preceding table, it is evident that the friction loss in 2½-inch rubber-lined hose is only 0.0029 of that in ¾-inch booster hose for the same flow, or 0.074 of that in 1½-inch rubber-lined hose. On the other hand, the friction loss in 2½-inch fire hose is 2.6 times that of 3-inch hose (with 3-inch couplings) for the same flow.

The table also can be used to arrive at lengths of 2½-inch hose that are equivalent to hose of other sizes. For example, 1,000 feet of 2½-inch hose = 1,000 × 0.074 = 74 feet of 1½-inch hose.

Example: What is 100 feet of 1½-inch hose equivalent to in 2½-inch hose?

Answer: The factor for 1½-inch hose is 0.074.

$$\frac{100}{0.074} = 1{,}351 \text{ feet of } 2\text{½-inch hose}$$

Example: What is 1,000 feet of 3-inch hose with 2½-inch couplings equivalent to in 2½-inch hose?
Answer: The factor for 3-inch hose with 2½-inch couplings is 2.5.

$$\frac{1000}{2.5} = 400 \text{ feet of } 2\text{½-inch hose.}$$

Here are some rules of thumb for converting hose lengths from one size to another:

Multiply the length of ¾-inch hose by 340 to determine the equivalent length of 2½-inch hose.

Multiply the length of 1-inch hose by 86 to determine the equivalent length of 2½-inch hose.

Multiply the length of 1½-inch hose by 13 to determine the equivalent length of 2½-inch hose.

Multiply the length of 3-inch hose by 0.4 to determine the equivalent length of 2½-inch hose.

Multiply the length of 3½-inch hose by 0.17 to determine the equivalent length of 2½-inch hose.

We have been talking about "equivalent lengths." Just what do we mean? Actually, what we mean is that if we convert a length of one size hose to the equivalent length of 2½-inch hose, the friction losses in each size hose will be the same at the same flow rate.

This leads us to another conclusion. We can use the conversion table factors with friction losses for various gpm in 2½-inch hose to determine the friction losses in other hose sizes. And the way we use the factors for friction losses is the same way we use them for determining equivalent hose lengths.

Example: What is the friction loss in 3-inch hose (2½-inch couplings) with 300 gpm flowing?
Answer: The friction loss for 2½-inch hose with 300 gpm flowing is 21.2 psi. Then $\frac{21.2}{2.5} = 8.41$ psi.

Allowing for the slight increase in friction loss caused by the 2½-inch couplings, this is right in line with the 8.2 figure in the standard friction loss table for 3-inch hose with 3–inch couplings.

FIRE SERVICE HYDRAULICS

For rapid calculation on the fireground, the following table can be used to determine the friction losses of siamesed lines of equal length:

Siamesed Lines of Equal Length	Divide 2½-Inch Hose Friction Loss by
2-2½-Inch lines	¼
3-2½-Inch lines	⅛
1-2½-Inch and 1-3-inch line	⅙
2-3-Inch lines	⅑
3-3-Inch lines	1/20

Conversion Table To 2½-Inch Rubber-Lined Hose

Length in Ft.	3-in. (2.6)	3½-in. (5.8)	2-2½-in. (3.6)	3-2½-in. (7.75)	2-3-in. (9.35)	1-1½-in. (.074)
100	38.461	17.24	27.78	12.90	10.7	1351.35
200	76.923	34.48	55.56	25.81	21.39	2702.7
300	115.384	51.72	83.33	38.71	32.09	4054.05
400	153.846	68.96	111.11	51.61	42.78	5405.40
500	192.307	86.21	138.89	64.52	53.48	6756.75
600	230.769	103.45	166.67	77.42	64.17	8108.1
700	269.2305	120.69	194.44	90.32	74.87	9459.45
800	307.692	137.93	222.22	103.23	85.56	10810.8
900	346.153	155.17	250.00	116.13	96.26	12162.15
1000	384.614	172.41	277.78	129.03	106.95	13513.5
1200	461.538	206.9	333.33	154.84	128.34	16216.2
1400	538.461	241.38	388.89	180.64	149.73	18918.9
1500	576.921	258.62	416.67	193.55	160.43	20270.25
1600	615.384	275.86	444.44	206.45	171.12	21621.6
1800	692.306	310.34	500.00	232.26	192.51	24324.3
2000	769.23	344.83	555.56	258.06	213.9	27027.
2200	846.153	379.31	611.11	283.87	235.29	29730.
2500	961.535	431.04	694.44	322.58	267.38	33784.
2800	1076.922	482.76	777.78	361.29	299.4	37838.
3000	1153.84	517.24	833.33	387.1	320.85	40540.

Friction loss in small hose

To determine the friction loss for ¾ to 2-inch rubber-lined hose, the following formula is used:

$$FL = 40(2Q^2 + \tfrac{1}{6}Q)C$$

Q is the flow in gallons per minute divided by 100.

FRICTION LOSS CALCULATIONS

The coefficient C has the following values:

Hose Diameter	C
¾-inch	11.7
1-Inch	2.5
1¼-Inch	1.0
1½-Inch	0.4
2-Inch	0.1

This formula provides a method of determining friction losses in lines of less than 2½-inch diameter that is more accurate than using the factors to convert small hose to 2½-inch hose and then calculating the friction loss.

Example: What is the friction loss in 100 feet of 1½-inch rubber-lined hose with a flow of 100 gpm?
Answer: Q, the gpm divided by 100, is 1, and for 1½-inch hose, the value of C is 0.4. Substituting in the formula:

$$FL = 40(2Q^2 + \tfrac{1}{6}Q)C$$
$$FL = 40(2 + \tfrac{1}{6})0.4 = 40 \times 2.17 \times 0.4 = 34.72 \text{ psi}$$

Example: What is the friction loss in 100 feet of ¾-inch booster hose with a flow of 18 gpm?
Answer: Q is $18 \div 100$, or 0.18, and the value of C for ¾-inch hose is 11.7. Substituting in the formula:

$$FL = 40(2Q^2 + \tfrac{1}{6}Q)C$$
$$FL = 40\left(2 \times 0.18 \times 0.18 + \frac{0.18}{6}\right)11.7$$
$$FL = 40(2 \times 0.0324 + 0.03)11.7 =$$
$$FL = 40 \times 0.0948 \times 11.7 = 44.36 \text{ psi}$$

Unlined linen hose friction loss

So far, only rubber-lined hose has been considered in discussing friction loss calculations. However, once you are able to determine friction losses for rubber-lined hose, the problem of unlined linen hose is surprisingly simple. To determine the friction loss in unlined linen hose, first calculate the friction loss for rubber-lined hose of the same size and same flow and multiply that friction loss figure by 2.1. This rule applies to all sizes of unlined linen hose.

Finding flow from friction loss

When the friction loss for 100 feet is known, the flow of water through a 2½-inch hose line can be calculated with fairly accurate

FIRE SERVICE HYDRAULICS

results by using this formula:

$$Q = \tfrac{1}{2}\sqrt{2FL - 0.25} - 0.25$$

FL is the friction loss per 100 feet of 2½-inch hose.
Q is the gallons per minute of flow divided by 100.

By omitting the 0.25 under the radical sign, this formula can be simplified without serious loss of accuracy in the results. The revised formula then is:

$$Q = \tfrac{1}{2}\sqrt{2FL} - 0.25$$

Example: If the friction loss in 2½-inch hose is 12 psi, what is the flow in gpm?
Answer: Substituting 12 for FL in the formula:

$$Q = \tfrac{1}{2}\sqrt{2 \times 12} - 0.25 = \tfrac{1}{2}\sqrt{24} - 0.25$$
$$Q = .5 \times 4.9 - 0.25 = 2.45 - 0.25 = 2.2$$

As $Q = 2.2$ and Q is the flow in hundreds of gpm, the answer therefore is:
The flow is 220 gpm.

Flow advantages of big hose

In hose stretches of 1,000 to 2,000 feet, friction loss becomes a limiting factor in determining the maximum amount of water that can be pumped through a line. Parallel lines are not always a logical answer because many fire departments that commonly stretch 1,500 or more feet of hose in a single lay are limited in both hose and the apparatus to carry and stretch it. Therefore, these departments in areas without hydrants must depend on a single line between the water source and the fire because they lack the facilities for stretching parallel lines.

Increasing pump pressure is an unproductive way of increasing the flow in hose because to double the gallonage being delivered by a nozzle, the nozzle pressure has to be quadrupled. This quickly skyrockets engine pressures beyond reality.

However, lucrative dividends can be harvested by increasing hose sizes. In a single line, 1,500-foot relay, a pumper operating at 200 psi can put 220 gpm through 2½-inch hose and provide a 20-psi residual pressure at the second pumper. Under the same conditions but with 3-inch hose, the flow will be increased by 145 gpm (66 percent) to 365 gpm. And if the only change is to use 3½-inch hose, the delivery of water will be increased 157 percent (345 gpm more than in 2½-inch hose) to a total of 565 gpm.

FRICTION LOSS CALCULATIONS

The following tables illustrate the flow rate advantages of larger diameter hose:

Flow Capabilities of Fire Hose

Single Line, 1,000 Feet

Engine Pressure (psi)	2½-Inch Hose (gpm)	3-Inch Hose (gpm)	3½-Inch Hose (gpm)
100	175	295	450
150	230	385	580
200	275	455	705
250	315	525	805

Single Line, 1,500 Feet

Engine Pressure (psi)	2½-Inch Hose (gpm)	3-Inch Hose (gpm)	3½-Inch Hose (gpm)
100	140	240	365
150	185	305	470
200	220	365	565
250	250	405	645

Single Line, 2,000 Feet

Engine Pressure (psi)	2½-Inch Hose (gpm)	3-Inch Hose (gpm)	3½-Inch Hose (gpm)
100	125	200	310
150	160	260	405
200	190	315	485
250	215	360	550

NOTE: Engine pressure includes allowance for providing approximately 20 psi residual pressure at the pumper receiving the water supply. Friction losses were taken from the AIA tables. Experience with new types of hose may indicate lower friction losses.

With the use of larger diameter hose, big flows can be supplied at lower engine pressures. This is important because more than friction loss is involved. When large flows are required on long lays, engine pressures rise to levels that reduce the capacity of the pump.

A pump in good condition will supply its rated volume at 150 psi plus the extra capacity built into the pump by design. At 200 psi, a pump is required to pump 70 percent of its rated volume; and at 250 psi, only 50 percent. Most pumps will do better than the required performance.

But the point is that as the pump pressure increases, the potential volume of the pump decreases. Therefore, when maximum volumes of water are required, engine pressures for hose layouts should be kept as low as practical. The larger the diameter of the hose, the more likely this is to be accomplished on the fireground.

3-Inch hose with 2½-inch couplings

The preceding tables emphasize the advantages of using 3-inch hose instead of 2½-inch hose to reduce friction losses and more easily provide larger flows. However, once these advantages have been acknowledged, the question arises as to whether 3-inch hose should have 3-inch or 2½-inch couplings.

Some of the metropolitan fire departments that have used 3-inch hose for many years use 3-inch couplings and all their master stream appliances have 3-inch threads. But in most departments, master stream appliances have 2½-inch threads. This makes the use of adapters mandatory if 3-inch couplings are used and this raises the constant fear of not having an adapter handy when it is needed.

The simple answer to this problem is to use 2½-inch couplings on 3-inch hose. The fear that this will cause an excessive rise in friction loss is unfounded. With properly beveled 2½-inch couplings, the increased friction loss in 3-inch hose can be expected to be less than 5 percent of the friction loss saved by going from 2½ to 3-inch hose.

For a specific example, let's look at 300 gpm flowing through a 2½-inch hose with a friction loss of 21.2 psi. In 3-inch hose with 3-inch couplings, the friction loss for the same flow is 8.2 psi, but in 3-inch hose with 2½-inch couplings, the friction loss will be about 8.5 psi. At this flow, the increase in friction caused by the 2½-inch couplings is much less than 5 percent of the friction loss saved by going from 2½ to 3-inch hose.

Another view of how much 3-inch hose friction loss is affected by the size of the couplings can be gained by recalling the conversion factors. To determine friction losses of 3-inch hose the loss figures for 2½-inch hose at various flow rates are divided by 2.6 when 3-inch couplings are used and 2.5 when 2½-inch couplings are used on 3-inch hose.

CHAPTER SIX

Engine and nozzle pressures

Like good cooking, good pump operation is the result of following certain rules seasoned with judgment. The pump operator essentially is striving to provide specific nozzle pressures to satisfy various fireground requirements.

Theoretically, he can do this by applying various rules of hydraulics to known conditions and doing some mathematical computations. But when he is standing on the pavement with one hand on the throttle and his eyes on the gages, he often finds that he has to make educated guesses about some of the conditions and trim his mathematics to basic rule-of-thumb reckoning.

In discussing engine and nozzle pressures, this chapter will look at the problems from the viewpoint of the working pump operator as well as that of the theorist.

Behind every engine and nozzle pressure problem is friction loss, and when the answer to any problem of this nature is determined, we will have specific figures for friction loss, nozzle pressure and engine pressure. To solve any problem, we must know two of these factors to obtain the third.

If a hose line is laid on level ground, the engine pressure is equal to the nozzle pressure plus the friction loss. The equation is:

$$EP = NP + FL$$

However, hose is not always on level ground and often it goes to an upper story of a building. Then the pull of gravity on water in that part of the hose line higher than the pump level creates what is termed back pressure. Back pressure will be discussed more extensively later in this chapter. At present, it is enough to know that it exists and therefore creates the need to modify the engine pressure formula to read as follows:

$$EP = NP + FL + BP$$

If a nozzle pressure of 45 psi results in a water flow creating 10 psi friction loss and if there are 350 feet of 2½-inch hose out, and if the nozzle is on the fifth floor of a building, causing a back pressure of 20 psi, then the engine pressure is figured as follows:

First, a 10-psi friction loss for 350 feet of hose makes a total friction loss of 35 psi. Substituting in the previous formula:

$$EP = 45 + 35 + 20$$
$$EP = 100$$

On the fireground, a pump operator is intent only on determining the required engine pressure. However, there are non-fireground occasions when the engine pressure is known and the unknown is one of the other factors. If the hose line is on level ground, which eliminates back pressure, then the basic formula can be written to solve either friction loss:

$$FL = EP - NP,$$

or nozzle pressure:

$$NP = EP - FL$$

If back pressure is involved, it too should be subtracted from engine pressure.

On the fireground, standard nozzle pressures are established by department rules for pump operation, so the pump operator knows the pressure he must supply at the nozzle. From the size nozzle and the nozzle pressure, he knows the water flow rate, which controls the friction loss per 100 feet of hose. If the officer orders a specific number of lengths of hose stretched, the pump operator then knows the exact length of the line. Otherwise, he estimates the length of the stretch. This is all the information he needs to determine the required engine pressure, and the result will be accurate enough for the fireground.

Underwriters' formula

In desk-top calculations, we can enjoy the luxury of having two unknowns: engine pressure and friction loss or nozzle pressure and

ENGINE AND NOZZLE PRESSURES

Table 1 Values of K for Large Hose Lines and Nozzles

Size Nozzle Inches	K For					
	Single Line 2½" Hose	Single Line 3" Hose	Single Line 3½" Hose	Two 2½" Lines Siamesed *	Two 3" Lines Siamesed *	3 Lines 2½" Hose Siamesed *
1	.105	.038	....	.025	....	....
1⅛	.167	.062	....	.043	....	....
1¼	.248	.092	.039	.066	.023	.028
1⅜	.341	.137	.059	.096	.034	.043
1½	.505	.192	.084	.135	.051	.061
1⅝	.680	.266	.113	.184	.068	.084
1¾	.907	.351	.152	.242	.093	.115
2	1.550	.605	.250	.418	.157	.190

Courtesy American Insurance Association.
* Allowance is made for loss in deluge set; these values will also give approximately correct figures for turret nozzles and ladder pipes, except that in the latter, pressure equal to 0.434 times the ladder pipe height must be subtracted from the engine pressure before solving for nozzle pressure.

friction loss. Then it is necessary to use what is known as the Underwriters' formula, which is:

$$EP = NP(1.1 + KL)$$

for rubber-lined hose of 2½-inch or greater diameter with a 1-inch or larger nozzle.

This is the formula for solving for engine pressure.

If we wish to solve for nozzle pressure, the formula is rearranged:

$$NP = \frac{EP}{1.1 + KL}$$

In the above formulas:

EP is engine pressure in pounds per square inch.
NP is nozzle pressure in pounds per square inch.
K is a factor depending on hose and nozzle diameters. Values of K for different sizes of hose and nozzles are given in Table 1.
L is the number of 50-foot lengths of hose in the line.

Small lines

Modified formulas are used for rubber-lined hose of 2½ inches or less diameter with a ⅞-inch or smaller nozzle. The modified formulas are:

$$EP = NP(1 + KL)$$

FIRE SERVICE HYDRAULICS

and
$$NP = \frac{EP}{1 + KL}$$

Unlined linen hose

For unlined linen hose of 2½ inches or greater diameter with a 1-inch or larger nozzle, the formulas for engine pressure and nozzle pressure are:
$$EP = NP(1.21 + 2.1KL)$$
$$NP = \frac{EP}{1.21 + 2.1KL}$$

For unlined linen hose of 2 inches or less diameter with a ⅞-inch or smaller nozzle, the formulas for engine pressure and nozzle pressure are:
$$EP = NP(1 + 2.1KL)$$
$$NP = \frac{EP}{1 + 2.1KL}$$

The K values for unlined linen hose are taken from either the table for large rubber-lined hose and nozzles or the table for small rubber-lined hose and nozzles, according to whether the problem involves large hose and nozzles or small hose and nozzles.

The values of K are shown in Tables 1 and 2. Care must be taken to make certain that the value of K in a specific problem is taken from the proper table for the size hose and nozzle.

Table 2 Values of K for Small Hose Lines and Nozzles

Size Nozzle Inches	K For				
	Single Line ¾" Hose	Single Line 1" Hose	Single Line 1¼" Hose	Single Line 1½" Hose	Single Line 2½" Hose
¼	0.235	.060	.025	.0084	.....
⅜	1.035	.256	.088	.036	.....
½	.....	.748	.258	.104	.0115
⅝	.....	.....	.578	.234	.023
¾	.....	.....	.....	.455	.042
⅞	.....	.....	.....	.....	.068

ENGINE AND NOZZLE PRESSURES

Solving for values of K

Sometimes it is necessary to work out parallel line problems that require the use of the Underwriters' formula. If the parallel lines are of equal length and size, it is a simple matter to convert them to a single equivalent 2½-inch line by using the appropriate conversion factor. If the lines are of unequal length you must first average them and then apply the factor.

If the lines are of unequal size, for example, one 2½-inch and one 3½-inch, convert the 3½ to 2½-inch before taking the steps given above. The same method applies to any combination of lines.

In addition, the two or more nozzles on the lines must be converted into a single equivalent nozzle. In the following example we are combining a 1-inch nozzle with an 1⅛-inch nozzle. To do this, we add the squares of the diameters and extract the square root of the sum to give us an equivalent nozzle:

$$1 \times 1 = 1$$
$$1\tfrac{1}{8} \times 1\tfrac{1}{8} = 1.265$$
$$1 + 1.265 = 2.265$$
$$\sqrt{2.265} = 1.505$$

If the nozzles are all the same size, Table 6, "Approximate Comparison of Nozzles," can be used to determine the equivalent nozzle size.

For both large and small nozzle sizes not in the table, K values can be obtained by using the following formula:

$$K = \frac{D^4}{10}$$

where D is the nozzle diameter in inches.

For example, for a 2¼-inch nozzle on 2½-inch hose:

$$K = \frac{(2\tfrac{1}{4})^4}{10} = \frac{(2.25)^4}{10}$$

$$K = \frac{2.25 \times 2.25 \times 2.25 \times 2.25}{10} = 2.563$$

Effects of changing nozzle diameters on a line

If a nozzle on a line is replaced with a smaller diameter nozzle and the engine pressure remains the same, the nozzle pressure will increase. Because the smaller nozzle discharges less water, friction loss in the

FIRE SERVICE HYDRAULICS

hose will decrease with the smaller flow. Because energy supplied by the pump is never lost, the energy formerly used in the larger flow for overcoming the higher friction loss will be diverted to increasing the nozzle pressure.

Conversely, if a nozzle on a line is replaced by a larger diameter nozzle, which can discharge more water, the increase in flow and friction loss will result in a lower nozzle pressure if the engine pressure remains the same.

Let's see what happens to the nozzle pressure and the flow when the engine pressure remains the same and a 1-inch tip is substituted for a $1\frac{1}{4}$-inch tip. This will provide an opportunity to use the Underwriters' formula.

Example: What engine pressure is needed to provide a 50-psi nozzle pressure for a $1\frac{1}{4}$-inch tip on 600 feet of lined $2\frac{1}{2}$-inch hose?

Answer:

$$EP = NP(1.1 + KL)$$
$$K \text{ (from Table 1)} = 0.248$$
$$L = 600 \div 50 = 12$$
$$EP = 50(1.1 + .248 \times 12)$$
$$EP = 50 \times 4.076$$
$$EP = 203.8 \text{ psi}$$

Now we know that an engine pressure of 203.8 psi will provide a nozzle pressure of 50 psi under the conditions stated in the problem for a $1\frac{1}{4}$-inch tip. But the engine pressure of 203.8 is unrealistic for practical pump operation, so we will round off the figure to 205 psi. Now let's restate our problem to determine nozzle pressure.

Example: What is the nozzle pressure when an engine is pumping at 205 psi through 600 feet of $2\frac{1}{2}$-inch hose with a $1\frac{1}{4}$-inch tip?

Answer:

$$NP = \frac{EP}{1.1 + KL}$$
$$NP = \frac{205}{1.1 + .248 \times 12}$$
$$NP = \frac{205}{4.076} = 50.2 = 50 \text{ psi}$$

ENGINE AND NOZZLE PRESSURES

A 1¼-inch tip with a nozzle pressure of 50 psi will flow 326 gpm. Now let's keep the engine pressure and the hose layout the same but change to a 1-inch tip and solve for the nozzle pressure.

Answer:

$$NP = \frac{EP}{1.1 + KL}$$

K (from Table 1) = 0.105

$$NP = \frac{205}{1.1 + .105 \times 12}$$

$$NP = \frac{205}{2.36} = 86.86 = 87 \text{ psi}$$

A 1-inch tip with a nozzle pressure of 87 psi will flow 275 gpm.

By keeping the engine pressure at 205 psi but changing from a 1¼-inch to a 1-inch tip, we get the following results:

Tip Size	Nozzle Pressure	Flow
1¼-in	50 psi	326 gpm
1-in	87 psi	275 gpm

By changing from a 1¼ to a 1-inch tip, the nozzle pressure increased 37 psi, or 74 percent. At the same time, the flow decreased 51 gpm, or 15.6 percent.

What happens when we keep the same engine pressure (205 psi) and the same hose layout but change to a ¾-inch tip? To solve this nozzle pressure problem, we have to use the Underwriter's formula for tips of ⅞-inch or less diameter on rubber-lined hose of 2½ inches or less diameter.

Answer:

$$NP = \frac{EP}{1 + KL}$$

K (from Table 2) = 0.042

$$NP = \frac{205}{1 + .042 \times 12}$$

$$NP = \frac{205}{1.504} = 136.3 = 136 \text{ psi}$$

FIRE SERVICE HYDRAULICS

A ¾-inch tip with a nozzle pressure of 136 psi will flow 195 gpm.

From this computation, we find that with an engine pressure of 205 psi, a ¾-inch tip shows a nozzle pressure increase of 86 psi, or 172 percent, over that of a 1¼-inch tip but a decrease in flow of 131 gpm, or 40.18 percent.

Now let's try the engine pressure formula for tips of ⅞-inch or smaller diameter with hose of 2½-inch or less diameter.

Example: What is the engine pressure needed to supply a 50-psi nozzle pressure for a ¾-inch tip on 600 feet of 2½-inch rubber-lined hose.
Answer:

$$EP = NP(1 + KL)$$
$$K \text{ (from Table 2)} = 0.042$$
$$L = 600 \div 50 = 12$$
$$EP = 50(1 + .042 \times 12)$$
$$EP = 50 \times 1.504 = 75.2 = 75 \text{ psi}$$

Open hose butt calculations

Once in a while an engine pumps through an open butt of a hose line either to flood an area or to pump water out of something. Calculations of engine and nozzle pressures are sometimes helpful in planning the job.

The Underwriters' formula can be used, and the calculations for engine and nozzle pressures are the same as when a nozzle is on the line.

The value of K for a 2½-inch open butt on a 2½-inch line is 3.35. The value of K for a 3-inch open butt on a 3-inch line is 2.4.

Because formulas are not always accurate for all conditions and layouts, the friction loss for a specific flow indicated by the butt pressure will not always agree with the engine pressure determined by formula, using the K values in the preceding two paragraphs.

Empirical formulas based upon experimental results unfortunately have this fault, and diverse results from two different methods of calculating have to be tolerated.

The discharge from an open butt on a hose line may be found by using the formula.

$$GPM = 29.7d^2 \sqrt{P} \ (0.90)$$

If the discharge from a 2½-inch open butt has a pressure of 9 psi, then:

$$GPM = 29.7(2.5 \times 2.5) \sqrt{9} \times 0.90$$
$$GPM = 29.7 \times 6.25 \times 3 \times 0.90 = 501.18$$

ENGINE AND NOZZLE PRESSURES

To save a little time on the arithmetic, the formula can be worked out without using the 0.90, which is the coefficient of discharge, but subtracting 10 percent from the answer.

Back pressure

Whenever a nozzle is higher than the pump level, then back pressure exists. This is the pressure exerted by a column of water as a result of gravity and is independent of the configuration of a hose line or the shape of the cylinder or plumbing containing the water.

Back pressure has a constant measurement of 0.434 psi per foot of height. If the surface of the water in a water tank or the nozzle of a hose line is 100 feet above the ground, then in each case the back pressure is $100 \times 0.434 = 43.4$ psi.

The size of the waterway makes no difference to back pressure because it is exactly that—a pressure and not a force. That is why back pressure is expressed in pounds per square inch. A column of water 50 feet high will have a back pressure of $50 \times 0.434 = 21.7$ psi whether the column is $\frac{1}{8}$ inch or 18 feet in diameter or width, if the column is rectangular.

On the fireground, because 0.434 is an awkward figure to handle in mental arithmetic, back pressure is arbitrarily defined as $\frac{1}{2}$ psi per foot of height or 5 psi per story in buildings. The latter figure is based on an average height of 12 feet per story in buildings constructed years ago. The height is strictly a vertical measurement. The actual length of a hose line has nothing to do with back pressure. If a nozzle is 50 feet high and the hose is 700 feet long, the back pressure is figured only on the basis of the 50 feet of height.

Back pressure must be considered in calculating engine pressures because it acts directly against the pump and must be neutralized by increasing the pump pressure (engine pressure). If the back pressure is 20 psi, then the engine pressure needed to provide nozzle pressure and overcome friction loss must be increased 20 psi to neutralize the back pressure.

For example, if the friction loss in a hose is 95 psi, the nozzle pressure is 50 psi, and the nozzle is 30 feet above the ground, we determine the engine pressure by first figuring the back pressure:

$$30 \times 0.434 = 13.02 = 13 \text{ psi}$$

Then we use the formula:

$$EP = FL + NP + BP$$
$$EP = 95 + 50 + 13 = 158 \text{ psi}$$

FIRE SERVICE HYDRAULICS

Or in the same problem, we could use the ½ psi per foot figure for back pressure. This would give us a back pressure of 15 psi for 30 feet of height. As can be readily seen, this would add 2 psi to the previous answer and give us an engine pressure of 160 psi. Actually, this would make no difference on the fireground because pump gages are not read closer than the nearest 5 pounds. With an engine pressure of 158 psi needed, the pump operator would keep the needle of the pressure gage hovering around the 160 psi line. The effect of a few psi one way or the other in engine pressure cannot be noticed by the man on the nozzle.

If the hose is on the roof of a four-story building, instead of estimating the height we can use the rule-of-thumb 5 psi per story for back pressure. This gives us 20 psi back pressure.

On the other hand, if the nozzle is working on the fourth floor of this same building, the back pressure would be $3 \times 5 = 15$ psi because when the nozzle is on the fourth floor, it is only three stories above the ground. In buildings that have the first floor a substantial number of feet above the ground, the pump operator can use his judgment as to whether almost a story of height has already been reached by the first floor.

From a practical standpoint, it is not necessary to consider back pressure until the nozzle is up to the third floor.

Nozzle lower than pump

We have been discussing back pressure when the hose line makes a steady rise upward from the pump. What happens if the hose line descends from the pump and the nozzle is lower than the pump? Then we have the reverse effect of back pressure.

If the nozzle is 30 feet lower than the pump, then the effect is to help the pump. Therefore, the 13-psi reverse back pressure, which can be regarded as "forward pressure," is subtracted from the engine pressure required for friction loss and nozzle pressure.

In the problem previously examined in which friction loss was 95 psi and the nozzle pressure was 50 psi, a nozzle 30 feet below the pump would make the engine pressure equation look like this.

$$EP = 95 + 50 - 13 = 132 \text{ psi}$$

If the hose line goes up a grade, then descends and finally rises so that the nozzle is higher than the pump, the back pressure is figured only for the height the nozzle is above the pump. The descent below the pump cancels out the rise back to the pump level.

ENGINE AND NOZZLE PRESSURES

Rule-of-thumb calculations

The pump operator has no time on the fireground to work out extensive calculations to determine the proper pump pressure. He must depend on some simple friction loss figures that are easily remembered and do all his arithmetic in his head.

Initial nozzle pressures, at least, should be standard throughout the department. For straight tips on hand lines, a 50-psi nozzle pressure works well. It provides a water delivery rate that is liberally proportional to the size of the tip and yet keeps the line within the limits of reasonable handling.

Table 3   Friction Loss in psi per 100 Feet of Hose

Tip Size	1½-Inch Hose	2½-Inch Hose	2 Lines 2½-Inch Hose Siamesed	3-Inch Hose	GPM Flow	Nozzle Pressure
¼	1				12	
⅜	3			½ that of a single 2½-inch line –or– Multiply 2½-inch friction loss by 0.4	25	
½	10				50	
⅝	15	1			75	
¾	30	3			100	50 psi
⅞		5			150	
1		10	¼ that of a single 2½-inch line		200	
1⅛		15			250	
1¼		25		10	325	
1⅜			15	20	500	Master stream straight tips, 80 psi
1½			20	30	600	
1⅝			30	40	700	
1¾			35	50	800	Fog nozzles, 100 psi for rated gpm
2			55	75	1000	

Note: Friction losses for one 2½ and one 3-inch line, siamesed, are about ¼ less than the friction loss for two 2½-inch lines siamesed.

FIRE SERVICE HYDRAULICS

For fog nozzles on 1½-inch and smaller lines, 100-psi nozzle pressures are recommended. On 2½-inch lines, fog nozzles rated at 200 to 250 gpm at 100 psi will be much easier to handle and will still break up the stream into a good fog at nozzle pressures as low as 70 psi.

Master stream nozzle pressures are generally 80 psi for straight tips and 100 psi for fog tips.

Table 3 shows the rule-of-thumb friction losses applicable to 1½ to 3-inch hose. Pump operators should memorize the figures in this table.

Friction losses in appliances

Friction losses in appliances vary according to the rate of water flow. A deluge set that has a friction loss of 3 or 4 psi at 400 gpm might show a friction loss of 8 or 10 psi with 800 gpm flowing. When using the rule-of-thumb method of calculating engine pressures, a 10-psi friction loss for old-style deluge sets and deck guns should be added to the computations. For newer design deluge sets, 5 psi is enough to allow for friction loss.

For most ladder pipes, 5 psi will compensate for the friction loss within the appliance.

The friction losses for siameses and wyes of modern design are insignificant and can be disregarded without any noticeable effect on the fire streams. With the older siameses and wyes, in which friction losses ranged from 3 to 5 psi, it is customary to use 5 psi as a standard friction loss figure.

How rule-of-thumb method is used

To see how the rule-of-thumb method works, let's first look at the problem of the pump operator who must supply a 1⅛-inch tip on a 350-foot line of 2½-inch hose operating on the fourth floor of a building.

From Table 3, the friction loss for a 1⅛-inch tip with a defined 250-gpm flow is 15 psi per 100 feet. The line is 350 feet long, so we multiply 15 by 3½, which gives a friction loss of 52½ psi for the entire line. That's an unwieldy figure to handle in your head, so we round it out to 55 psi. The hose is on the fourth floor, three stories above the ground, so we include a back pressure of 3 times 5 psi (the back pressure per story) which is 15 psi. Adding 55 psi for friction loss and 15 psi for back pressure, the total is 70 psi. Then we add the desired nozzle pressure, 50 psi, to obtain the 120-psi engine pressure that the pump operator must maintain to properly supply this hose line.

Using this same procedure, let's consider the problem of the pump operator who has to supply a ladder pipe with a 1¾-inch tip 60 feet

ENGINE AND NOZZLE PRESSURES

above the ground. Two 2½-inch parallel lines, each 250 feet long, go to a siamese, and there is 100 feet of 3-inch hose to the ladder pipe.

Using the rule-of-thumb Table 3, here is how the friction losses are totaled:

2½-in parallel lines, 250 ft, 800 gpm = 2½ × 35 = 87½ =	90 psi
3-in hose, 100 ft, 800 gpm =	50 psi
Back pressure, 60 ft =	30 psi
Total friction loss and back pressure =	**170 psi**

Note that no friction loss was calculated for the siamese. The 87½-psi friction loss in the parallel lines was rounded off to 90 psi. This adequately compensates for friction loss in the siamese. If the three friction losses computed had all been in multiples of 5 so that none had to be rounded off, the failure to compensate in this way for the friction loss in the siamese would have no noticeable effect on the ladder pipe stream.

In rule-of-thumb computations for fire streams, all figures are rounded off to the nearest multiple of 5 for ease in adding mentally and for practical setting of pump pressures. Pump gages are not calibrated any finer than 5-psi intervals, and it is impractical to try to read a pump gage any closer than that because of the slight needle flutter, minor fluctuations in engine speed (rpm) and the slight variations in the angles at which the operator reads the gage.

PARALLEL LINES

When using the Underwriters' formula to solve for engine or nozzle pressures for parallel lines of equal length siamesed into deluge sets, the value of L is obtained by dividing the number of feet in one of the parallel lines by 50. In giving the values of K for parallel lines, the table includes an allowance for friction loss in deluge sets.

Example: An engine is pumping through two parallel lines, 400 feet long, to a deluge set with a 1½-inch tip. What engine pressure is required to obtain a nozzle pressure of 80 psi?

Answer: In determining the value of K from Table 1, select the value for a 1½-inch tip with two 2½-inch lines siamesed. This K value includes an allowance for the friction loss in the deluge set.

$$EP = NP(1.1 + KL)$$
$$K = 0.135$$
$$L = 400 \div 50 = 8$$
$$EP = 80(1.1 + .135 \times 8)$$
$$EP = 80 \times 2.18 = 174.4 = 175 \text{ psi}$$

FIRE SERVICE HYDRAULICS

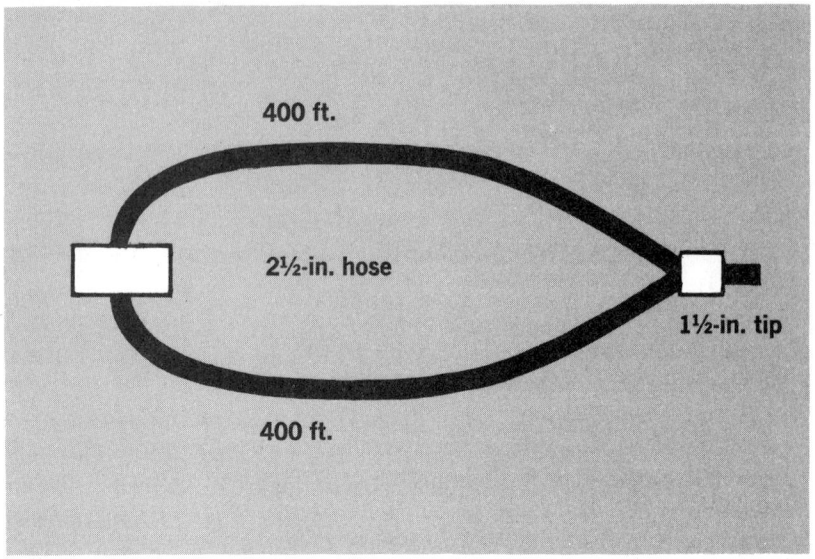

For theoretical reasons off the fireground, the question of what nozzle pressure will be attained with a specific pump pressure through parallel lines might arise.

Example: An engine is pumping at 190 psi through two 400-foot, 2½-inch parallel lines to a deluge set with a 1½-inch tip. What is the nozzle pressure?
Answer:

$$NP = \frac{EP}{1.1 + KL}$$
$$K = 0.135$$
$$L = 400 \div 50 = 8$$
$$NP = \frac{190}{1.1 + .135 \times 8}$$
$$NP = \frac{190}{2.18} = 87.1 = 87 \text{ psi}$$

Now let's determine the engine pressure for the same layout by using the rule-of-thumb figures in Table 3.

First look up the friction loss for a 1½-inch tip supplied by two 2½-inch lines of the same length. The friction loss per 100 feet of parallel lines is 20 psi. The length of the parallel lines in this problem is 400. Therefore:

FL in parallel lines = 4 × 20 = 80
NP = 80
FL in old-style deluge set = 10
EP = 170 psi

Simple calculation for parallel lines

When parallel lines of the same diameter are of equal length, the friction losses can be computed quite easily if the pump operator remembers the friction losses for a few key flow rates.

If the flow through two parallel lines totals 600 gpm, then it follows that 300 gpm are flowing through each line. In parallel lines of equal length and diameter, the friction loss in one line is the only friction loss figure needed to compute the engine pressure. In this case, the friction loss for 300 gpm in a 2½-inch line is 21 psi, so 21 psi is the friction loss per 100 feet of parallel 2½-inch hose flowing a total of 600 gpm.

If there were three parallel lines in the layout and the total flow was 600 gpm, then each line would carry one third the water or 200 gpm. The friction loss for 200 gpm in 2½-inch hose is 10 psi, and therefore 10 psi is the friction loss in three parallel lines flowing a total of 600 gpm.

If four parallel lines were being used, the friction loss per 100 feet for the layout would be the friction loss for one quarter of the total gpm being delivered.

Table 4 illustrates what has been discussed:

Table 4 Friction Loss Table

gpm	Single 2½" Line	2 Siamesed 2½" Lines
200	10	3
250	15	4
300	21	6
350	28	8
400	36	10
500	55	15
600	78	21
700	105	28
800	136	36
900	171	45
1000	210	55

Notice how the friction loss figures 10, 15, 21, 28, 36 and 55 appear in both columns. Note also that the gallonage for the two siamesed

FIRE SERVICE HYDRAULICS

2½-inch lines is twice that of the single 2½-inch line with the same friction loss.

This characteristic holds true for friction losses for parallel lines of any size.

The friction loss for parallel lines can be determined by dividing the total gpm being delivered by the number of lines. The friction loss for the parallel line layout will be the friction loss for one line flowing its equal share of the total flow.

If the pump operator can remember friction losses for single 2½-inch lines flowing up to 500 gpm, he can handle any 2½-inch parallel line problem he is apt to encounter. If he doesn't remember the friction loss for a specific gallonage, he can resort to the friction loss formula, $2Q^2 + Q$, to figure the friction loss for the single line carrying its equal share of the total layout flow. And that figure, of course, will be the friction loss for the layout, per 100 feet of the length of the layout—not each line in it.

If the pump operator is supplying 3-inch lines and can't remember friction losses for those lines, he can work the problem as though it involved only 2½-inch lines. Then he multiplies his answer by 0.4 to get the friction loss for 3-inch parallel lines.

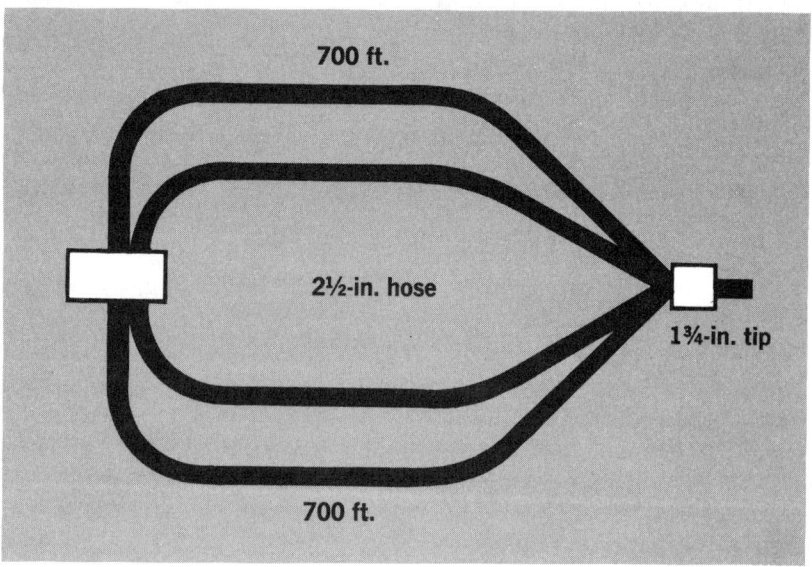

ENGINE AND NOZZLE PRESSURES

He would have to multiply the answer for 2½-inch lines by 0.17 for 3½-inch lines.

Once the friction loss figure for the parallel lines is determined, multiply it by the number of hundreds of feet in the length of the layout (not the total length of the individual lines) to compute the total hose friction loss. Adding friction loss for parallel lines, nozzle pressure, any back pressure and a friction loss for the appliance or a 20-psi residual pressure for the receiving pumper in a relay totals up to the required engine pressure.

Example: An engine supplies a 1¾-inch tip with a nozzle pressure of 80 psi on a deluge set through four lines of 2½-inch hose, each 700 feet long. What is the engine pressure?
Answer:

Flow of 1¾-inch tip at 80 psi is 800 gpm
Each of 4 lines is flowing 200 gpm
Friction loss of 2½-inch line flowing 200 gpm is 10 psi

FL of parallel lines (700 feet) = 7 × 10 = 70 psi
FL in deluge set (old style) = 10 psi
NP = 80 psi
EP = 160 psi

Complicated siamese layouts

So far, the discussion of parallel lines has been confined to parallel lines of equal length supplied by the same pumper. Eight variations of parallel line problems exist. They are:

Diameters of Parallel Lines	Lengths of Parallel Lines	Engine Pressures
1. Equal	Equal	Equal
2. Equal	Equal	Unequal
3. Unequal	Unequal	Equal
4. Unequal	Equal	Equal
5. Equal	Unequal	Equal
6. Equal	Unequal	Unequal
7. Unequal	Unequal	Unequal
8. Unequal	Equal	Unequal

To solve the problems of such layouts, first change the layout to a single line of 2½-inch hose and then calculate the engine pressure or the nozzle pressure, whichever is required.

FIRE SERVICE HYDRAULICS

To change other sizes of hose to 2½-inch hose, conversion factors are used. The factors are listed in Table 5:

Table 5 Hose Conversion Factors (Changing to 2½-Inch Hose)

Hose Size	Factor
3-Inch:	
2½-Inch couplings	2.5
3-Inch couplings	2.6
3½-Inch	5.8
4-Inch	11.0
4½-Inch	19.5
5-Inch	32.0

The larger diameter hose length is divided by the factor to find the equivalent length of 2½-inch hose.

Example: What is the length of 2½-inch hose that is equivalent to 850 feet of 3-inch hose with 2½-inch couplings?
Answer: The factor for 3-inch hose with 2½-inch couplings is 2.5. Therefore:

$$850 \div 2.5 = 340 \text{ ft}$$

Or, in the case of 3-inch hose with 2½-inch couplings, you can multiply the length of the 3-inch line by 0.4. Working the above problem this way, the result is:

$$850 \times 0.4 = 340 \text{ ft}$$

Example: What is the length of 2½-inch hose that is equivalent to 750 feet of 3½-inch hose?
Answer: The factor for 3½-inch hose is 5.8. Therefore:

$$750 \div 5.8 = 129.3 \text{ ft}$$

Various parallel line layouts

All parallel line layout problems can be solved by converting the entire layout, including any single line at the start or end of the layout, into a single 2½-inch line. To convert two parallel 2½-inch lines of equal length to an equivalent single 2½-inch line, divide the length of one line by 3.6.

When one engine is pumping three 2½-inch siamesed lines of equal length, divide the length of one line by 7.75 to convert the layout to a single 2½-inch line. Then use the appropriate form of the Underwriters' formula to calculate engine or nozzle pressure.

If there are four 2½-inch parallel lines of equal length, divide the length of one line by 12.4.

For reducing other combinations of siamesed parallel lines of equal length to a single 2½-inch line, see the 2½-Inch Hose Friction Loss Conversion Table on page 232.

When parallel lines have the same diameter but different lengths, average the lengths and then reduce to a single 2½-inch line.

Parallel lines of different diameters and lengths are converted to 2½-inch lines and the equivalent lengths are averaged before the entire layout is reduced to a single 2½-inch line for final solution of the engine or nozzle pressure by using the appropriate form of the Underwriters' formula.

When two or more engines are pumping through one or more parallel lines apiece, average the engine pressures and change the layout to a single 2½-inch line to find the nozzle pressure. When the nozzle pressure is known and the engine pressure is determined by use of the Underwriters' formula, the result will be the average pressure for all engines.

Note that parallel lines of different diameters and/or lengths must be reduced to a single 2½-inch line to use the Underwriters' formula.

When parallel lines are the same size, they are not reduced to a 2½-inch line to use the rule-of-thumb method of calculating nozzle pressure or engine pressure. However, if they are different lengths, then the lengths are averaged.

Clappers in siameses

When two engines are each pumping a line to a siamese, the question of what happens to the clappers often arises. Ideally, each engine should be pumping at a pressure that will get its water to the siamese at the same pressure the other engine is providing to that point. Then the clappers for each line would remain open during the pumping operation, and each engine would be doing its share of the work.

But sometimes one, or both, of the pump operators comes up with the wrong engine pressure and the pressures at each inlet of the siamese differ considerably.

Let's assume that the pressure needed at the siamese is 160 psi. Engine 1 is providing 170 psi at the siamese after overcoming a friction loss of 65 psi with an engine pressure of 235 psi. Engine 2 has the same hose layout, but the operator made an error in calculating and decided to use an engine pressure of 205 psi. As a result, Engine 2 cannot provide the desired 160 psi at the siamese.

As both engines pump, the higher pressure from Engine 1 closes the

FIRE SERVICE HYDRAULICS

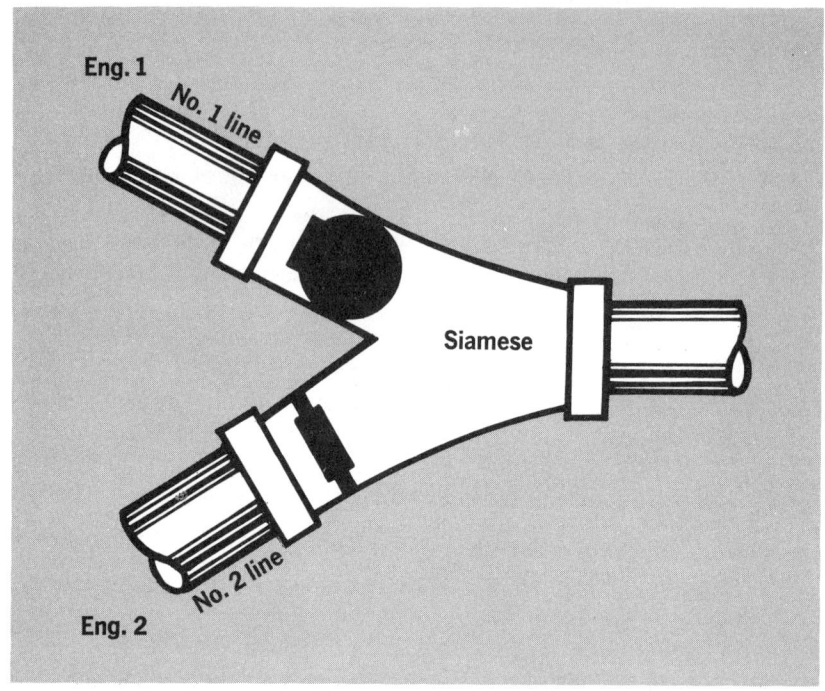

clapper against the line from Engine 2, which was to be expected. But once the clapper closes and halts all movement of water from Engine 2, the pressure throughout No. 2 line becomes the same as the engine pressure, 205 psi. That is, the pressure at the siamese inlet would be 205 psi if the clapper stayed closed. Now when the pressure in Engine 2's static line rises enough above the 170 psi provided at the siamese by Engine 1, the clapper for No. 2 line will begin to open.

The pressure in No. 2 line has to exceed 170 psi to open the clapper because there are more square inches of area on the inner than the outer surface of the clapper. Therefore, it takes a little less pressure to keep a clapper closed than it does to open it when pressures are applied at the same time to both sides.

As soon as the clapper on No. 2 line begins to open, the pressure begins to drop. When the clapper has opened wide enough and long enough for the pressure to drop below about 170 psi, the clapper will shut again. Then the entire process will be repeated. The clapper will flutter, slightly or greatly, as long as both engines are pumping at different pressures.

ENGINE AND NOZZLE PRESSURES

If the pressures provided by two engines are fairly even at the siamese, both clappers will remain open, but they will flutter slightly because of the modest pressure difference. However, when the pressures from each line are practically the same, both lines provide their full share of the water to the siamese.

For a clapper to remain shut during a pumping operation, the pump supplying the line to that clapper would have to be operated at an engine pressure less than the pressure provided at the siamese by the other engine's line.

WYED LINES

When a hose line is divided into two or more lines, these lines are known as wyed lines. The four more common classes of wyed line problems are:

	Diameters of Wyed Lines	Lengths of Wyed Lines	Nozzle Diameters
1.	Equal	Equal	Equal
2.	Equal	Unequal	Equal
3.	Equal	Equal	Unequal
4.	Equal	Unequal	Unequal

Just as in doing compound problems with siamesed lines, the first step in using the Underwriters' formula for solving wyed line problems is to reduce the layout to an equivalent single line of $2\frac{1}{2}$-inch hose. The second step is to combine the nozzles into a single nozzle of equivalent orifice area. Then the problem can be solved for engine or nozzle pressure.

Class 1

These problems involve equal diameters and lengths of wyed lines and equal nozzle diameters.

Example: An engine is pumping through 500 feet of $2\frac{1}{2}$-inch hose which is wyed off to two $2\frac{1}{2}$-inch lines each 200 feet long and each with a 1-inch tip and a 45-psi nozzle pressure. What is the engine pressure?

Answer: Combine the 200-foot wyed lines into a single $2\frac{1}{2}$-inch line:

$$200 \div 3.6 = 55.6 \text{ ft of } 2\frac{1}{2}\text{-in hose}$$

FIRE SERVICE HYDRAULICS

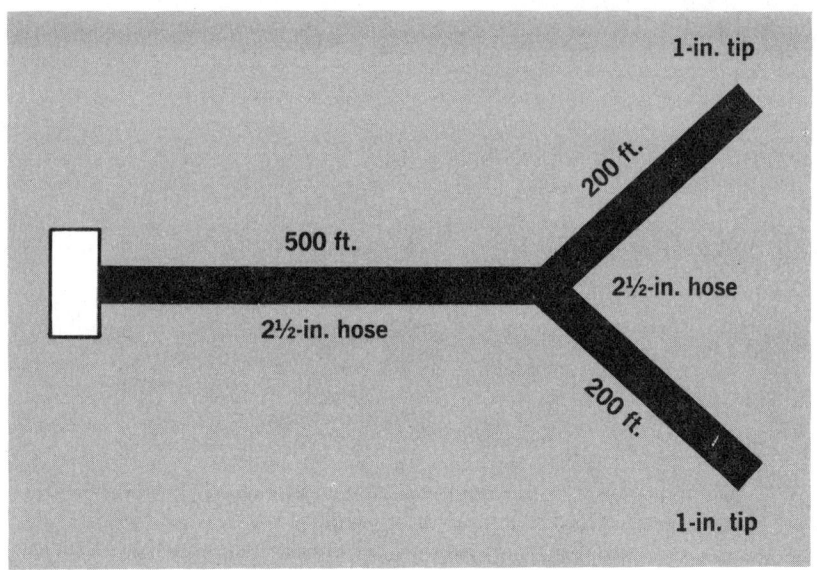

55.6 + 500 = 555.6 feet of a single 2½-inch line, the equivalent of the hose layout.

From Table 6, two 1-inch nozzles are equivalent to one 1⅜-inch nozzle.

K for a 1⅜-inch-nozzle on 2½-inch hose is 0.341

$$L = 555.6 \div 50 = 11.1$$
$$EP = NP(1.1 + KL)$$
$$EP = 45(1.1 + .341 \times 11.1)$$
$$EP = 45 \times 4.885 = 219.8 = 220 \text{ psi}$$

The Underwriters' formula also can be used to solve for nozzle pressure (NP) instead of engine pressure. However, the nozzle pressure so obtained will be an average nozzle pressure, not the nozzle pressure for each tip, except when the branch lines and tips are identical.

Example: If an engine pumps through the layout in the previous example, what is the nozzle pressure if the engine pressure is 220 psi?

Answer: The formula is changed to solve for nozzle pressure:

$$NP = \frac{EP}{1.1 + KL}$$

$$NP = \frac{220}{1.1 + .341 \times 11.1}$$

$$NP = \frac{220}{4.885} = 45.03 = 45 \text{ psi}$$

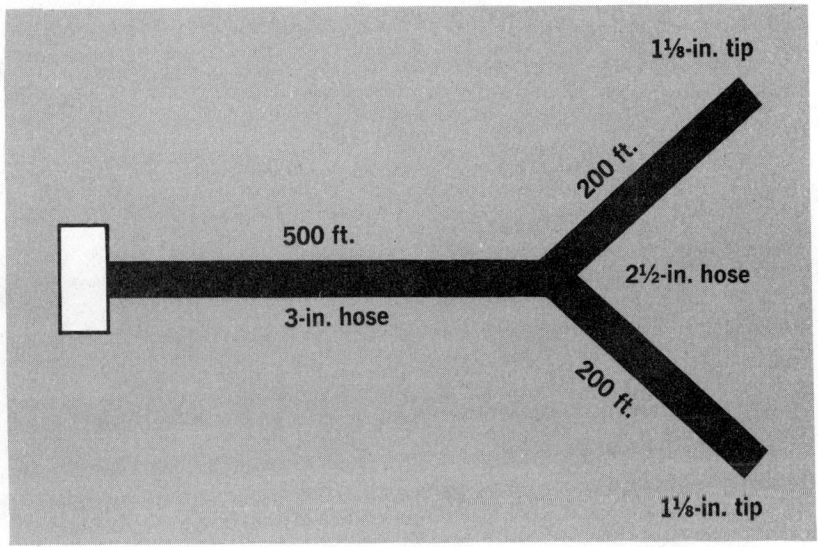

Example: An engine is pumping through 500 feet of 3-inch hose with 2½-inch couplings which is wyed off to two 200-foot lines of 2½-inch hose, each with 1⅛-inch nozzles with a nozzle pressure of 50 psi. What is the engine pressure?

Answer: Combine the 200-foot wyed lines into a single 2½-inch line:

$$200 \div 3.6 = 55.6 \text{ ft of 2½-in hose}$$

Change the 3-inch hose to an equivalent length of 2½-inch hose:

$$500 \div 2.5 = 200 \text{ ft of 2½-in hose}$$

200 + 55.6 = 255.6 feet, the equivalent length of 2½-inch hose for the layout

FIRE SERVICE HYDRAULICS

From Table 6, two 1⅛-inch nozzles are equivalent to one 1⅝-inch nozzle.

K for a 1⅝-inch tip is 0.680

$$L = 255.6 \div 50 = 5.1$$
$$EP = NP(1.1 + KL)$$
$$EP = 50(1.1 + .68 \times 5.1)$$
$$EP = 50 \times 4.568 = 228.4 = 228 \text{ psi}$$

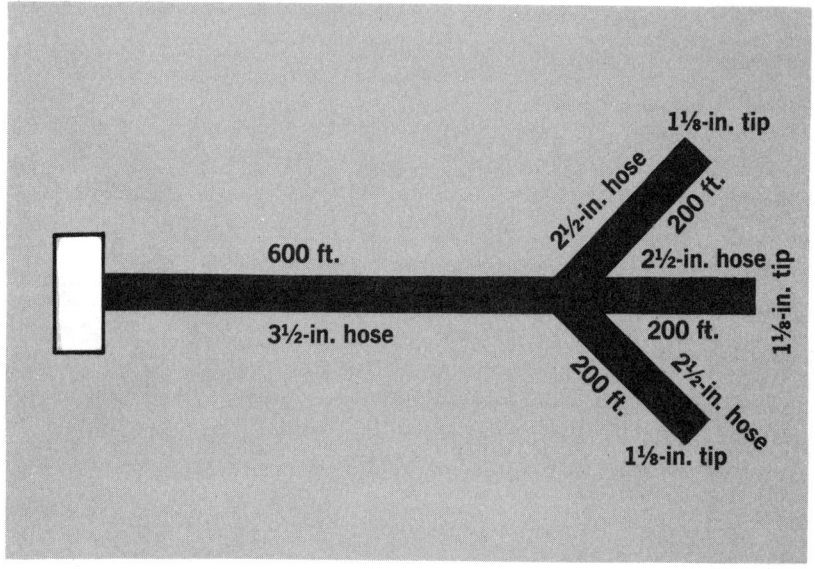

Example: An engine is pumping through 600 feet of 3½-inch hose which is wyed off to three 2½-inch lines, each 200 feet long and each with a 1⅛-inch tip. The nozzle pressure at each tip is 45 psi. What is the engine pressure?

Answer: First combine the wyed lines into an equivalent, single, 2½-inch line:

$$200 \div 7.75 = 25.8 \text{ ft}$$

The equivalent 2½-inch line for the 600 feet of 3½-inch hose is:

$$600 \div 5.8 = 103.4 \text{ ft}$$
$$103.4 + 25.8 = 129.2 \text{ ft}$$

ENGINE AND NOZZLE PRESSURES

Three 1⅛-inch nozzles are equivalent to one 1¹⁵⁄₁₆-inch nozzle on a 2½-inch hose line.

$$K = \frac{D^4}{10} = \frac{(1^{15}/_{16})^4}{10} = \frac{(1.9375)^4}{10} = \frac{(1.94)^4}{10} = 1.414$$

$L = 129.2 \div 50 = 2.6$
$EP = NP(1.1 + KL)$
$EP = 45(1.1 + 1.414 \times 2.6)$
$EP = 45 \times 4.78 = 215.1 = 215$ psi

For rapid calculation of engine pressure on the fireground, we must know the size tips on the wyed lines, the nozzle pressure desired and the diameters and lengths of the wyed lines and the feeder line.

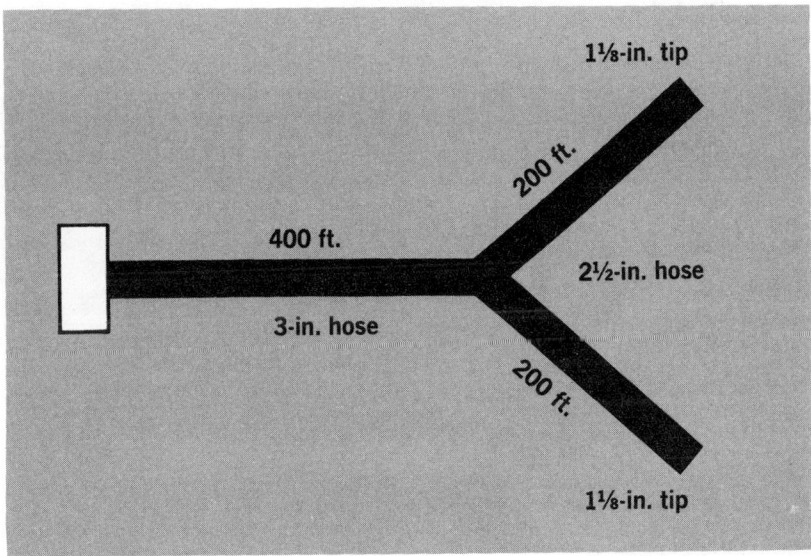

Example: An engine is to pump through 400 feet of 3-inch hose which is wyed into two 200-foot 2½-inch lines with 1⅛-inch nozzles. What is the engine pressure?

Answer: The two 1⅛-inch tips at 45-psi nozzle pressure will deliver 250 gpm each. The friction loss in 2½-inch hose for 250-gpm is 15 psi. With 200 feet in each wyed line, the friction loss in the wyed lines is 30 psi. The total delivery of the two nozzles is 500 gpm, so that is what must move through the 3-inch feeder line. The friction loss in 3-inch hose for 500 gpm is 21, so the friction loss in the supply line is $4 \times 21 = 84$. Adding, we get:

$$EP = 45 \text{ psi } NP + 30 \text{ psi } FL \text{ in wyed lines} + 84 \text{ psi } FL \text{ in the 3-in line} = 159 \text{ psi}$$

The operator would pump at 160 psi. Pumping is not accurate to the pound for several reasons, so pumping at the nearest 5-psi mark on the gage is satisfactory because even this is not precise.

In this example, the friction loss in the wye has been ignored because of the reasons previously detailed in this chapter. Also, figuring by rule of thumb is not so exact that ignoring the wye makes any perceptible difference in the fire streams.

Approximate nozzle comparisons

To expedite solving wyed line problems, the approximate comparison of nozzles is made in Table 6. Some of the combined nozzles in

Table 6 Approximate Comparison of Nozzles

Number of Nozzles												
1	5/8″	3/4″	7/8″	1″	1 1/8″	1 1/4″	1 3/8″	1 1/2″	1 5/8″	1 3/4″	1 7/8″	2″
2	7/8″	1 1/16″	1 1/4″	1 3/8″	1 5/8″	1 3/4″	1 15/16″	2 1/8″	2 5/16″	2 1/2″	2 5/8″	2 13/16″
3	1 1/16″	1 5/16″	1 1/2″	1 3/4″	1 15/16″	2 3/16″	2 3/8″	2 5/8″	2 13/16″	3″	3 1/4″	3 1/2″
4	1 1/4″	1 1/2″	1 3/4″	2″	2 1/4″	2 1/2″	2 3/4″	3″	3 1/4″	3 1/2″	3 3/4″	4″
5	1 3/8″	1 11/16″	2″	2 1/4″	2 1/2″	2 3/4″	3 1/6″	3 3/8″	3 5/8″	3 15/16″	4 3/16″	
6	1 1/2″	1 13/16″	2 1/8″	2 7/16″	2 3/4″	3 1/16″	3 3/8″	3 11/16″	4″			
7	1 5/8″	2″	2 5/16″	2 5/8″	3″	3 5/16″	3 5/8″	4″				
8	1 3/4″	2 1/8″	2 1/2″	2 13/16″	3 3/16″	3 9/16″	3 7/8″					
9	1 7/8″	2 1/4″	2 5/8″	3″	3 3/8″	3 3/4″	4 1/8″					
10	2″	2 3/8″	2 3/4″	3 3/16″	3 9/16″	4″						
11	2 1/16″	2 1/2″	2 7/8″	3 5/16″	3 3/4″							
12	2 3/16″	2 5/8″	3″	3 1/2″	3 15/16″							

Courtesy American Insurance Association.

ENGINE AND NOZZLE PRESSURES

the table are not commonly used sizes, so there is no value of *K* for them in the table of *K* values. For these nozzle sizes, the *K* value will have to be solved by the method previously described.

A more accurate method of handling the nozzles on wyed lines is to compute the exact size of the nozzle equal to those of the wyed lines and solve for *K*.

The table's value is limited by the fact that it is restricted to comparing nozzles of the same size. To find the nozzle size equivalent to three $1\frac{1}{8}$-inch tips, select the line numbered 3 at the left and then follow that line to the figure under $1\frac{1}{8}$ inch on the top line. The figure $1\frac{15}{16}$ inch at the junction of the vertical and horizontal columns is the size nozzle equivalent to three $1\frac{1}{8}$-inch nozzles.

Class 2

Wyed lines of unequal lengths but with equal hose and nozzle diameters.

Example: An engine is pumping through 200 feet of $2\frac{1}{2}$-inch hose which is wyed off to two $2\frac{1}{2}$-inch lines. One branch of the wye is 200 feet long and the other is 300 feet long. The nozzles are both 1-inch and the average nozzle pressure is 50 psi. What is the engine pressure?

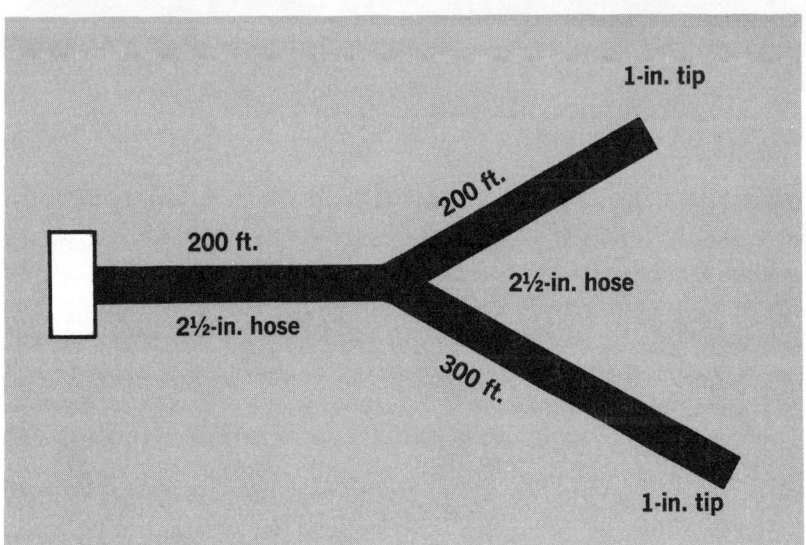

Answer: First average the wyed lines:

$$200 + 300 = 500 \div 2 = 250 \text{ ft}$$

Then change the layout to a single line of 2½-inch hose. Two 2½-inch parallel lines, each 250 feet long, are equivalent to a single 2½-inch line that is $250 \div 3.6 = 69.4$ feet.

The total length of 2½-inch hose is $200 + 69.4 = 269.4$ feet.

Two 1-inch tips are equivalent to one 1⅜-inch nozzle (Table 6). K for a 1⅜-inch tip (Table 1) is 0.341.

$$EP = NP(1.1 + KL)$$
$$K = 0.341$$
$$L = 269.4 \div 50 = 5.4$$
$$EP = 50(1.1 + 0.341 \times 5.4)$$
$$EP = 50 \times 2.9414 = 147.07 = 147 \text{ psi}$$

Now let's solve the same problem using the rule-of-thumb procedure.

Answer: Each 1-inch tip flows 200 gpm. Therefore the total flow through the line from the engine to the wye is 400 gpm. In wyed lines of equal length and size with nozzles of the same diameter, the friction loss for only one wyed line, regardless of how many are in the layout, is considered. If the wyed lines are the same size but different lengths, use the average length as a single line.

FL in wyed 2½-inch line = 2.5×10 = 25
FL in supply line = 2×36 = 72
NP = 50
EP = $\overline{147}$ psi

Instead of solving for engine pressure, let's consider the same basic problem but solve for the average nozzle pressure.

Note that only when wyed lines are equal in diameter and length and the nozzles are the same diameter can the nozzle pressure at each nozzle be the same. When there is any variation in the wyed lines, the nozzle pressure determined by formula is an average for the number of nozzles in the wyed line layout. In hose layouts with a wye and lines off the wye, the pressure at the wye can be considered to be the same as if it were at a pump. If instead of being wyed lines, these same lines of different characteristics (length, diameter, nozzle size) were individual lines off a pump, then the nozzle pressures would be different. The pressure at the wye creates the same results in the

ENGINE AND NOZZLE PRESSURES

wyed lines as this pressure would if it were at a pump with the same lines taken directly off the pump. Thus, the answer to the problem, as restated below, will be an average nozzle pressure.

Example: An engine is pumping at 150 psi through 200 feet of 2½-inch hose which is wyed off to two 2½-inch lines. One branch of the wye is 200 feet long and the other is 300 feet long. The nozzles are both 1-inch. What is the average nozzle pressure?
Answer- First average the wyed lines:

$$200 + 300 = 500 \div 2 = 250$$

Then change the layout to a single line of 2½-inch hose. Two 2½-inch parallel lines, each 250 feet long, are equivalent to a single 2½-inch line that is $250 \div 3.6 = 69.4$ feet.

The total length of 2½-inch line is $200 + 69.4 = 269.4$ feet.

Two 1-inch tips are equivalent to one 1⅜-inch nozzle (Table 6). K for a 1⅜-inch tip (Table 1) is 0.341.

$$NP = \frac{EP}{1.1 + KL}$$
$$L = 269.4 \div 50 = 5.4$$
$$NP = \frac{150}{1.1 + 0.341 \times 5.4}$$
$$NP = \frac{150}{2.9414} = 50.9 = 51 \text{ psi (average)}$$

Calculating individual nozzle pressure

To determine the nozzle pressure for each wyed line, first calculate the total gpm flow of the layout. With this figure, obtain the friction loss in the hose from the engine to the wye. Subtracting the total friction loss in the supply line from the engine pressure gives the pressure at the wye. Use the pressure at the wye as you would engine pressure and calculate each nozzle pressure separately.

To find the total flow for the layout, determine the flow from the equivalent nozzle size, in this problem 1⅜ inches, at the average pressure, in this problem 51 psi, by using the formula for flow:

$$\text{GPM} = 29.7 d^2 \sqrt{P}$$
$$\text{GPM} = 29.7 (1⅜)^2 \sqrt{51}$$
$$\text{GPM} = 29.7 (1.375 \times 1.375) 7.1414$$
$$\text{GPM} = 29.7 \times 1.89 \times 7.1414$$
$$\text{GPM} = 400$$

FIRE SERVICE HYDRAULICS

To find the friction loss per hundred feet in the 2½-inch line from the engine to the wye, use the formula:

$$FL = 2Q^2 + Q$$
$$FL = 2(4)^2 + 4 = 2 \times 16 + 4 = 36 \text{ psi}$$

The friction loss for the 200 feet in the feeder line is:

$$2 \times 36 = 72 \text{ psi}$$

To find the pressure at the wye, subtract the feeder line friction loss from the engine pressure:

$$150 - 72 = 78 \text{ psi, the pressure at the wye}$$

The pressure at the wye, 78 psi, is used as engine pressure in solving for nozzle pressure for each branch of the wye with the Underwriters' formula:

$$NP = \frac{EP}{1.1 + KL}$$

K for a 1-inch tip on a 2½-inch line is 0.105. If the tips were different sizes, the appropriate K value for each size would be used.

Solving for the 200-foot line:

$$L = 200 \div 50 = 4$$
$$NP = \frac{78}{1.1 + .105 \times 4}$$
$$NP = \frac{78}{1.52} = 51.3 = 51 \text{ psi}$$

Solving for the 300-foot line:

$$L = 300 \div 50 = 6$$
$$NP = \frac{78}{1.1 + .105 \times 6}$$
$$NP = \frac{78}{1.73} = 45.08 = 45 \text{ psi}$$

Here is another problem with wyed lines of unequal length and equal hose and nozzle diameters but with a feeder line of larger diameter.

Example: An engine is pumping through 200 feet of 3-inch hose with 2½-inch couplings which is wyed off to two 2½-inch lines. One is 200 feet long and the other is 300 feet long. Each branch

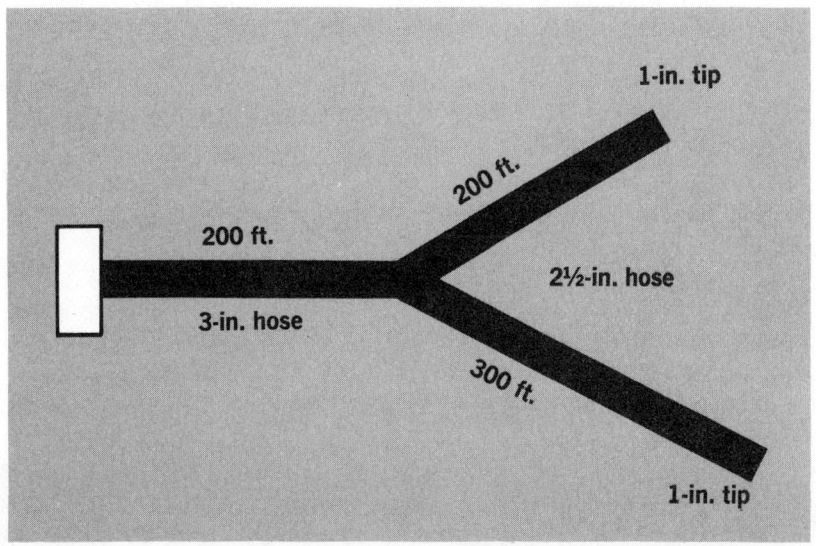

line has a 1-inch tip with an average nozzle pressure of 50 psi. What is the engine pressure?

Answer: First average the two wyed lines:

$$200 + 300 = 500 \div 2 = 250 \text{ ft}$$

Two 2½-inch parallel lines each 250 feet long are equivalent to a single 2½-inch line that is $250 \div 3.6 = 69.4$ feet.

To change the 200 feet of 3-inch hose with 2½-inch couplings to an equivalent single length of 2½-inch hose:

$$200 \div 2.5 = 80 \text{ ft}$$

The total length of equivalent 2½-inch hose is $80 + 69.4 = 149.4$ feet.

Two 1-inch nozzles are equivalent to one 1⅜-inch nozzle (Table 6).

K for a 1⅜-inch nozzle on 2½-inch hose is 0.341.
$L = 149.4 \div 50 = 2.988$.

$$EP = NP(1.1 + KL)$$
$$EP = 50(1.1 + .341 \times 2.988)$$
$$EP = 50 \times 2.1189 = 105.9 = 106 \text{ psi}$$

FIRE SERVICE HYDRAULICS

Using rule-of-thumb figures to calculate the engine pressure in this same problem, remember that a 1-inch tip flows 200 gpm. There are two wyed lines, so the total flow for the layout is 400 gpm. In a 2½-inch line, the friction loss for 400 gpm is:

$$2Q^2 + Q = 2(4)^2 + 4 = 2 \times 16 + 4 = 36$$

To find the friction loss for 3-inch hose, multiply:

$$36 \times 0.4 = 14.4 = 14 \text{ psi for 100 ft}$$

FL for 200 feet of 3-inch hose = 28 psi
FL for 250 feet (average length) of wyed line = 25 psi
NP = 50 psi
EP = 103 = 105 psi

Class 3

Wyed lines of equal size and length but with nozzles of different sizes.

Example: An engine is pumping through 400 feet of 3½-inch hose, which is wyed off to two 300-foot lines of 2½-inch hose. One branch of the wye has a 1-inch nozzle and the other has a 1¼-inch nozzle. The nozzle pressures average 50 psi. What is the engine pressure?

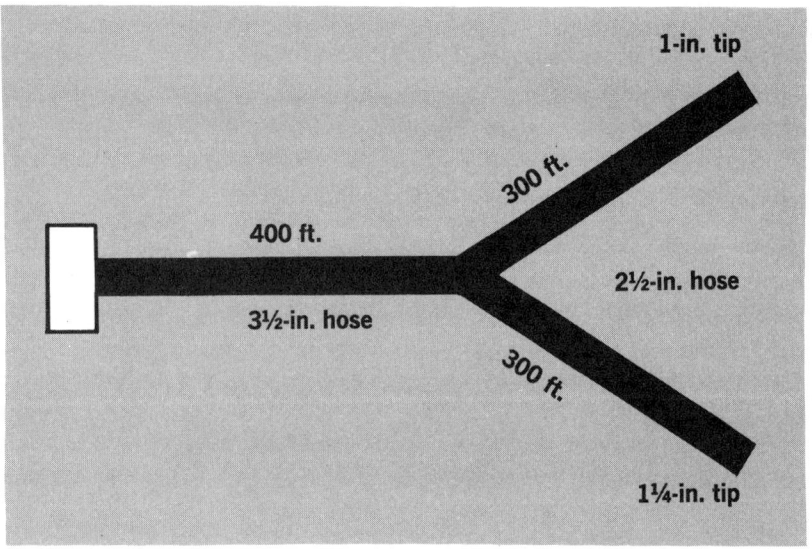

ENGINE AND NOZZLE PRESSURES

Answer: Combine the two 300-foot, 2½-inch, wyed lines into a single 2½-inch line:

$$300 \div 3.6 = 83.3 \text{ ft}$$

Convert the 400 feet of 3½-inch hose to an equivalent length of 2½-inch hose:

$$400 \div 5.8 = 68.9 \text{ ft}$$

The total single length of equivalent 2½-inch hose is:

$$83.3 + 68.9 = 152.2 \text{ ft}$$

To find the size nozzle equivalent to one 1-inch and one 1¼-inch nozzle, first square and then add their diameters:

$$1 \times 1 = 1$$
$$1.25 \times 1.25 = 1.5625$$
$$\text{The sum of the squares} = 2.5625$$

Find the square root of this sum:

$\sqrt{2.5625} = 1.6$, the size nozzle equivalent in area to one 1-inch and one 1¼-inch nozzle

K for a 1.6-inch nozzle on a 2½-inch hose is determined by the formula:

$$K = \frac{D^4}{10} = (1.6 \times 1.6 \times 1.6 \times 1.6) \div 10 = 0.655$$

$$L = 152.2 \div 50 = 3.04$$

Substituting in the formula:

$$EP = NP(1.1 + KL)$$
$$EP = 50(1.1 + .655 \times 3.04)$$
$$EP = 50 \times 3.09 = 154.5 = 155 \text{ psi}$$

If we wish to solve this same problem by rule of thumb, the first step is to estimate the total flow for the layout. At 50-psi nozzle pressures, the 1-inch tip would flow 200 gpm and the 1¼-inch tip would flow 325 gpm. However, if the 1-inch tip has a nozzle pressure of 50 psi, we know that the pressure at the 1¼-inch tip will be less because the higher gpm it flows creates greater friction loss. Therefore, the flow will not be 325 gpm for the 1¼-inch tip nor 525 gpm for the entire layout.

In estimating a gpm for the entire layout, we look for a figure that is less than 525 and is easy to handle, and 500 gpm fills the bill nicely.

FIRE SERVICE HYDRAULICS

The 400-foot, 3½-inch line will handle 500 gpm. The friction loss in a 2½-inch line is 55 psi for 500 gpm ($2Q^2 + Q$). Instead of dividing 55 by 5.8 to determine the friction loss for 3½-inch hose, let's make the mental arithmetic easy by dividing 55 by 6 (rounding out the 5.8), which gives us 9 psi as the friction loss for a 3½-inch line flowing 500 gpm. (Actually, we're 0.5 psi shy of the friction loss table figure.) Multiplying 4 (the number of hundreds of feet in the 3½-inch line) by 9 gives us 36 psi for the friction loss in the feeder line.

The friction loss in the 300-foot, 1-inch tip line can be estimated at 3 × 10 psi, or 30 psi. The friction loss in the 1¼-inch tip line, also 300 feet, can be estimated at 3 × 25 psi, or 75 psi. Averaging these friction losses in round numbers, we get 55 psi. Now the problem adds up like this:

FL for feeder line = 36 = 35 (rounding off for convenience)
FL for wyed line layout = 55
NP = 50
EP = 140 psi

The rule-of-thumb engine pressure is 15 psi less than the engine pressure obtained through the use of the Underwriters' formula. We'll restate the problem to determine how much difference the lower engine pressure makes.

Example: An engine is pumping through 400 feet of 3½-inch hose which is wyed off to two 300-foot lines of 2½-inch hose. One branch of the wye has a 1-inch nozzle and the other has a 1¼-inch nozzle. The engine pressure is 140 psi. What are the pressures at each nozzle?

Answer: First we have to find the pressure at the equivalent nozzle, 1.6 inches, determined in solving the original problem.

$$NP = \frac{EP}{1.1 + KL}$$

From the original problem solution:

$$K = 0.655$$
$$L = 3.04$$

Substituting in the formula:

$$NP = \frac{140}{1.1 + .655 \times 3.04}$$

$$NP = \frac{140}{3.09} = 45.3 = 45 \text{ psi}$$

ENGINE AND NOZZLE PRESSURES

From the nozzle pressure, we can determine the flow by the formula:

$$GPM = 29.7d^2 \sqrt{P}$$
$$GPM = 29.7(1.6)^2 \sqrt{45}$$
$$GPM = 29.7 \times 2.56 \times 6.7$$
$$GPM = 509.4 = 509$$

To find the friction loss for 509 gpm in the 3½-inch hose, we use the formula for 2½-inch hose:

$$FL = 2Q^2 + Q$$
$$FL = 2(5.09)^2 + 5.09$$
$$FL = 56.9 \text{ psi}$$

In the original problem, we found that the 400 feet of 3½-inch hose was equivalent, as far as friction loss is concerned, to 68.9 feet of 2½-inch hose. Since 68.9 is 0.689 of 100 feet, then 0.689 × 56.9 = 39.2 psi, the total friction loss in the 400 feet of 3½-inch hose.

At this point, it is interesting to see what we could have done to reach the same answer. Instead of converting the 3½-inch hose to 2½-inch hose, we could have multiplied the friction loss figure for 509 gpm in 2½-inch hose, 56.9, by 4 as the first step to obtain the total friction loss in the 400 feet of 3½-inch hose:

$$4 \times 56.9 = 227.6 \text{ psi}$$

Then we could have done either of two things:
1. Divide by the factor, 5.8, for converting 3½ to 2½-inch hose:

$$227.6 \div 5.8 = 39.24 \text{ psi}$$

2. Multiply by the factor, 0.17, for the same conversion:

$$227.6 \times 0.17 = 38.69 \text{ psi}$$

The three methods give answers that vary only slightly—39.2, 39.24 and 38.69. Now note that each of these figures rounds off to 39 psi, which is more than accurate enough for fireground pumping.

Now we know that the friction loss in the 400 feet of 3½-inch hose is 39 psi, so we subtract this from the engine pressure:

$$140 - 39 = 101 \text{ psi, the pressure at the wye}$$

The pressure at the wye, 101 psi, is treated as though it were engine pressure, and the nozzle pressure for each line off the wye is determined by calculating the pressure at each nozzle separately.

FIRE SERVICE HYDRAULICS

For the 300-foot branch line with the 1-inch tip:

$$NP = \frac{EP}{1.1 + KL}$$
$$K = 0.105$$
$$L = 300 \div 50 = 6$$
$$EP = 101, \text{ the pressure at the wye}$$
$$NP = \frac{101}{1.1 + .105 \times 6}$$
$$NP = \frac{101}{1.73} = 58.38 = 58 \text{ psi}$$

For the 300-foot branch line with the 1¼-inch tip:

$$NP = \frac{EP}{1.1 + KL}$$
$$K = 0.248$$
$$L = 300 \div 50 = 6$$
$$EP = 101, \text{ the pressure at the wye}$$
$$NP = \frac{101}{1.1 + .248 \times 6}$$
$$NP = \frac{101}{2.588} = 39.02 = 39 \text{ psi}$$

Class 4

Wyed lines of equal diameters but unequal lengths and with nozzles of different sizes.

Example: An engine is pumping through 400 feet of 3-inch hose with 2½-inch couplings which is wyed off to two lines of 2½-inch hose. One branch is 200 feet long with a 1⅛-inch nozzle and the other is 300 feet long with a 1¼-inch nozzle. The average nozzle pressure is 50 psi. What is the engine pressure?

Answer: First average the two wyed lines:

$$200 + 300 = 500 \div 2 = 250 \text{ ft}$$

Two 2½-inch parallel lines each 250 feet long are equivalent to a single 2½-inch line that is:

$$250 \div 3.6 = 69.4 \text{ ft}$$

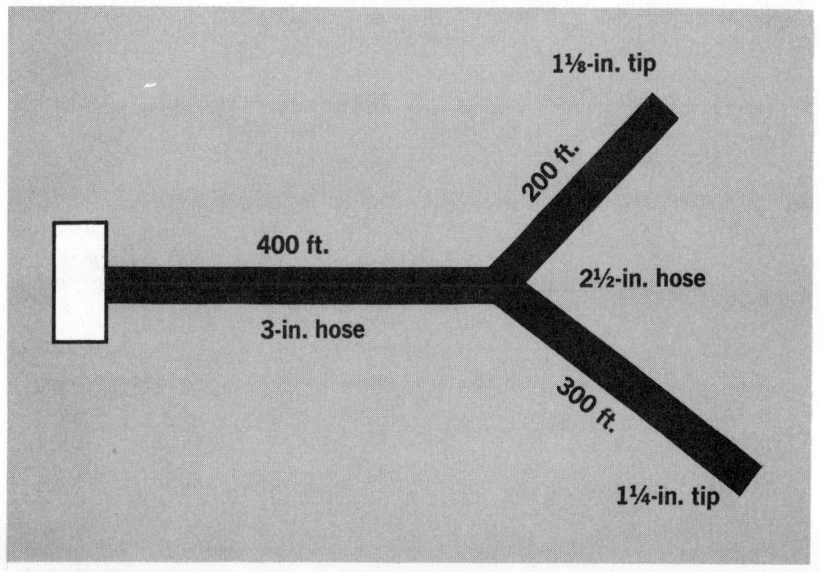

The equivalent length of 2½-inch hose for the 400 feet of 3-inch hose with 2½-inch couplings is:

$$400 \div 2.5 = 160 \text{ ft}$$

The total length of equivalent 2½-inch hose is:

$$69.4 + 160 = 229.4 \text{ ft}$$

The size nozzle equivalent to one 1⅛ and one 1¼-inch nozzle is determined by first squaring and adding the nozzle diameters:

$$1.125 \times 1.125 = 1.265625$$
$$1.25 \times 1.25 = \underline{1.5625}$$
$$\text{Sum of squares} = 2.828125$$

The square root of this sum, $\sqrt{2.828125} = 1.68$, is the diameter of the nozzle size equivalent to the two nozzles in the problem. The K value for the equivalent size nozzle is determined by the formula:

$$K = \frac{D^4}{10} = (1.68 \times 1.68 \times 1.68 \times 1.68) \div 10 = 0.796$$
$$L = 229.4 \div 50 = 4.58$$
$$EP = NP(1.1 + KL)$$
$$EP = 50(1.1 + .796 \times 4.58)$$
$$EP = 50 \times 4.746 = 237.3 = 237 \text{ psi}$$

The rule-of-thumb method can be used to solve the same problem as follows:

First determine the total flow for the layout. The flow from a 1⅛-inch tip is considered, by rule of thumb, to be 250 gpm, and the flow from a 1¼-inch tip, 325 gpm. Therefore, the total flow for the layout is 250 + 325 = 575 gpm. Round this off to 600 gpm.

In a 2½-inch hose, the friction loss for 600 gpm is:

$$FL = 2Q^2 + Q$$
$$FL = 2(6)^2 + 6 = 2 \times 36 + 6 = 78 \text{ psi}$$

For 3-inch hose, multiply the friction loss of 2½-inch hose by 0.4.

First round off 78 to 80 and then multiply: $0.4 \times 80 = 32$

There is 400 feet of 3-inch hose, so the total friction loss is: $4 \times 32 = 128$ psi. Round this off to 130 psi.

Remember this figure and calculate the friction losses for the wyed lines.

For the 200-foot line with the 1⅛-inch tip, $FL = 2 \times 15 = 30$
For the 300-foot line with the 1¼-inch tip, $FL = 3 \times 25 = 75$

Find the average friction loss for the two lines and add the 50-psi nozzle pressure:

$$30 + 75 = 105 \div 2 = 52.5 = 53 + 50\ NP = 103 \text{ psi}$$

Round off the 103 to 105 and add the friction loss in the feeder line to get the engine pressure:

$$EP = 105 + 130 = 235 \text{ psi}$$

In this example, this is only 2 psi less than the answer obtained by solving the problem with the Underwriters' formula.

If fireground demands make it necessary to favor one nozzle at the expense of the other, the engine pressure can be computed by figuring the friction loss and nozzle pressure total for each branch line and

ENGINE AND NOZZLE PRESSURES

adding either the higher or lower branch total to the friction loss for the feeder line.

For example, if the 1⅛-inch tip could use a 50-psi nozzle pressure but the 1¼-inch nozzle was in a precarious location for control, then the engine pressure could be based on the lower engine pressure required to supply the 1⅛-inch tip:

$$\begin{aligned} 1\tfrac{1}{8}\text{-inch tip line } FL &= 30 \\ NP &= 50 \\ \text{Feeder line } FL &= 130 \\ \hline EP &= 210 \text{ psi} \end{aligned}$$

But if the line with the 1¼-inch tip needed a full flow and the other line could handle a little extra pressure:

$$\begin{aligned} 1\tfrac{1}{4}\text{-inch tip line } FL &= 75 \\ NP &= 50 \\ \text{Feeder line } FL &= 130 \\ \hline EP &= 255 \text{ psi} \end{aligned}$$

Or the company officer could order a compromise between the 210 and 255-psi engine pressures to provide adequate fire streams and safe pressure on both lines.

Rule-of-thumb method choices

In solving any wyed line problems involving any unequal features, much time can be saved by turning to the rule-of-thumb method. First determine the total flow of all nozzles. This total is needed to find out the friction loss in the feeder line from the pump to the wye. Multiply the friction loss figure by the number of hundreds of feet in the feeder line to obtain the friction loss total for that line. Then compute the total friction loss and nozzle pressure in each wyed line.

Now the pump operator must exercise his judgment and do one of four things:

1. He can average the wyed line friction loss and nozzle pressure totals and add the average obtained to the supply line friction loss to determine the engine pressure. As a result, one line will have too much nozzle pressure and the other will have too little. In some cases the differences between the wyed lines are not enough to cause any trouble with nozzle pressures after averaging.

2. The pump operator can use the friction and nozzle pressure total for the line that requires the most. This will provide good nozzle pressure for the line that entered into the final figuring and too much pressure for the other nozzle.

FIRE SERVICE HYDRAULICS

3. The operator can use the friction and nozzle pressure figures for the wyed line demanding the lowest total. This will still give this line a good nozzle pressure, but the other line (or lines) will have insufficient nozzle pressure.

4. The good operator will use good judgment. If neither line has any special problems and the difference in the required pressures is not too great, averaging the needs of the wyed lines will be a practical answer in calculating engine pressure. On the other hand, if the line that requires the most pressure is in serious need of the extra water, then the operator might consider the second choice above. But he has to make certain that when he gives the most demanding line everything it needs, the line that needs less pressure is not overwhelmed. If the men on this line are on a slippery roof in the winter, then it would be inadvisable to overburden this hose crew with a beefed-up line. Fire fighting conditions will influence the operator in compromising the different needs of the wyed lines. This procedure can be used for any number of wyed lines.

Mental calculation for engine pressure for wyed lines

Example: An engine is pumping through 500 feet of 3-inch hose with 2½-inch couplings. This line is wyed off to two 2½-inch lines, one of them 200 feet long with a 1⅛-inch nozzle operating

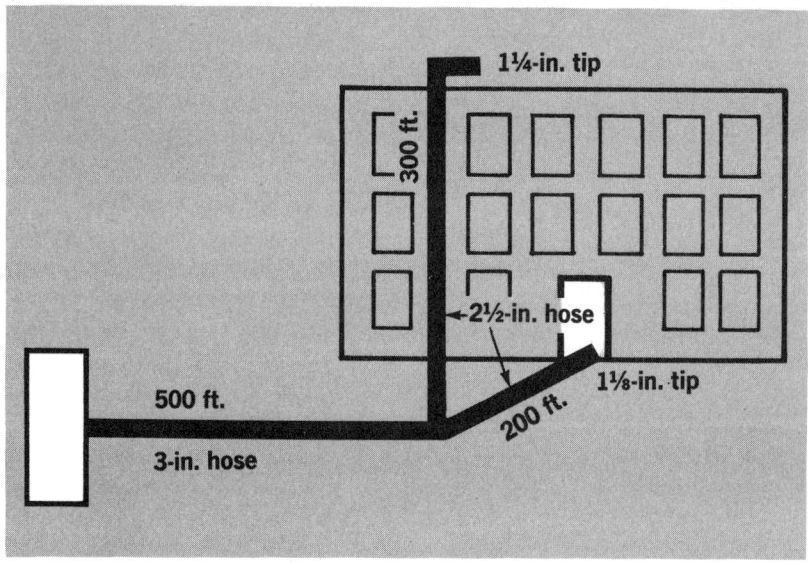

ENGINE AND NOZZLE PRESSURES

on the first floor of a building. The other branch is 300 feet long with a 1¼-inch nozzle and is operating on the roof of the four-story building. What is the engine pressure?

Answer: The 1⅛-inch nozzle requires 250 gpm and the 1¼-inch tip demands 325 gpm at a nozzle pressure of 50 psi. Therefore, the 3-inch hose feeding the wye must carry 575 gpm. Now let's follow the process of mental arithmetic and judicious reasoning that a pump operator follows to determine the proper engine pressure.

The total flow is close enough to 600 to figure friction loss at that gpm for the 3-inch hose. For 2½-inch hose, $FL = 2Q^2 + Q = 2 \times 6 \times 6 + 6 = 78$ psi. But that is the friction loss for 2½-inch hose. The friction loss for 3-inch hose can be approximated by multiplying the friction loss for 2½-inch hose by 0.4. Again, let's go to round numbers and instead of 78, multiply 80 by 0.4, which gives a friction loss of 32 psi for each 100 feet of 3-inch hose. But we took larger figures for flow and friction loss originally, so let's cut the 32 to 30 psi. (It's also easier to multiply in your head.)

FL for 500 feet of 3-inch hose (575+ gpm) = 5 × 30 = 150 psi
FL for 200 feet, 2½-inch hose, 1⅛-inch tip = 30 psi
NP = 50 psi
EP = 230 psi

This is the engine pressure for the shortest line and the one requiring the lesser flow, so this engine pressure would not satisfy the longer line with the 1¼-inch tip on the roof. But because of this, the pump operator knows that it is safe to charge his hose layout. Actually, both lines will fail to get the nozzle pressures they should have, but neither line will have excessive pressure that might lead to injuring a fire fighter.

If water is required at this point, the pump operator can supply the hose layout at once with an engine pressure of 230 psi while he figures the engine pressure that would properly supply the 1¼-inch tip on the roof of the four-story building. His mental calculations are as follows:

FL for 3-inch hose (as before) = 150 psi
FL for 300 feet, 2½-inch hose with 1¼-inch tip = 75 psi
Back pressure at roof (4 × 5) = 20 psi
NP = 50 psi
EP = 295 psi

The pump operator now has figures for his minimum and maximum

FIRE SERVICE HYDRAULICS

pumping range for this layout—an engine pressure of 230 psi, which will satisfy the 1⅛-inch tip, and 295 psi, which will satisfy the 1¼-inch tip. What does he do to keep both hose crews safe and reasonably happy?

The operator's problem beyond the wye is actually no different than the problem presented by two single lines off a pump when each line is a different length and has a different size tip and the engine does not have individual hose gate gages. This is where the pump operator puts his judgment into the engine pressure equation.

The difference between the minimum and maximum engine pressures is 65 psi, so an engine pressure midway between the extremes, 260 or 265 psi, would service both nozzles reasonably well. The smaller tip would have a little too much pressure and the larger tip would not have quite enough pressure.

But the pump operator also must consider the conditions under which the hose lines are operating. If the roof is icy or the crew is working off a roof ladder, then a nozzle pressure that is less than 50 psi is advisable. In this case, if the operator pumped at 230 psi, the pressure at the 1¼-inch tip would be in the neighborhood of 40 psi, which would still provide a respectable fire stream of about 290 gpm.

This problem was selected because it also introduces another problem—the problem of pump capacity. At 250 psi, a fire apparatus pump is required to pump only 50 percent of its rated volume, although most pumps will do more than that even after years of service. However, this requirement is based on the net pump pressure at draft. If the residual hydrant pressure remains high enough so that the intake pressure is at least 45 psi, then an engine pumping at 295 psi would be operating at a net pressure of 250 psi. Under these conditions, a 1,250-gpm pump should have no difficulty providing the 575 gpm required by our problem, and many 1,000-gpm pumps could meet this demand.

However, if the residual hydrant pressure were not sufficient, then the pump operator would be influenced by necessity to pump at a lower pressure, possibly even 230 psi.

AERIAL STREAM CALCULATIONS

Operating an engine to supply an aerial stream is much like supplying a deluge set except for one thing—back pressure, which is always present. In reckoning the engine pressure for a ladder pipe operation, the pump operator must add up:

1. Friction loss in the supply line from the pump to the siamese,

ENGINE AND NOZZLE PRESSURES

2. Friction loss of 5 psi in the siameses,
3. Friction loss in the line up the ladder,
4. Back pressure, ½ psi per foot of height,
5. Friction loss in the ladder pipe, 5 psi,
6. Nozzle pressure, 80 psi recommended for straight tips, 100 psi for fog tips.

Although the supply lines may be 2½-inch hose, the two lengths up the ladder should be 3-inch hose to keep down the friction loss in the single line that runs from the siamese to the ladder pipe.

Calculations on the fireground can be reduced substantially if a specific pressure figure is selected for the siamese. This means that the height of the ladder pipe for all operations must be taken as a single figure to eliminate a variable in the computations. In Table 7, the pressures at the siamese range from 140 to 205 psi for streams delivering 350 to 1,000 gpm at an average height of 60 feet. Raising the pipe to 80 feet would cause an additional back pressure of about 10 psi (8.6 psi, to be exact). The nozzle pressure, incidentally, would drop but a fraction of 10 psi, so a good fire stream would still be maintained at the higher elevation. At elevations lower than 60 feet, the stream would gain pressure to a slight extent. From Table 7, you can select a standard siamese pressure for the tip normally carried on your ladder pipe.

Table 7 Ladder Pipe Pressures

Average height of pipe—60 feet.
Three-inch hose, 100 feet, from siamese to ladder pipe.

GPM	Tip Size	FL (psi) in 3-Inch Hose	Nozzle Pressure	Back Pressure	PSI at Siamese
350	Fog	11	100	30	140
400	1¼	14	80	30	125
500	1⅜	21	80	30	130
500	Fog	21	100	30	150
600	1½	30	80	30	140
700	1⅝	40	80	30	150
750	Fog	45	100	30	175
800	1¾	50	80	30	160
1000	2	76	80	30	185
1000	Fog	76	100	30	205

NOTE: Pressures at siamese are rounded off to nearest multiple of 5 psi.

Elevating platform stream calculations

For elevating platforms, an average working height that is slightly on the high side should be selected as the basis for the back pressure figure.

Friction loss in the pipe inside the articulated or telescoping boom will vary with the gallonage, so the figure used should be for the tip generally carried on the monitor gun. With 4-inch pipe, friction loss for a 75 to 80-foot elevating platform will range from about 10 psi at 500 gpm to about 35 psi at 1,000 gpm. With a 6-inch pipe, friction loss will range from 1.5 psi at 500 gpm to about 5 psi at 1,000 gpm.

If you add a reasonable back pressure figure, the friction loss in the piping in the boom to the monitor, nozzle pressure (100 psi for fog tips, 80 psi for straight tips), and about 5 psi for loss in the gun, you will have a figure for the pressure needed at the intake manifold. Generally, if a pressure of 125 to 150 psi is maintained at the manifold, depending on the particular apparatus, the monitor in the basket will produce a good fire stream.

Engine pressure can be established just as it is for supplying a hose layout for a deluge set. The pressure at the manifold can be regarded as the "nozzle pressure."

Deck guns

The hydraulic problems of deck guns and deluge sets operated from their mounting bracket atop pumpers are minimal. A deck gun is rarely more than 5 or 6 feet higher than the pump, and deluge sets are generally mounted no higher than that even when on the roof of the cab. So the back pressure for either appliance usually is not more than $6 \times 0.434 = 2.6$ psi, hardly anything to worry about.

Friction loss in a deck gun and its connections may range between 5 and 10 psi, according to whether the flow is light or heavy. A rule-of-thumb figure of 10 psi for friction loss is considered to be reasonable for deck guns.

When two 10-foot lengths of 3-inch hose with 2½-inch couplings are used to feed a deluge set from the pump of the engine that the set is mounted on, the friction loss in the hose lines will not exceed 3 psi for 1,000 gpm, and the loss in the set itself may range from 3 to 10 psi, depending on the volume and the design of the unit. Again, a friction loss of 10 psi is a reasonable constant working figure for deluge sets mounted on an engine and being fed by that engine's pump.

When these master stream devices are not fed directly from the pump of the engine carrying them, add 5 or 10 psi for friction loss in new or old-style guns when computing the engine pressure for supplying feeder lines.

RELAY PUMPING

The paradox about relay pumping is that the harder you work to perfect it, the less likely you are to be successful. If the engines are

ENGINE AND NOZZLE PRESSURES

spaced so that each is required to do the same amount of work in terms of engine pressure, that should be regarded as a coincidence and not as a virtue.

The key word in the previous sentence about engine pressure is "required." It is efficient and practical to have each engine, except the one at the fire, start a relay with the same engine pressure. This eliminates mental calculations by the pump operators, who have only to provide the previously chosen pressure for all relay work in their department or mutual aid system.

The standard engine pressure for relays should be selected in relation to the diameter and amount of hose carried on apparatus and the amount of water the relay is expected to supply in gallons per minute. Most relays are set up with a single line because the amount of hose available is limited and the distance between the fire and the water source is often several thousand feet.

When 2½-inch hose is used, the flow is generally in the 200 to 250-gpm range. The gpm can range up to 400 gpm with 3-inch hose, but with 3½-inch hose, 500 gpm can be supplied at about the same engine pressures and distances between pumpers as 200 gpm with 2½-inch hose. Or to look at it the other way, an engine can supply a 3½-inch line nearly six times the length of a 2½-inch line with the same gallons per minute.

The virtues of hose for relay work multiply as the diameter increases. Four-inch hose is about 11 times as efficient as 2½-inch hose in terms of waterway capacities, and 5 and 6-inch hose, about 32 and 80 times as efficient, respectively.

When the size hose and the average amount carried on apparatus for relays is known, then a standard engine pressure can be established for all pumpers in relay except the one at the fire.

If no apparatus carries more than 1,800 feet of 2½-inch hose and a delivery of 200 gpm to the fireground is acceptable, then a starting engine pressure of 200 psi can be established as standard operating procedure for all engines in the relay except the one at the fire. The latter will operate at the proper pressure for the lines it supplies as though it were operating from a hydrant. The flow, of course is limited to 200 gpm, but this can be handled through one or more lines as long as the total demand of the nozzles does not exceed the supply from the relay.

Each pump receiving water in a relay should have an intake (residual) pressure of 10 to 20 psi. If suction relief values are used, they are set to operate and relieve intake pressures in excess of 10 psi. Without the suction relief valves, which are generally used on larger-diameter,

FIRE SERVICE HYDRAULICS

single-jacket hose, a minimum intake, or residual, pressure of 20 psi is desirable to ensure a constant supply of water to each pump.

With this information in mind, consider a relay with engines spaced not more than 1,800 feet apart supplying 200 gpm. The friction loss will be about 180 psi with approximately 20 psi left for intake pressure at each pump. If the fireground engine has 1,800 feet of hose out in a single 2½-inch line with a 1-inch tip, then this engine must pump at 230 psi (180 psi for friction loss in the line and 50 psi for nozzle pressure).

Engine pressures should not exceed the annual hose test pressure of the department as a matter of reliability and safety in the relay.

Other standard engine pressures for relays can be determined in a similar manner for other hose sizes and water supply demands.

Now let's see how a relay with a single 2½-inch line to supply 200 gpm might work out.

First-in Engine 3 drops a 2½-inch line with a 1-inch nozzle at the fire and stretches all the 2½-inch hose in its bed, 1,700 feet. Engine 2 lays its bed of 1,800 feet of 2½-inch hose, and Engine 1 stretches 900 feet of 2½-inch hose to the static water source.

With a nozzle pressure of 50 psi at the 1-inch tip, the relay flow will

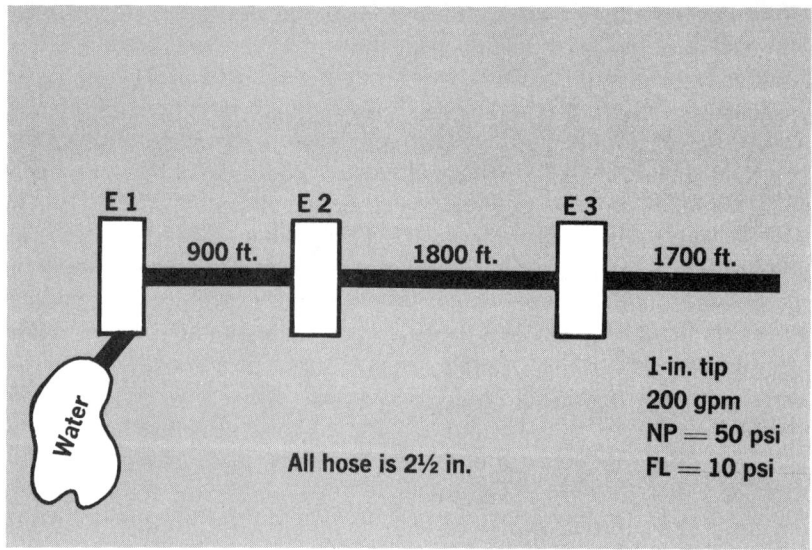

be 200 gpm at a friction loss of 10 psi. If each engine pumps at 200 psi, the intake pressure will be 110 psi at Engine 2 and 20 psi at Engine 3. With 1,700 feet of hose to pump through, Engine 3 has to allot 170 psi for friction loss and 50 psi for nozzle pressure, making the engine pressure 220 psi.

This is the quickest method of setting up a relay. If an engine carries more than 1,800 feet of 2½-inch hose, then the operator will have to use a higher engine pressure to compensate for additional friction loss. If all engines expected to participate in relay work in an area, for example, carry 2,000 feet of 2½-inch hose, then a starting pressure for each engine in the line will be 220 psi. The engine at the fire, of course, will pump at the pressure required by the line, or lines, working the fire.

Note that the intake pressure at Engine 2 in our original problem is 110 psi. There are hydrants in some municipalities with better than 100-psi static pressure, so the 110 psi at Engine 2 is no problem. It can be compared to hooking up to a hydrant with that as a residual pressure. However, once a relay is under way, an assigned water officer can supervise the refinement of engine pressures so that no pump has to supply more pressure than required.

An alert operator would notice that a sizable portion of the hose load was still in the bed. Then he would estimate the amount of hose he stretched, add a couple of hundred feet more to this figure for safety and reduce his initial 200-psi engine pressure closer to the pressure actually required. In this example, it would be 110 psi (90 psi for friction loss and 20 psi for intake pressure at Engine 2).

Parallel lines make larger volumes feasible in relays. When each engine stretches parallel lines, the distances between engines are half as great as when single lines are laid. Then up to 300 gpm can be supplied through each 2½-inch line for a total of 600 gpm at a friction loss of about 21.2 psi per hundred feet. If an engine carried 1,800 feet of 2½-inch hose, there will be 900 feet in parallel lines with a total friction loss of about 191 psi. Adding 20 psi for intake pressure at the next pump in the line, the engine pressure would be about 210 psi.

Because distance is the major problem to be solved in areas where hydrants are nonexistent, parallel lines are rarely laid in these areas. In hydrant areas, parallel line relays generally are two-engine relays, which actually are simple problems in supplying parallel lines by the engine at the water source.

However, let's see how a parallel line relay with 2½-inch hose might work out.

FIRE SERVICE HYDRAULICS

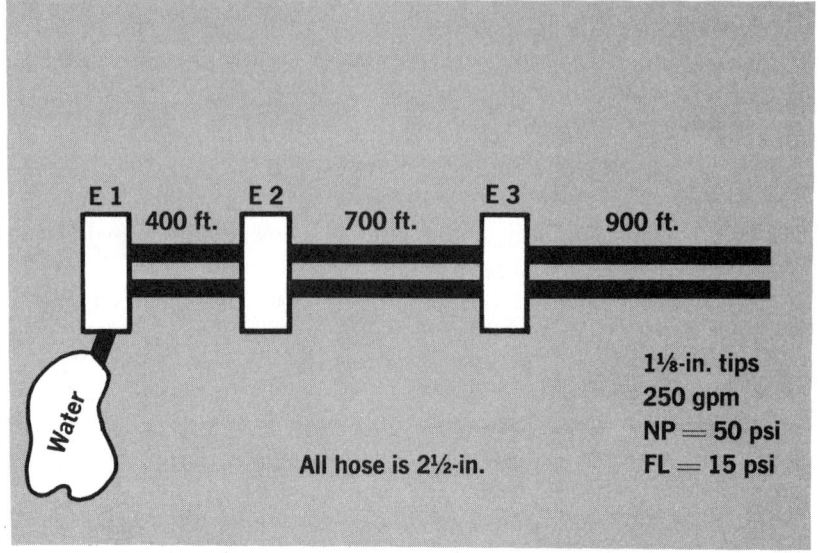

Example: In stretching from fire to water, Engine 3 drops parallel 2½-inch lines at the fire with a 1⅛-inch tip on each. The stretch is 900 feet. Engine 2 stretches 700 feet with parallel 2½-inch lines and Engine 1 completes the relay by laying parallel 2½-inch lines 400 feet to a pond. The desired nozzle pressure is 50 psi. What should the engine pressures be?

Answer: With 50 psi tip pressures the flow through each line (rule of thumb) will be 250 gpm for a total delivery of 500 gpm to the fireground. The friction loss for 250 gpm in 2½-inch hose is 15 psi. Engine 1 has a friction loss of 4 × 15 = 60 psi to overcome. An intake, or residual, pressure of 20 psi is desired at Engine 2. Therefore, Engine 1 will pump at 80 psi. The friction loss for Engine 2 is 7 × 15 = 105 psi. Adding intake pressure for Engine 3, Engine 2 will pump at 105 + 20 = 125 psi. The friction loss for Engine 3 is 9 × 15 = 135 psi and 50 psi for nozzle pressure must be added to make an engine pressure of 185 psi.

There is a method of determining distances between pumpers in a relay so that each pump will work at the same pressure. However, while this method is interesting from a theoretical standpoint, it has little value on the fireground where time is vital and it may be necessary to utilize most of the available hose.

For those who like to calculate hydraulic problems with a pencil, this is the way the method is used:

1. Determine the total length of the relay.
2. Add enough feet to the relay length to make the friction loss in this "phantom hose footage" equivalent to the desired nozzle pressure less 20 psi. (Add only hundreds of feet.)
3. Divide the total of the relay length and the "phantom hose footage" by the number of engines in the relay except the one at the fire. Figure the engine pressure for this last pumper in the normal way.
4. Find the friction loss for this average hose lay length between engines.
5. Add 20 psi to the friction loss obtained in step 4 to obtain the engine pressure for each pumper in the relay.

This method will provide an intake pressure of 20 psi at each pumper. When the average length is stretched between pumpers, starting at the source pumper, it will be found that the fireground pumper will have less hose to supply. The friction loss thereby saved is what provides the nozzle pressure.

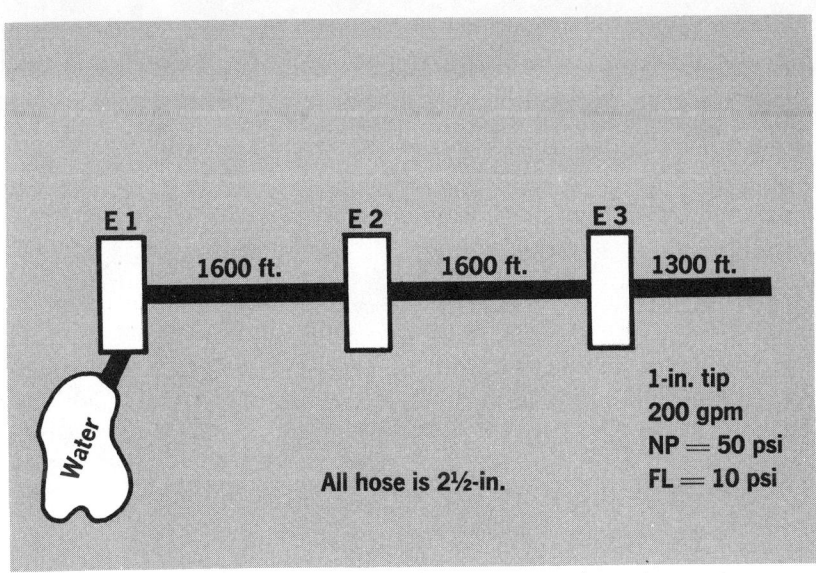

FIRE SERVICE HYDRAULICS

Example: Three engines are to pump a 4,500-foot relay with 2½-inch hose and a 1-inch tip with 50 psi nozzle pressure. What are the distances between engines and what is the engine pressure for each pump?

Answer: By rule of thumb, a 1-inch tip flows 200 gpm at 50 psi nozzle pressure with a friction loss in 2½-inch hose of 10 psi. The 50 psi nozzle pressure less 20 psi is 30 psi, which is equivalent in this hose layout to 300 feet of hose. Therefore, 300 is theoretically added to the relay: 300 + 4,500 = 4,800 feet.

Now divide by the number of engines (3) in the relay:

$$4,800 \div 3 = 1,600 \text{ ft.}$$

There should be 1,600 feet of hose from Engine 1 to Engine 2 and from Engine 2 to Engine 3. This adds up to 3,200 feet of hose, leaving 1,300 feet of hose for Engine 3 to supply at the fire scene. With a 10 psi friction loss, here is how each engine pressure is figured:

Engine 1—10 × 16 + 20 for intake pressure = 180 psi
Engine 2—10 × 16 + 20 for intake pressure = 180 psi
Engine 3—10 × 13 + 50 for nozzle pressure = 180 psi

Thus, if each engine pumps at 180 psi, the requirements of this relay layout will be satisfied.

This time, let's figure a three-engine relay with 2½-inch hose and a 1⅛-inch tip.

Example: A 3,700-foot relay is to be handled by three engines. There is a single line of 2½-inch hose and the tip to be supplied is 1⅛-inch. What is the engine pressure for each pumper and what is the distance between pumpers?

Answer: A 1⅛-inch tip at 50 psi nozzle pressure flows, by rule of thumb, 250 gpm with a friction loss of 15 psi. The 50 psi nozzle pressure less 20 psi is 30 psi, which is equal to the friction loss of 200 feet of hose under the conditions in this problem. Therefore, add 200 feet to the relay length:

$$200 + 3,700 = 3,900 \text{ ft.}$$

Now divide by the number of engines in the relay:

$$3,900 \div 3 = 1,300 \text{ ft.}$$

ENGINE AND NOZZLE PRESSURES

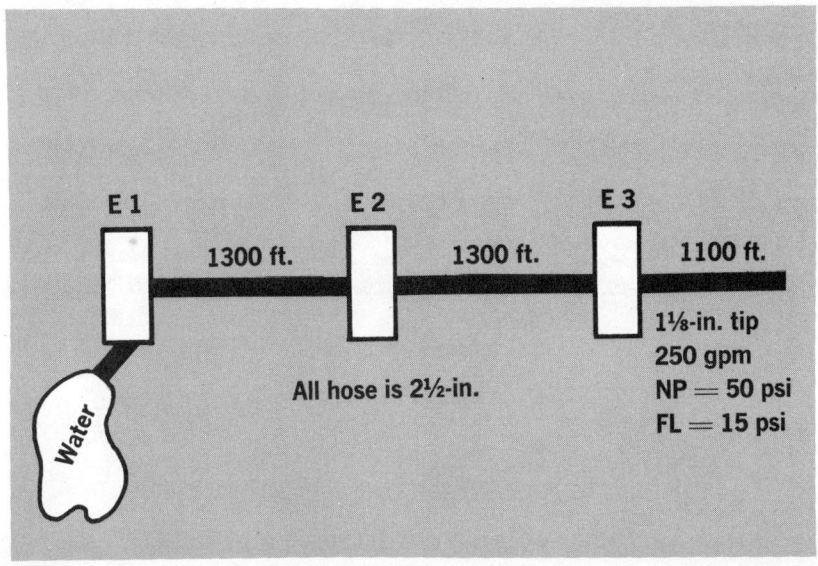

There should be 1,300 feet of hose between Engines 1 and 2 and between Engines 2 and 3. That makes a total of 2,600 feet and leaves 1,100 feet of hose for Engine 3 to supply. With a 15 psi friction loss, the engine pressures are figured as follows:

Engine 1—15 × 13 + 20 for intake pressure = 215 psi
Engine 2—15 × 13 + 20 for intake pressure = 215 psi
Engine 3—15 × 11 + 50 for nozzle pressure = 215 psi

If each engine pumps at 215 psi, the requirements of this relay will be satisfied. Note that Engines 2 and 3 will have an intake (residual) pressure of 20 psi, as desired.

Now let's look at the use of 3½-inch hose in a relay.

Example: Three engines are placed in a relay with a single line of 4,800 feet of 3½-inch hose to supply a deluge set with a 1⅜-inch tip at 80 psi nozzle pressure. What is the distance between engines and at what pressure should they pump?

Answer: With a 1⅜-inch tip at 80 psi, the deluge set requires 500 gpm with a rule-of-thumb friction loss in 3½-inch hose of

FIRE SERVICE HYDRAULICS

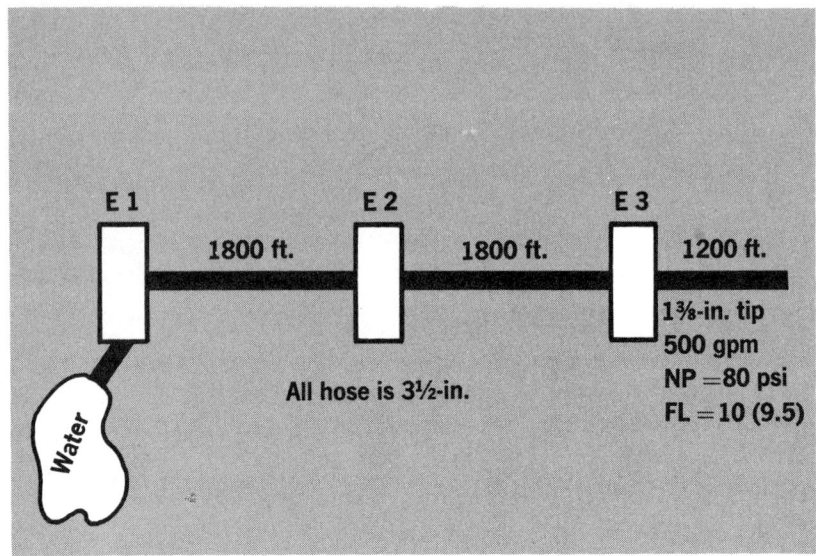

10 psi (actually 9.5 psi). The 80 psi nozzle pressure less 20 psi is 60 psi, which is equal to the friction loss of 600 feet of 3½-inch hose with 500 gpm flowing. Therefore, add 600 to the relay length:

$$600 + 4{,}800 = 5{,}400$$

Divide by the number of engines in the relay:

$$5{,}400 \div 3 = 1{,}800 \text{ ft.}$$

There should be 1,800 feet of hose between Engines 1 and 2 and between Engines 2 and 3. The total of 3,600 feet leaves 1,200 feet of hose for Engine 3 to supply. With a 10-psi friction loss, the engine pressures are figured as follows:

Engine 1—10 × 18 + 20 for intake pressure = 200 psi
Engine 2—10 × 18 + 20 for intake pressure = 200 psi
Engine 3—10 × 12 + 80 for nozzle pressure = 200 psi

By pumping at 200 psi, each engine will fulfill its share of the work in the relay. Both Engines 2 and 3 will have an intake (residual) pressure of 20 psi.

ENGINE AND NOZZLE PRESSURES

LEEWAY IN CALCULATIONS

How precise hydraulic computations must be to attain good engine and nozzle pressures is a problem that often bothers new pump operators. Actually, the rules of thumb work remarkably well because there is a healthy margin for error built into hydraulics for fireground use.

Many times, errors we make on the fireground cancel themselves out quite accurately, or the minor ones don't really make that much difference.

For example, there is a comfortable latitude in the engine pressure calculated by a pump operator. An error of 5 or 10 psi in the engine pressure means nothing at the nozzle. And variations in engine pressures have less effect on the nozzles of longer lines than on the tips of shorter lines.

The engine pressure ranges that provide nozzle pressures of 40 to 52 psi in different length hose stretches and with different size tips are shown in Table 8.

When working with 300 to 900-foot lines having 1 to 1¼-inch nozzles, a change of 30 psi in the engine pressure will be reflected in an average change of 10 psi at the tip. These figures, of course, are averages that are reflected in Table 9. When the engine pressure was changed 10 psi, the average change reflected by the tip was 3.63.

Table 8 Engine Pressure and Nozzle Pressure Ranges

Tip	Length of 2½-Inch Hose Line		
	300 Feet	600 Feet	900 Feet
1″	EP: 70–90 psi NP: 40–52 psi	EP: 100–120 psi NP: 42–51 psi	EP: 120–150 psi NP: 40–50 psi
1⅛″	EP: 90–110 psi NP: 43–52 psi	EP: 130–160 psi NP: 42–52 psi	EP: 170–210 psi NP: 41–51 psi
1¼″	EP: 110–130 psi NP: 43–50 psi	EP: 170–210 psi NP: 42–51 psi	EP: 220–280 psi NP: 40–50 psi

Note: In working out the nozzle pressures, decimals were rounded out to the nearest full psi.

FIRE SERVICE HYDRAULICS

Table 9   Average psi change in tip pressure, using 300 to 900-foot stretches of $2\frac{1}{2}$-inch hose, when engine pressure is changed 10 psi in 50 to 250-psi range.

Tip Size	300 Feet	600 Feet	900 Feet
1 in.	5.78	4.24	3.35
$1\frac{1}{8}$ in.	4.76	3.23	2.44
$1\frac{1}{4}$ in.	3.86	3.25	1.80
Average	4.80	3.57	2.53

These figures are meant to reassure the pump operator who fears rules of thumb may be too inaccurate. They also should bolster the faith of those who have used hydraulics as a working tool on the fireground by adding judgment to make the equations adaptable to the needs.

CHAPTER SEVEN

Fire streams

Simply stated, a good fire stream is one that will extinguish a fire in the shortest period of time with a minimum amount of water. A good fire stream has also been described as one that has adequate reach and volume to penetrate to the seat of a fire and cool burning materials to the point where they cease to give off flammable vapors.

The point to remember is that the burning material must be cooled to a point below its ignition temperature. Extinguishing flame is just not enough.

A good fire stream therefore can be one from a garden hose or one from a 1000-gpm master stream appliance. It depends mainly on the quantity (fuel loading) of the material that is burning.

The most complete and reliable series of experiments on the hydraulics of fire streams were conducted in the years 1888 and 1890 by John R. Freeman, C. E. These experiments took place in Lawrence, Mass., on the premises of the Washington Mills Company, which had recently installed a new fire pump and hydrant system.

Later experiments were conducted by Marston, Fleming, and the Associated Factory Mutual Fire Insurance Company. These experiments, including Freeman's though extensive, covered only nozzle pressures up to 100 psi, and tips up to $3\frac{1}{2}$ inches in diameter. The National Board of Fire Underwriters in conjunction with the Chicago Fire Department made additional tests in 1942 that extended the pressure range. However, Freeman's findings were reinforced by the later experiments with but minor changes. And his definition of a good fire stream still holds as true today as it did in 1888.

An effective solid fire stream is one that:

1. At the limit named has not lost continuity of stream by breaking into showers of spray;

2. Up to the limit named appears to discharge nine-tenths of its volume of water inside a circle 15 inches in diameter and three-quarters of it inside a 10-inch circle;

3. Is stiff enough to attain in a fair condition the height or distance named even though a fresh breeze is blowing;

4. At a limit named will, with no wind, enter a room through a window opening and just barely strike the ceiling with force enough to spatter well.

Freeman's choice of these conditions was arbitrary, but the choice in general was good and it established a standard by which streams can be compared and tied in to his experiments. Later experimenters tended to modify Freeman's statements as to what constitutes a good stream and accept in its place the distance from the tip where the stream appears to break into slugs of water which are still closely grouped and effective for fighting fires. This effective reach apparently is what Freeman and Marston took as the reach of a fair stream.

In a paper published in FIRE ENGINEERING, H. A. Musham adjusted Freeman's data with other experiments and plotted them on logarithm paper. He extended them with a degree of accuracy within 2 to 3 percent of the original experiments to cover all ranges and tips in common use, or which might be used in the future.

Air resistance

Very little is known about the resistance encountered by a fire stream as it passes through air. However, the analytical work that has been done indicates that resistance increases at an accelerated rate as the pressure is raised with the same tip. And at a high, though probably lower rate, if the pressure is kept constant while the tip size is increased. This rapid increase in resistance accounts for the decelerating increase in horizontal effective reach attained by raising the pressure and increasing the tip. It is probable that resistance can be broken down into three components:

That due to friction; that caused by the piling up of air in front of the stream, which eventually breaks it up into spray; and that caused by the displacement of surrounding air necessary to permit the passage of the water. This third component probably causes some form of wave disturbance in the air which is accentuated by the pulsations of the stream itself.

Because of the numerous elements which enter into air resistance, it is not likely that such resistance will ever be accurately known. These elements include the temperatures of both air and water, atmospheric pressure, density of the air and its relative humidity, viscosity and surface tension of both air and water.

Local air currents set up by the stream between it and the ground also play an important part as can be deduced from the fact that the greatest horizontal effective reach occurs at elevations of from 30 to 34 degrees for the tips and pressures in common use, instead of 45 degrees as would normally be expected. The length of the curved path of the stream itself remains approximately the same regardless of the elevation of the nozzle.

If the nozzle were placed but a few feet off the ground, then the stream would strike the ground a short distance from the tip. As the elevation is increased, the distance is increased until all the effective length of the stream is clear of the ground. Nearness of the stream to the ground drags the air away from over the ground. This creates a decrease in air pressure under the stream, and air above it presses down and reduces reach.

As the elevation increases, the reach increases until the maximum is reached at 30 to 34 degrees. Beyond 34 degrees, the horizontal effective reach decreases as the length of the curve's path remains the same or approximately so.

Effect of wind

Little experimental work has been done to determine the effect of the wind on fire streams. Such experiments as have been performed indicate that a faint tail breeze may increase horizontal distances by 10 percent. A moderate tail breeze may lower effective vertical reach from 10 to 15 percent. Head breeze will raise the vertical reach and shorten the horizontal. A high-velocity head wind will destroy the stream at the tip by turning it into spray that is then driven to the base of the nozzle or pipe. Similarly, a strong tail wind will, while carrying the water forward, break it up into spray.

The wind also has a serious effect on the air resistance. Tail winds reduce it and head winds increase it. In both cases the stream wastage is increased.

Limits of reach

Any given tip has a pressure limit, beyond which its effective reach decreases. As an example, small tips at high pressure have very little reach. Increase of pressure beyond proper limits increases wastage and such increase will progressively destroy the stream until its reach is nil. In general, increase of pressure increases reach, but the reach progressively decreases at a rapid rate as excessive pressure builds up. This is due to the acceleration of air resistance, and the coning of the stream at the nozzle.

As the water leaves the tip, gravity takes hold and it begins to drop. The stream takes on a curved conical form, increasing the surface area exposed to friction. The water shreds off in wisps of mist which with drops of water form a water curtain under the stream. If the coning becomes excessive, it is probably because of a defect in the nozzle's interior shape, or a roughness of the interior due to lack of proper maintenance.

The interior of nozzle barrels should be kept highly polished to insure good streams. Excessive coning also indicates that the pressure is too high and the discharge too much for the nozzle to handle.

NOZZLE PERFORMANCE

The source of water supply, the operating rate of the apparatus, the length of hose lay, system losses, the size of nozzle and its elevation all contribute to the performance of a nozzle. With the exception of water supply, which is here considered adequate, nozzle performance is directly controlled by the pump operator and the company officer or nozzleman.

Range of streams

In considering the range of fire streams and the proper size of stream to be employed, many points have to be taken into consideration. For instance, a stream that has broken into a spray may be quite effective and satisfactory on small fires, but when directed at a hot blaze may be useless. Again a stream which would be ideal in calm weather could be absolutely worthless in a gale.

Horizontal range

Knowledge of the horizontal range of fire streams is particularly valuable when streams are to be thrown from one building to another across the street, or from pier to pier at waterfront fires, or from fireboats to waterfront property. For instance, knowing that the distance required to span a street and hit in on a floor is 60 feet can make for efficient fire stream operation right from the start.

Theoretically a fire stream has its greatest horizontal range when the center of the stream as it leaves the nozzle makes a 45-degree angle from the horizontal. However, Freeman found that an angle of

FIRE STREAMS

32 degrees gave the greatest actual horizontal range. The formula derived empirically from his experiments was therefore based on this 32-degree angle.

The formula, $S = \frac{1}{2}NP + 26$, is applicable to nozzle pressures over 30 psi and for a $\frac{3}{4}$-inch nozzle where:

S = horizontal distance in feet
NP = nozzle pressure in pounds per square inch

For nozzle diameters in excess of $\frac{3}{4}$-inch, add 5 to the 26 for each $\frac{1}{8}$-inch increase in nozzle diameter. This formula gives satisfactory results within usual working pressures and nozzle sizes.

Vertical reach

The maximum vertical reach is attained when the nozzle is placed perpendicular to the ground. However, such a nozzle position is rarely found in actual fire fighting.

The angle of stream commonly used for maximum effective reach varies between 60 and 75 degrees (Figure 1). Stream A making a 45-degree angle does not require calculations to determine the maximum range. It would be difficult in an ordinary street to get the nozzle sufficiently far from the building while at an angle of 45 degrees to make the vertical height the important factor. An ordinary stream at 45 degrees and 50 feet from a building strikes the building only 40 feet from street level

Streams B and C are directed at an angle of 75 degrees and 60 degrees, respectively, and the determination of their effective reach requires a different formula. This formula was worked out (also empirically) to give values of vertical distances of fire streams when nozzles are inclined at or around an angle of 70 degrees.

The formula gives the height of an effective stream using a 1-inch tip and is applicable up to a nozzle pressure of 100 pounds:

$H = \sqrt{240p - p^2 - 1900} - 15$, where
H = effective height in feet
p = nozzle pressure in pounds per square inch

For pressures of 50 psi or less, subtract 1 from the 15 for each $\frac{1}{8}$-inch increase in nozzle diameter. Above 50 psi, subtract 2 from the 15 for each $\frac{1}{8}$-inch diameter increase.

Both formulas give approximate agreement with the table drawn by Freeman (Table 1). And from a practical standpoint they are satis-

FIRE SERVICE HYDRAULICS

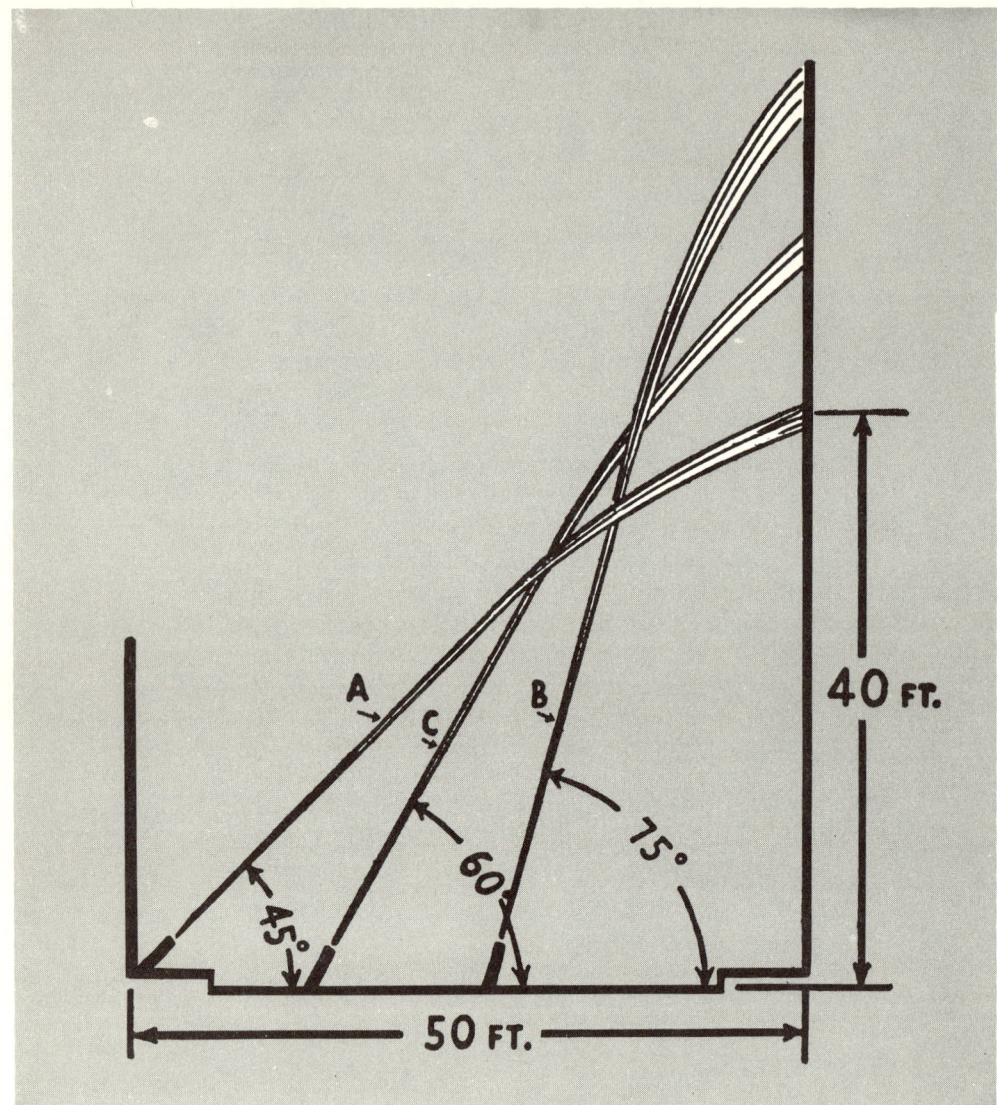

Figure 1

FIRE STREAMS

Table 1—Effective Reach of Fire Streams

Showing the distance in feet from the nozzle at which streams will do effective work with a moderate wind blowing. With a strong wind, the reach is greatly reduced.

Pressure at Nozzle	Size of Nozzle									
	1-Inch		1⅛-Inch		1¼-Inch		1⅜-Inch		1½-Inch	
	Vertical Distance, Feet	Horizontal Distance, Feet	Vertical Distance, Feet	Horizontal Distance, Feet	Vertical Distance, Feet	Horizontal Distance, Feet	Vertical Distance, Feet	Horizontal Distance, Feet	Vertical Distance, Feet	Horizontal Distance, Feet
20	35	37	36	38	36	39	36	40	37	42
25	43	42	44	44	45	46	45	47	46	49
30	51	47	52	50	52	52	53	54	54	56
35	58	51	59	54	59	58	60	59	62	62
40	64	55	65	59	65	62	66	64	69	66
45	69	58	70	63	70	66	72	68	74	71
50	73	61	75	66	75	69	77	72	79	75
55	76	64	79	69	80	72	81	75	83	78
60	79	67	83	72	84	75	85	77	87	80
65	82	70	86	75	87	78	88	79	90	82
70	85	72	88	77	90	80	91	82	92	84
75	87	74	90	79	92	82	93	84	94	86
80	89	76	92	81	94	84	95	86	96	88
85	91	78	94	83	96	87	97	88	98	90
90	92	80	96	85	98	89	99	90	100	91

Note.—Nozzle pressures are as indicated by Pitot tube. The horizontal and vertical distances are based on experiments by Mr. John R. Freeman, Transactions, Am. Soc. C. E., Vol. XXI.

factory since in fire fighting an exact measure of reach cannot be determined.

Approximate calculation

For quick calculation, but with less accuracy, fire fighters in the field can allow a pound of pressure for each foot of effective range of stream either vertically or horizontally. This rule is most accurate at low pressures, but for high-pressure streams where effective reach is a problem, it is still fairly accurate.

Furthermore, this rough rule gives results which in most cases are too large, but which can be desirable.

The ranges usually given in tables for fire streams are for those operating in comparatively still air. With an appreciable wind blowing, this range is materially reduced. Therefore, greater pressure than that actually needed provides a margin of safety.

The above formula and approximate calculation apply only to solid streams. The reach of fog streams varies from nozzle to nozzle and manufacturer to manufacturer. In effect, the reach is built into the particular nozzle in question. It can only be determined by observance of the stream at various nozzle pressures. Reach of straight streams (from combination fog and straight stream nozzles) will be approximately the same as from standard nozzles.

The quantity flowing from a fog nozzle has also been built in by the manufacturer. It would be well therefore to consult the manufacturer's specifications to determine what can or cannot be done with each nozzle.

Directing the stream

As mentioned previously, the curvature of the stream causes it to drop so as to strike the building at a point lower than that at which the nozzle is pointed. This is due to gravity, which pulls it downward so that the water takes a curved path from the instant it leaves the nozzle until it strikes the building.

Figure 2 shows diagrammatically a calculated stream, based on the

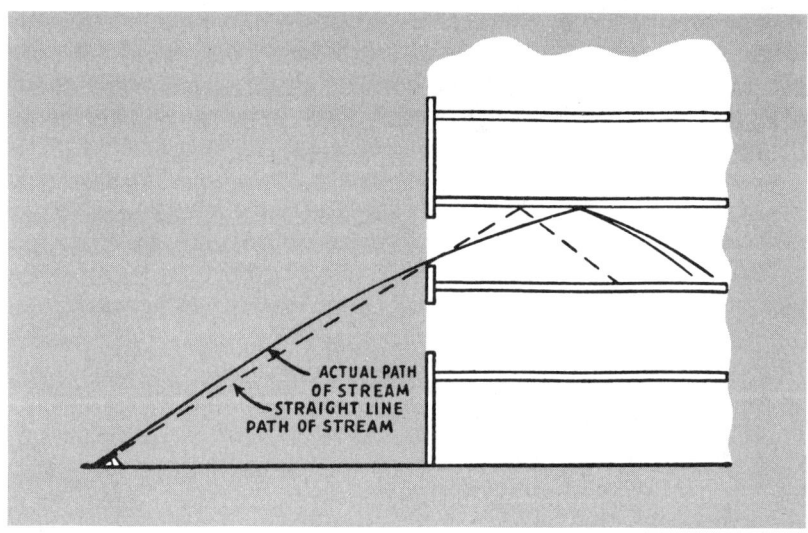

Figure 2

assumption that the stream travels in a straight line, and also an actual path of a stream.

The dotted line indicates the straight line path of a stream, while the solid line gives the actual path.

The curvature of the stream is an advantage in one respect: A stream from the nozzle to an upper floor of a building, instead of going straight on over the sill and striking the ceiling at a point directly in a straight line from the nozzle, curves as it crosses the sill and tends to strike farther in on the ceiling than calculated. Not only does it strike farther in on the ceiling, but the deflection of water is also at a smaller angle, which means that the water will spread much farther than it would if it traveled in a straight line.

Where it has been mentioned that the 50-degree angle is the maximum angle at which to direct a stream, the 50 degrees refers to the angle of the stream where it leaves the nozzle and not the angle of the stream where it crosses the sill or after it enters the building.

Thus while a sketch drawn with straight lines showing the probable penetration of the stream might lead one to believe that penetration was really ineffective or useless, an actual stream might prove to be effective. This is, as noted above, accounted for by the tendency of the stream to curve or take a curved path due to the action of gravity.

From the street: The third floor may be said to be the highest floor to which streams may be thrown effectively from the street level.

Theoretically, it might be possible to reach a higher floor than the third, but this would necessitate moving the nozzle farther away from the building to keep the angle of the stream small. But moving the nozzle away from the building means that the stream will have to travel farther, and the possibility of its breaking into spray is thereby increased. Then, too, increasing the range of the stream necessitates a higher nozzle pressure to do effective work.

Street streams are ordinarily directed with full effectiveness into the second floor of a building. But they may, as noted above, be used into the third floor. In this case, it is good practice to place the nozzle at least as far from the building as the point, where it enters the building, is above the street level.

Refer to Figure 3. While the greatest effective range is secured by sending the stream from the position as placed through the second-floor window (at C), still satisfactory penetration can be secured at the third-floor window (at point B). A stream crossing the sill at point B passes into the building quite a distance before it strikes the ceiling, and the water is deflected downward almost as far again, increasing the penetration to double that which is represented by the distance

FIRE SERVICE HYDRAULICS

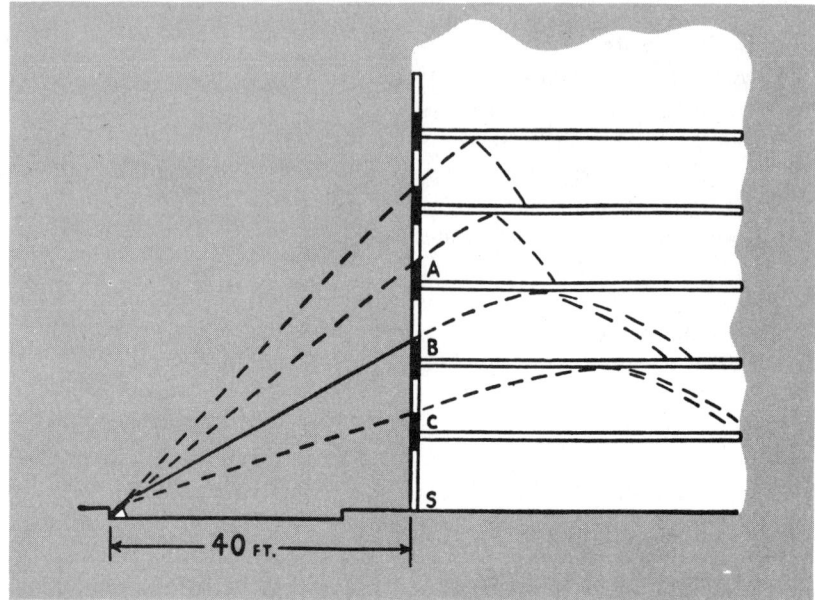

Figure 3

from the window sill to the point where the stream strikes the ceiling.

The distance B-S in the diagram should not be greater than the distance from S out to the nozzle. This means that an angle of 45 degrees is the maximum which can be considered effective for any street stream or other stream use.

A lesser angle than this is of course much more effective.

The diagram shows the nozzle placed 40 feet from the building. This is a little more than should actually be used in practice, unless apparatus is between the nozzle and the building.

When the stream is thrown through the third-floor window (Figure 3), the angle it makes with the ground is 32 degrees. This angle gives the greatest effective horizontal reach to fire streams.

To sum up, the third-floor window should be considered the maximum height to which a stream can be operated successfully from the street.

Under no conditions should the angle of the stream be greater than 50 degrees at the point where it leaves the nozzle. Allowing for the natural and unavoidable sag of the stream from the nozzle to where it strikes the building, the stream would likely pass, under these conditions, through the third-floor window instead of through the fourth-

FIRE STREAMS

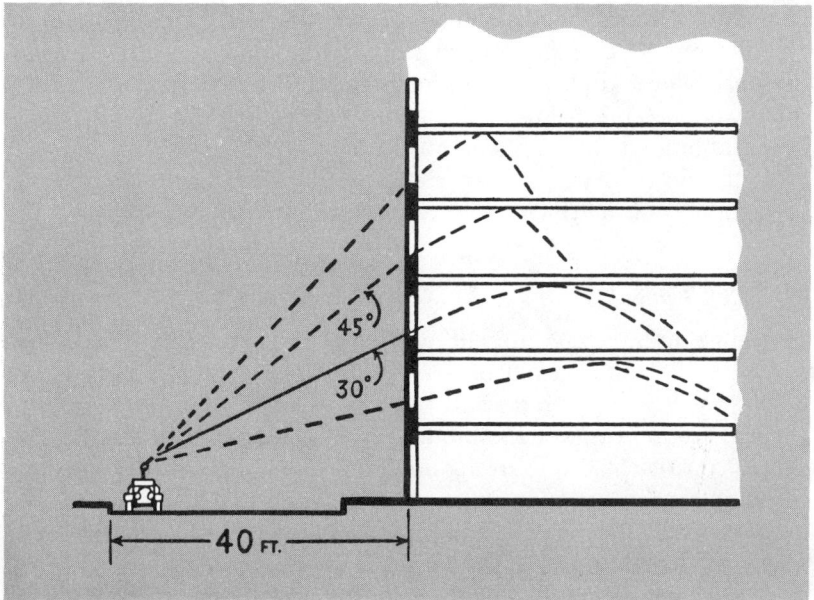

Figure 4

floor window as might be assumed from theoretical calculation and straight line path of stream travel.

Turret: The turret pipe mounted on a pumper or hose wagon possesses two advantages over the street stream. First, the nozzle is up to 9 feet above the street, and second, the heavy-caliber stream is easily manipulated.

The logical position for a turret pipe is at the opposite curb from the fire (Figure 4). This gives the maximum distance away from the building, which in turn means a maximum vertical reach, provided the stream is not broken up before it strikes the building. But in this case again, the stream is not considered effective above the third floor.

As in the case of the street line, a 30-degree angle for throwing water to the third floor gives the most effective results, and above the third floor the effectiveness decreases. This decrease in effectiveness is directly due to the decrease in penetration of the stream, or in other words, the decrease in distance from the front wall to the point at which the stream strikes the ceiling. The higher the angle at which the stream is directed, the nearer to the window it strikes the ceiling.

Also, as in the case of the street line, when the stream leaves the nozzle at a 50-degree angle, the natural curvature of the stream brings

FIRE SERVICE HYDRAULICS

the point (at which the stream might be calculated to strike) down considerably on the building and instead of the stream passing through the fourth-floor window to the building, the stream strikes into the window of the floor below. This is the case unless usually high pressures and large-diameter tips are used on the pipe.

To sum up the case of the turret pipe, the third floor is considered the highest floor at which effective work can be done.

Heavy streams

One-line fires rarely present any problems to a fire department. However, the multiple-alarm fire raging fiercely in a warehouse or lumber yard is another matter. Heavy streams, also called master streams, are in order for the big fires, where reach and penetration of great quantities of water are required.

Heavy streams range in size from a $1\frac{1}{4}$-inch nozzle up to 2 inches. Fireboats and some special land apparatus may have nozzles up to 4 inches. Nozzle pressures required for such streams range from 60 to 90 psi. Delivery from such streams runs from 400 gpm on an $1\frac{1}{4}$-inch nozzle to over 1000 gpm on a 2-inch tip such as may be found on an elevating platform.

Such tremendous streams require a fixed appliance for operation. A $2\frac{1}{2}$-inch hose line flowing 400 gpm is too dangerous for men to hold and a pipe holder of some sort is indicated. Flow beyond this requires appliances such as portable turrets, turrets mounted on apparatus, or nozzle assemblies rigidly fixed to an apparatus as are found on aerial ladders and platforms.

Uses of heavy streams

Heavy streams have definite uses which are as marked from small streams as are aerial ladders from portable ladders. And heavy streams are not confined entirely to outside use. A brief summary of the uses of heavy streams follows:

1. For fires involving large floor areas.
2. Fires in buildings which have involved entire floors and made it necessary for lines to operate from fire escapes.
3. Fires on second, third, or even fourth floors which have to be handled by street lines.
4. For operating across streets, air shafts, alleys.
5. For operating in buildings of large area and height such as sheds, armories, churches, railroad stations, etc.

It is not only the quantity of water which the large stream discharges that makes it more effective, but also its greater range.

An example of increased range by using a large nozzle may be seen from Table 1, which gives the vertical reach of streams from nozzles 1 inch to $1\frac{1}{2}$ inches in diameter at pressures of 20 to 90 psi.

Hand lines may be described as those lines stretched to the seat of a fire and operated or manipulated by fire fighters at this point. A hand line may be used through a shutoff nozzle, a cellar pipe or a fog nozzle which may be a combination fog and straight-stream nozzle.

A shutoff nozzle should be used at all times when working in a building or at other locations where it is apparent that there will be considerable shifting of the line. Before entering the floor of a fire building fire fighters should understand that the nozzle should be opened and closed judiciously. Where there is a great deal of smoke, but no great amount of heat, water should not be used until the fire is located. By waiting until fire shows, a considerable amount of unnecessary water damage may be avoided.

A delay of a few seconds may give the smoke time enough to break and disclose the seat of the fire. Shut off the line as quickly as possible when not needed. Do not keep it playing after the fire no longer requires it. Minimize the use of water as much as possible and thereby minimize water damage. Use reducing tips as early as conditions permit.

Hand lines and multiple lines should not be stretched from the same pumper. Two lines supplying a turret can produce a discharge at the tip of 600 gpm or better, which is close to the capacity of a 750-gpm pump. Putting a hand line into the same pump can reduce the efficiency of the turret while producing only another ineffective stream. At the same time, it complicates the job of the pump operator who will have to make frequent adjustments of his controls, particularly if the handline supplies a shutoff nozzle that is used intermittently.

Supplying the nozzle

Assuming that water supply is adequate, the final factor contributing to an effective stream is the pump (or more likely the pump operator). On level ground the hydraulic gradient runs in a straight, descending line from the pump to the nozzle. Losses are represented as friction loss. The pump must also supply sufficient pressure to overcome head loss if any, and provide proper nozzle pressure for adequate reach and effectiveness of the stream. Pump operators should therefore have a good knowledge of what constitutes an effective fire stream and how to achieve it.

Nozzle size and pressure

For interior fire fighting in occupancies with a heavy fuel load, shut-off nozzles in sizes up to $1\frac{1}{4}$-inch with pressures ranging from 40 to 60 psi will supply lines that are considered good. Apartment houses, dwellings, and small interior fires call for smaller nozzles ranging from $1\frac{1}{8}$-inch downward. In fact, today's fire fighters generally carry a combination straight stream and fog nozzle to fit either $1\frac{1}{2}$-inch or $2\frac{1}{2}$-inch hose for interior fire fighting.

As a general rule, nozzle diameter should not exceed one-half the diameter of the hose used and should be reduced depending upon the length of stretch. A short stretch is considered to be about six lengths; a medium stretch about 12 lengths; and a long stretch, 18 lengths or over.

A nozzle which is too large for a particular diameter hose gives a weak ineffective stream. A large nozzle results in a large flow of water and a large flow of water produces large friction loss. With a long hose stretch and a large flow of water, the nozzle pressure may be weak because of excessive friction loss. In such case, an increase of nozzle pressure requires the use of a smaller nozzle, assuming engine pressure, hose diameter and length of stretch remain constant. Since the velocity of flow remains the same in the hose, the water must flow faster through the smaller nozzle, thereby increasing pressure and reach at the tip.

If the nozzle is too small, and the engine pressure high, the stream will form a spray pattern on leaving the nozzle and its effectiveness is lost.

The chief causes of defective streams might be listed as follows:

1. Insufficient pressure
2. Too mucn pressure
3. Defective tip (nozzle)
4. Air in hose line
5. Hose twisted near nozzle or pipe

Insufficient pressure: This may be caused by too little pressure at the engine or hydrant, or it may be caused by too lengthy hose stretches and consequent heavy friction loss. It may also be due to using tips that are too large on lines of excessive length. The latter practice increases the friction loss in the hose.

Too much pressure: High pressure can be used on large nozzles without excess spraying, whereas with small tips high pressures often cause the stream to break up within a few feet of the tip, with immediate loss of effectiveness.

Damaged tip: This is occasionally the cause of a poor stream. It results from a burred or dented waterway brought on by rough handling. A nozzle with a small burr at the edge where the water leaves will break up the stream and cut its range down greatly.

Air in the line: This is usually caused by operating pumps under negative pressure. Air leaks into the line and is carried along with the water to the nozzle. It can be detected by the sputtering sound and broken condition of the stream. Air, when imprisoned in a line under high pressure, is compressed and on reaching the nozzle is free to expand in all directions, with the result that it breaks the stream.

Twist in hose near pipe: A double twist in the line near the nozzle causes a rotary motion of the water in the line, which in turn, results in a sprayed stream as it leaves the nozzle. The remedy is to make sure that the line of hose leading up to the nozzle has but one bend in it: the bend which leads from the horizontal stretch up to the nozzle.

CHAPTER EIGHT

Standpipe systems

A standpipe is in effect a vertical water main that extends through a building from cellar to roof and is used exclusively for fire fighting. A standpipe system is one that incorporates a standpipe, to which hose outlets have been connected on each floor, and which has its own positive water supply.

Next to the sprinkler system it is one the most important tools in fire fighting operations. It is essential, therefore, that the fire fighter be familiar with the various standpipe systems, their components and operations.

Standpipe systems are most commonly installed in tall buildings and those of large floor areas—also in hotels, schools, theaters, factories, high-hazard occupancies, places of public assembly where large number of people congregate and similar occupancies.

The systems may be efficiently used by building occupants (employees, watchmen, etc.) to control or extinguish small fires pending the arrival of the fire department. Use of the standpipe system eliminates the necessity of an exhausting and time-consuming hose stretch by fire personnel.

Standpipe systems can be efficiently used to prevent an exposure fire from entering a building. Water damage is minimized by the judicious use of shutoff nozzles instead of the open nozzles which are commonly found on house lines. In some cases, standpipe lines from the upper floors of a noninvolved building may be the only means of

STANDPIPE SYSTEMS

directly hitting a fire in an adjacent fire building. Fog nozzles on standpipe lines may be used to advantage from points above the fire to cool the atmosphere where men are working under particularly adverse and hot conditions in the street below. These fog streams are directed downward to provide relief.

Under unusual fire conditions (hydrants too far distant or failure of public water supply) the standpipe water supply in one building may be used to fight a fire in another building by stretching lines from the ground floor outlet to the intake side of pumpers. These then discharge for direct application on the fire or through the outside siamese for extinguishment of fire in the adjacent building.

Standard first-aid systems for use by building occupants have 1½-inch hose with ⅜-inch or ½-inch open nozzles. Systems for fire department use have 2½-inch outlets. A convenient arrangement is to use reducers to connect 1½-inch hose to 2½-inch outlets.

Standpipe classifications

Wet systems with the supply valve open and water pressure maintained at all times.

Systems so arranged through use of approved devices as to admit water to the system automatically by opening a hose valve.

Systems so arranged as to admit water to the system through manual operation of approved remote control devices located at each hose station.

Dry standpipes for which water is supplied through fire department pumper connections (siameses).

System components

Risers: The size of standpipe risers is governed by the size and number of fire streams that may be needed simultaneously and by the distance of the floor outlets from the source of water supply. Both of these factors will be reflected in friction loss in the piping. The number of standpipe risers and arrangement or distribution of equipment for proper protection are governed by local conditions such as occupancy, character and construction of building, exterior exposures and accessibility.

The number of risers (single or multiple system) as well as their location is further dictated by the fact that all portions of each story must be within reach of a stream from a nozzle attached to not over 100 feet of hose. Standpipes for small hose should be located so that all portions can be reached by a stream from not over 75 feet of hose.

The outlets (hose stations) on risers are located near or in stairway enclosures or near fire escapes for ready availability to firemen.

FIRE SERVICE HYDRAULICS

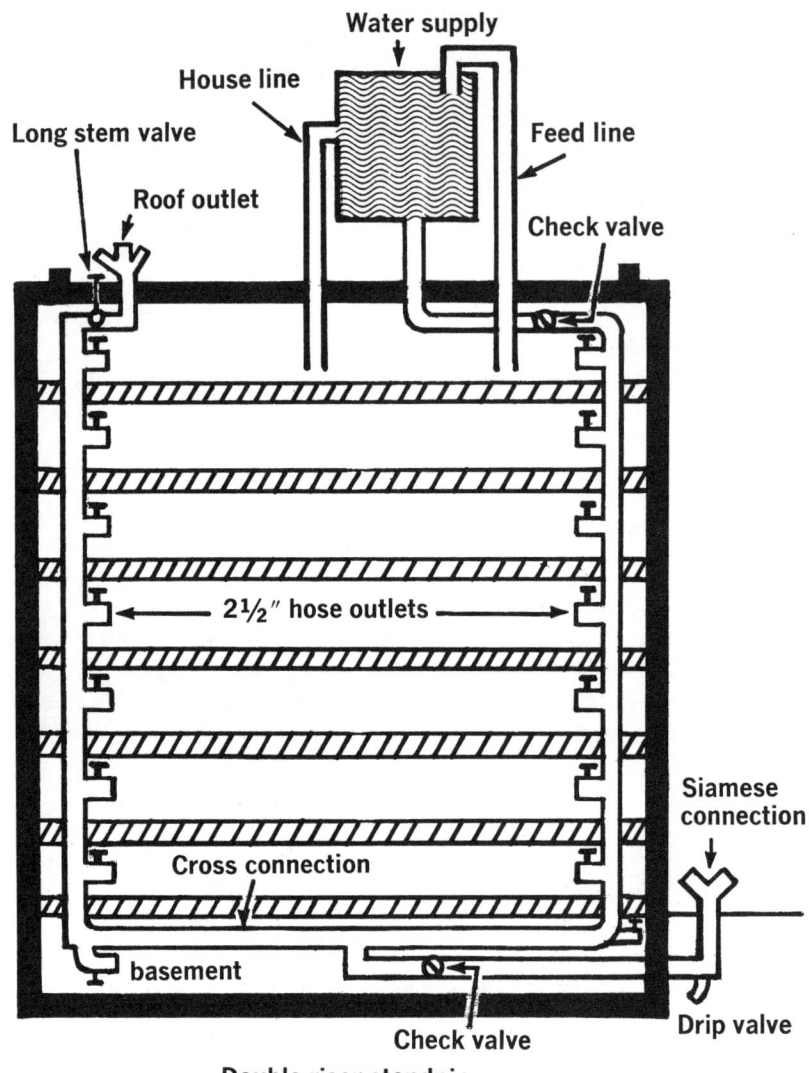

Double riser standpipe

Figure 1

Control valves: These valves, known also as indicating, stop, outside stem and yoke (O.S. & Y.) or gate valves, are installed throughout the system in order to meet emergency and other conditions such as repair, breaks in the standpipe lines, and for testing purposes. Globe valves are used at the hose outlets off the riser on each floor and also at the roof manifold. In the case of a multiple-riser system, sufficient

STANDPIPE SYSTEMS

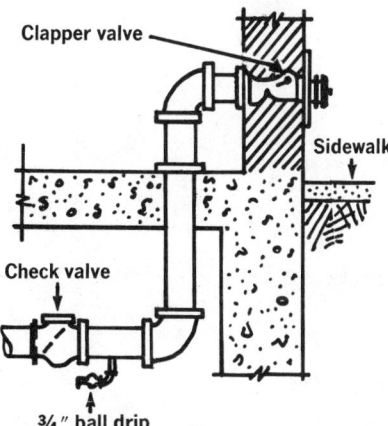

Figure 2. Check valves are installed throughout standpipe systems. The one shown here is just inside the building and prevents water from the system from leaving the building.

controls should be provided so that one riser may be cut off without interrupting the supply to other risers from the same source of supply. Sections of risers may also be similarly isolated.

Check valves: These are installed at various locations to prevent the flow of water past a check point in a standpipe or sprinkler system. The check valve near the siamese connection is normally kept on its seat (closed position) by the pressure of the water in the riser above it. This valve is unseated and permits water to be fed into the system when pressure on the pumper side exceeds that on the riser side. A drip valve is installed between the check valve and the siamese connection to prevent an accumulation of water which may seep past the check valve in the line between these two points.

The check valve in the line beneath the roof tank is normally in the open position (off its seat) to allow the flow of water. It is closed when flowing pressure on the pumper side exceeds the pressure on the tank side of the valve. This prevents the water supply from below from entering and overflowing the roof tank. Therefore, the roof tank cannot be refilled by this means.

In case the tank supply runs low and the automatic fill pump is inoperative, water may be fed into the tank through its inspection hatch by supplying water through the street siamese and taking a line off the roof manifold.

Roof manifold: These devices are installed to connect hose lines for roof operation. No house line is ordinarily found connected to this manifold. Since this part of the standpipe system is subject to freezing, it is normally maintained dry and water is supplied to it from the roof tank by operation of a long-stem valve reaching down

FIRE SERVICE HYDRAULICS

Figure 3. Roof manifold of standpipe system is normally kept dry. Water is supplied to it by operation of long-stem valve (left) that connects to top-floor piping.

to the top-floor piping. Because of the limited elevation of the tank, pressures at the manifold are often insufficient for effective streams and should be built up from another source.

Individual control of the pressure in each line off the manifold is exercised by manual operation of the shut-off handle. Lines from the roof manifold may be used to advantage in fighting roof fires, for stream application on an adjacent building, or to cover exposures. The threads on the outlets are protected by a cap on a chain.

STANDPIPE SYSTEMS

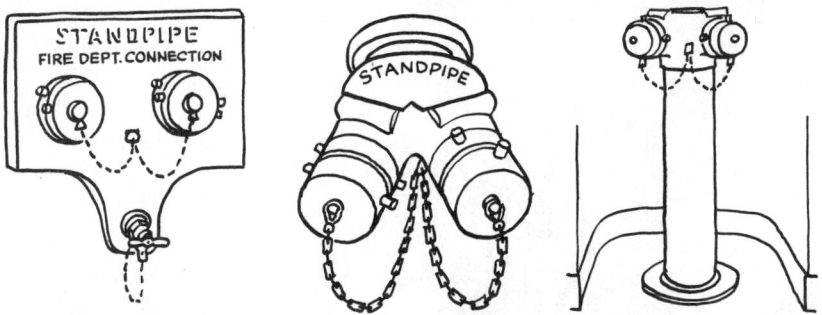

Figure 4. Typical fire department standpipe connections: flush type (left), wye type siamese and free standing type that is generally found in wall recess.

Siamese connection: The number of siamese connections found on a building depends on the size and number of risers, as well as the street frontage of the building. More than one such connection may be installed on the same street, or additional connections may be found at the sides or rear of the building if it has street frontage there. One type has a sill cock for connecting a garden hose to flush sidewalks. The wye type of siamese connection resembles a siamese on a pumper or deluge set. The free-standing type is generally found in a wall recess.

Before connecting a line to it, the inlets of the siamese should be checked for the presence of gaskets. It is recommended that the left

Figure 5. When attaching hose lines to siamese it is recommended that the left outlet be connected first. This is because the female outlets on the siamese connection rotate counterclockwise. Attaching the left line first permits freer manipulation of the hands when the right connection is made.

313

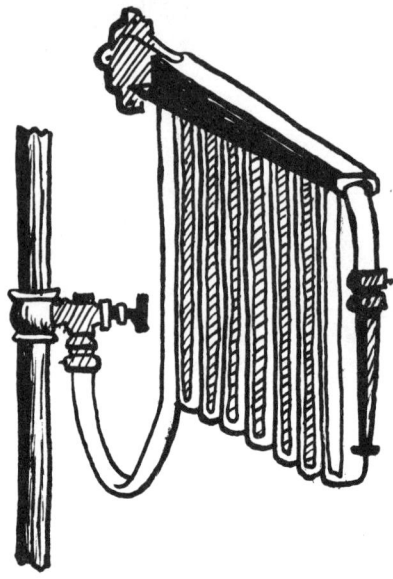

Figure 6. Hose outlets and drip valves are taken off the standpipe riser at each floor and provided with sufficient hose to reach distant points on the floor.

inlet be supplied first and that the reverse procedure be followed when disconnecting the line.

Hose outlets and drip valves: These are taken off the standpipe riser at each floor and are provided with sufficient linen hose to reach distant points on the floor for stream application. For fire department operations, the house line is removed and the stronger fire department hose substituted. The house line may be either 2½ or 1½-inch, depending on the occupancy. In the latter case, a reducer (2½-inch × 1½-inch) will have to be removed before connecting fire department 2½-inch hose.

To guard against wetting and resultant deterioration in the linen hose used on house lines, a drip connection is often provided between the control valve and the first hose coupling. If the drip valve is of the pet cock type, it should be closed before opening the control valve. Nozzles provided are usually of the open type.

Pressure reducers: To guard against excessive pressures, a pressure reducer is often installed between the valve outlet and the hose coupling. One common method of obtaining the pressure reduction is by use of an orifice disk.

Where hydrostatic pressure at any outlet for small hose exceeds 100 psi, an approved device is installed at the outlet to reduce the pressure so that the nozzle pressure will be approximately 80 psi. Where hydrostatic pressure on any outlet for 2½-inch hose exceeds 55 psi, and 2½-inch hose is provided for use by building occupants, a pressure-reducing device should be installed.

The pressure-reducing device should be removed before connecting a fire department hose line and the pressure should be controlled by a member who remains at the control valve.

Water supplies

The quantity of the water supply for a standpipe system is dependent on the size and number of streams likely to be required and the length of time such streams may have to be operated. This standpipe supply is calculated over and above that required for the simultaneous operation of automatic sprinklers. In some installations, it will be found that both types of systems are supplied from the same source, e.g., a gravity tank with a total capacity equal to the needs of both a standpipe and a sprinkler system.

In buildings in the course of construction and in other occasional instances where standpipes are installed exclusively for fire department use, the standpipes are normally dry and the only means of supply is through a fire department connection (siamese).

The sources of water supply for standpipe system may be: gravity tanks, pressure tanks, automatic fire pumps, city or town water works systems, where domestic pressure is adequate, manually controlled fire pumps with pressure tanks, manually controlled fire pumps operated by remote control devices at each hose station, or siamese connections which are supplied by lines from a fire department pumper.

In buildings of unusual height, standpipes are often supplied by a series of fire pumps and tanks, usually about 20 stories apart.

Two independent sources of water supply are desirable. The primary source should be capable of supplying the streams first operated until the secondary source can be brought into action. The secondary supply should enable the streams to be kept in operation for a long time. To promote efficiency, the layout, controls and sources of water supply for each installation in the company district should be studied, mapped and carried on the apparatus.

Gravity tanks: These tanks, with a minimum recommended capacity of 5000 gallons for standpipe supply, are located on the roof or as a separate water tower, at least 20–25 feet above the highest hose outlet or top line of sprinklers to provide pressure by gravity.

If the tank is not located in a heated enclosure, special precautions are necessary to prevent freezing in the tank or its piping, unless the tank water is also used for domestic purposes and there is a continuous flow. Freezing is prevented by the installation of a tank heater, which should be started before cold weather sets in. The temperature of the water should never be permitted to fall below 40°F.

The frostproof casing around the riser leading from the gravity tank prevents heat loss. And care should be taken to see that the casing remains intact.

Access ladders to the inspection hatch are provided and may be used by fire department personnel on their inspection visits. Proper safety precautions should be observed on these inspections and the integrity of the ladder established before it is mounted.

At times it may be advisable to drain the tank by means of the emergency drain valve. This condition could result if the tank supports were weakened by fire, or if the heating unit failed.

The tank water is maintained at the proper level by the action of a ball float, which is connected to the controls of a fill pump, generally located in the basement. If the fill pump is not automatic, an audible alarm is sounded when the water level in the tank drops below normal. Although it is connected to a street main, the capacity of this fill pump (usually about 65 gpm) is not enough to replace the drain of a fire hose line, and so other sources of supply must be brought into action when the tank supply is depleted.

Pressure tanks: In some installations, standpipe or sprinkler supply may come from a tank filled with air and water under pressure in a ratio of 1:2. The minimum recommended capacity of these tanks for standpipe supply is 4500 gallons, although for light-hazard occupancies only, the tank may have a minimum capacity of 3000 gallons. Tank capacity is considered as the total contents, both air and water. These tanks are normally located on the top floor or in a heated enclosure on the roof.

Water level is maintained by the action of an automatic fill pump; air is maintained at the proper pressure (at least 75 psi) by air compressor equipment with suitable automatic controls. Audible alarms are sounded when either the water or air varies from normal levels or pressures.

A sight water gage and an air-pressure gage are installed so that the operative condition of the tank may be determined during a fire or maintenance inspection.

Automatic fire pumps: Standpipe risers may be supplied by a fire pump, which is commonly installed in a pump room at or below ground level. This pump should be capable of delivering a satisfactory water supply (250 gpm minimum) and pressures at the highest hose outlet.

STANDPIPE SYSTEMS

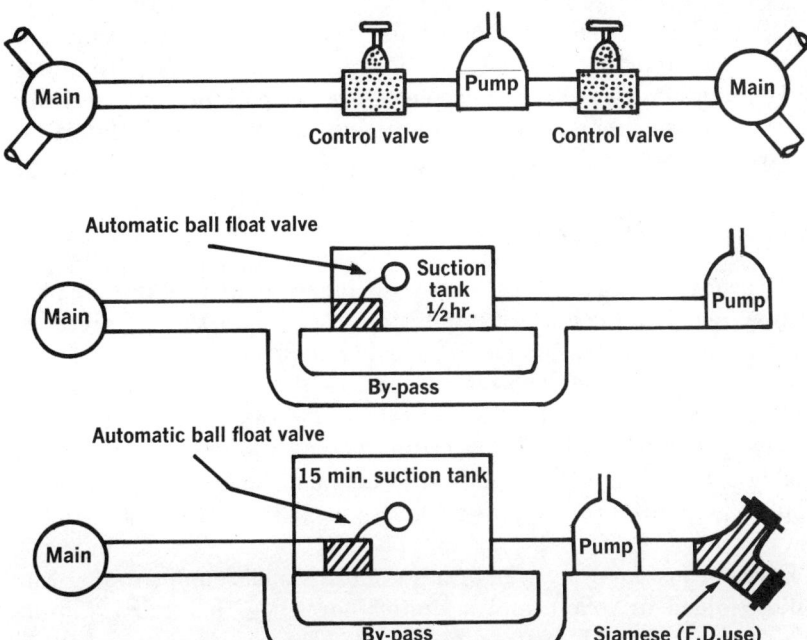

Figure 7. Fire pumps for standpipe risers can be fed by the above arrangements (top to bottom): by connections to two public water mains each fed two ways; from one main connected to suction tank that has sufficient capacity to furnish pump with at least one-half rated capacity for 15 minutes; by connection to either public water main or suction tank capable of supplying pump at rated capacity for 15 minutes.

The same fire pump or pumps which supply the standpipe system may be used for sprinkler supply, provided their capacity is adequate for both systems and relief valves are installed to prevent pressure in excess of the safe operating pressure for the sprinkler system. (See Chapter 4.)

The fire pump may be fed by any of the following arrangements:

1. By piping connected to two public water mains, each fed two ways, and with control valves provided on inlet and outlet sides of the pumps.

2. From one main, provided a suction tank (or tanks) is installed with sufficient capacity to furnish each pump with at least one-half hour of water supply at the rated capacity of the pump. A by-pass is also provided so that the pump is fed directly from the public water main in case of failure of the suction tank.

3. By a connection to either a public water main or a suction tank capable of supplying the pump at its rated capacity for 15 minutes. In conjunction with this arrangement, the fire pump supply may also be from one or more siamese connections reserved exclusively for fire department use and with this purpose clearly stated.

Operating at fires

It is considered good practice for all engine companies that respond to buildings equipped with standpipes to carry on the apparatus at least two lengths of $2\frac{1}{2}$-inch hose, folded or rolled.

Beyond the immediate need to accurately locate the fire, no hard and fast rule can be established as to when the standpipe system should or should not be used. This decision must be based on existing fire and other conditions. If the fire is located above the fourth floor, connect a pumper to the street siamese and stretch from the riser at the upper floors.

Carry at least two lengths of hose (more if sufficient personnel is at hand), folded or rolled, and a shutoff nozzle up the stairway *to the floor below the fire*. Elevators may be used for this purpose if they are not endangered by the fire.

If the building is of fireproof construction, it may be more efficient to operate from outlets in the enclosed stairways or fire towers on the fire floor. Forcible entry tools are also required.

On arrival at the selected outlet, detach the house line, as well as any pressure reducer or the reducing connection if the house line is $1\frac{1}{2}$-inch hose. The lengths of fire department hose are then coupled and attached to the hose outlet. The nozzle is put on and the hose is stretched up the stairway to the fire floor.

Surplus hose should be laid neatly down the stairway toward the floor below the outlet, or toward the floor above the fire if working from a fire tower. The hose should be free of kinks so that air may readily escape and a good stream be promptly assured at the nozzle when the door of the fire floor is opened. A fireman should remain at the control valve to regulate pressure on the hose line.

In high buildings where fire pumps are directly connected to the standpipe system, the officer in charge of the fire should immediately send orders to the building engineer to start the pumps and maintain sufficient pump pressures to provide effective nozzle streams. Necessary changes in pumper or fire pump pressures can be relayed by messenger, walkie-talkie radio or standpipe telephone system, if one is installed.

If a second company is ordered to stretch from the standpipe, the officer in charge of this unit should consider the point from which his unit is to operate so that enough hose is brought up to operate from this position.

If the officer in command orders a second company to operate from a fire escape, the company officer should order two extra lengths of 2½-inch hose to be carried into the building in addition to the usual pair of roll-ups.

The reason for this order may be noted from the following: The first-due company has connected to the outlet on the floor below the fire. It will probably require two lengths of hose to reach from this outlet to the rear or side of the building on that floor, and two additional lengths will be needed to make the stretch up the fire escape and into the fire floor.

In taking-up operations, the hose outlet control valve should be closed solidly, the drip pet cock opened, and the building line reconnected. To prevent unnecessary water damage, the line should be emptied by discharging the nozzle out a window or down a stairway at a point lower than the outlet.

In hotels, office buildings, and other light occupancies, 1½-inch hose is often found on standpipes and may under certain fire conditions be replaced with the same size hose by fire departments. However, in standpipe operations where riser water supply comes from a gravity tank or a weak water main, it is inadvisable to use 1½-inch lines on floors immediately below the roof. Here, the pressure on the riser outlet will not be sufficient to overcome friction losses in the small hose.

When siamese is out of order

Occasionally the outside siamese may be inoperable in a system that is otherwise serviceable. In such instances, and also where the check valve below the siamese connection is inoperative, the following procedure should be followed:

Stretch a line from the pumper to the gated outlet on the first floor. The house line, with reducing connection and pressure reducer, must be removed and the pumper line connected, using a double female coupling. If the supply line is 3-inch, a reducer (3 × 2½-inch) must be placed on the male before attaching it to the double female coupling. When the water is started in the supply line, the outlet valve is opened fully to allow water to flow into the riser. If necessary, additional lines can be similarly stretched to hose outlets on other floors.

Where the hose outlet extends at a right angle from the riser, the weight of the hose and fittings should be supported by a short length of rope. Tie a clove hitch and binder around the riser above the outlet valve. With the other end of the rope, tie a clove hitch and binder around the butt of the hose and double female connection. Push the clove hitch upward on the riser until the rope is taut.

Engine pressures needed

Pencil and paper, and long hydraulic formulas, have little place at an actual fire operation. From a practical standpoint, rule-of-thumb methods, giving approximate figures, will prove satisfactory.

Excessive pressure on the pumper may burst standpipe lines or hose lines, possibly losing the fire and endangering or injuring men in the building. Too little pressure may give unsatisfactory streams and have the same results. The average requirements call for a range of 150–200 psi on the pumper. If the engine pressure is excessive, surplus pressure on a fire line can be compensated for by partially closing the standpipe outlet valve.

The following information will serve as a guide in determining supply pumper pressures to provide effective streams:

It requires 0.434 psi to push up water 1 foot vertically, or 43.4 pounds per 100 feet. Assuming that the average height of a story is 12 feet, it will require 12 × 0.434, or about 5 pounds per story to overcome back pressure.

A standard figure of 25 pounds is taken to cover the friction losses in the riser, siamese connection and hose outlet valve.

A nozzle pressure of about 30 to 35 pounds is usually adequate for reach and maneuverability of inside lines.

The proper-size nozzle to be used depends on the policy of the department or the judgment of the officer in charge, but usually a 1, 1⅛, or 1¼-inch tip is used. Approximate friction loss allowances, equally applicable to a ground-level stretch, may be used for the hose line from the standpipe outlet to the nozzle. With 35 pounds, on a 1⅛-inch nozzle, about 220 gpm is discharged. For a flow of 220 gpm, there is a friction loss of about 12 psi per 100-feet of 2½-inch hose. The same nozzle pressure on the 1-inch and 1¼-inch nozzles gives a friction loss per 100 feet of about 8 and 18 pounds, respectively.

The friction loss from the engine to the siamese connection will depend on the number of lengths and the size of the hose. In the first stretch (35 pounds on a 1⅛-inch nozzle), the loss will be about 5 psi (4.6) per 100 feet, if a single supply line of 3-inch hose is used. From these explanations and figures, the required engine pressure in a typical case can readily be determined:

STANDPIPE SYSTEMS

Assume a fire occurs in a building 10 floors above street level, and the company is to operate from the standpipe:

10 floors at 5 psi back pressure per floor...................................	50
35 psi at the 1⅛-inch nozzle plus nearly 15 pounds friction loss in 100 feet of 2½-inch hose from outlet...................................	50
Friction loss for standpipe, siamese and outlet valve...................................	25
Friction loss in 200 feet of 3-inch hose from pumper...................................	10
Total (required engine pressure)...................................	135 pounds

From the above, an officer or pumper operator should be able to estimate the pump pressure necessary to give the proper nozzle pressure and volume on the fire floor. If an additional similar line is stretched from the standpipe riser to the fire floor, the original nozzle pressure and engine pressure will drop as the second nozzle is opened. Engine pressure must, therefore, be increased slightly to allow for the increased friction loss in the supply line, or lines, to the siamese. Friction losses in each hose line off the riser outlet, the required nozzle pressures and the back pressure would remain the same. The standard allowance of 25 pounds would still apply to cover the friction losses in the standpipe riser as well as the losses which exist at the siamese connection and the hose outlet valve.

Where the volume of fire requires an additional line or lines from a standpipe system, it is good policy to connect another pumper and feed another siamese in the system or an outlet valve on the ground floor.

On arriving at the fire building, the pumper operator should not wait for orders before stretching into the standpipe siamese, regardless of apparent fire conditions. One 3-inch line—and if fire conditions warrant it, a second one—should be stretched immediately to the siamese; but water should not be started until the order to do so is received. If only 2½-inch hose is available, two lines should be initially laid to the siamese connection.

Many authorities consider it good practice to stretch two lines of hose to the fire department connection even if one line can supply the required volume. If one line bursts, the other can continue the supply. Where men are in a dangerous position in a building, the failure of a single supply line can place their lives in jeopardy.

Where a pumper is using its full power to supply a standpipe and a number of streams are in operation, a second pumper should be connected to the system at another siamese, or connected to the same siamese by placing two portable siamese connections on the two branches of the standpipe siamese. This will permit four lines to be connected to the same street siamese connection.

In making a stretch from a pumper to a standpipe connection, the hose should be laid along the curb on the fire side of the street, if the pumper is on the same side. Otherwise, it should be stretched on the pumper side of the street to a point opposite the siamese connection and then carried across the street. This will reduce the crossing of hose lines by fire apparatus. When placing lines to a standpipe system, care should be taken not to obstruct entrances or interfere with the placement of ladders.

Occasions may arise when it is necessary to play a heavy stream from a standpipe in one building to another building involved in fire. When it is necessary to throw water in large volume a considerable distance, siamese lines by taking the supply from standpipe outlets on two different floors or from two outlets on the same floor. But when the volume and pressure secured by a "two-to-one" connection is insufficient, additional connections may be used.

If a building with standpipes is exposed to fire, the services of members of truck companies should be utilized to enter the building, stretch lines from standpipes and extinguish any incipient fire. By carefully wetting down material within the flame or heat radiation zone, they can prevent the fire from extending. These units can further help by closing all windows on the exposure side and removing portable stock near them.

At fires in theaters and other places of amusement which are equipped with standpipe systems, special precautions should be taken when stretching lines. Hose should be stretched from the standpipe on the side of the theater in which the fire occurs. For example, if a fire occurs in the smoking room on the west side of a theater, the proper stretch from the standpipe is from the outlet on that side—not the east side. This avoids laying a line of hose across the rear of the auditorium and obstructing front entrances and exits.

Inspection of standpipe systems

All portions of a standpipe system should be periodically examined by building maintenance men and fire department inspectors. A check should be made to see that gravity and pressure tanks are filled to the proper level and that at least 75 psi is maintained constantly in the pressure tanks. If the source of water supply is automatic, the control valves should be opened at all times.

Valves at the hose stations should be examined frequently for tightness. Leakage may be detected by inspecting the drip valves. Care should be taken to see that they are not clogged. If the system is normally dry, make sure that all hose valves are operable and closed.

Fire hose should be in good condition and properly positioned on the racks. Periodically, new gaskets should be installed in the couplings, at hose valves, and at the nozzles. More frequently, nozzles should be removed and examined for foreign objects which might adversely affect a hose stream.

Unlined linen hose should not be tested because it is difficult to dry thoroughly. It should, however, be carefully examined for cuts, loose couplings and deterioration, which may readily result from a leaky hose valve or from long standing in a damp atmosphere.

The outside siamese connection should be checked to see that its threads are compatible with those of the fire department, its usage is properly indicated, the couplings are not out of round, and the swing check valves and washers are present.

CHAPTER NINE

Automatic sprinkler systems

Mankind's constant search for better defenses against fire achieved a major victory in the 1880's with the development of the automatic sprinkler. An extension of the earlier non-automatic perforated pipes and open sprinkler, the new sprinkler not only extinguished fire but detected it.

The automatic sprinkler was immediately incorporated into an automatic sprinkler system that could protect all areas and all floors of a building. Ever since, it has been called man's first line of defense against fire.

An automatic sprinkler system consists of a grid of pipes which hangs from the ceiling of each floor of a building, and which is filled with water or compressed air. At a designated spacing, sprinkler heads are tapped into piping. These heads embody a mechanism whereby a rise in temperature to a predetermined level causes a head or heads to open and discharge water in the form of spray. The heads are so spaced that, if more than one head opens, the area sprayed by each overlaps the area sprayed by the others.

Sprinkler systems are required by law in some occupancies. They may also be installed voluntarily by the owner or occupant to protect a building, its contents and occupants. Such installation can also obtain a reduction of insurance premiums.

This form of protection may be found in the following types of occupancies: schools, institutions, theaters, factories, hospitals, hotels, flammables stored or used, oil cloth or linoleum manufacturing, garages, pyroxylin plastics stored or used, large undivided floor areas, transformer vaults, paint spray booths, basements of private homes, rooming houses, and many others. Except for chemicals which react violently with water, or for Class B fires, there are few fires in which water cannot be used from a sprinkler head.

The installation of sprinklers has a pronounced effect in reducing fire losses. A study of the most recent records shows that sprinklers either extinguished or held in check 96 percent of the fires in which they were involved. The 4 percent failure was due to a variety of causes, such as explosions that ruptured the piping, closed supply valves, freeze-ups, and failure of the water supply.

Sprinkler heads

Sprinkler heads are made of metal, are screwed into the piping at standard intervals, and generally have a ½-inch opening. The disk or seat is held in place (preventing the escape of water) by a strut, or two levers inserted between the seat and top of the yoke. Most frequently, the strut is of metal, consisting of two or more pieces held together with solder. Many types, however, use as a strut a quartz bulb which expands and breaks under heat, or a solid chemical, held in a cylinder, which disintegrates by heat action.

When constructed, the struts or levers are set in place under compression and are released when the fusible device operates from the heat of the fire. The disk is released from the opening and water flows out under pressure. The force of the water against the deflector creates a heavy spray which is directed outward and downward.

The sprinkler head is designed to withstand at least 500 psi without injury or leakage. If properly installed, there is little danger of the sprinkler breaking apart unless it is damaged.

Care must be taken to make certain that no part of an automatic sprinkler head is covered when the piping is painted or whitewashed. Such a coating may interfere with the free movement of parts and delay its opening, or render it inoperative. During the painting of piping or nearby areas, the heads should be protected by covering them with paper bags that are removed immediately on completion of the job.

Sprinkler heads of specific types are used for special purposes and in certain locations. When they fuse, care should be taken that their replacements are exact duplicates. In installing wax-coated sprinklers (used in corrosive atmospheres), precautions should be observed to avoid damaging the coating. Other types of corrosion-resistant coat-

FIRE SERVICE HYDRAULICS

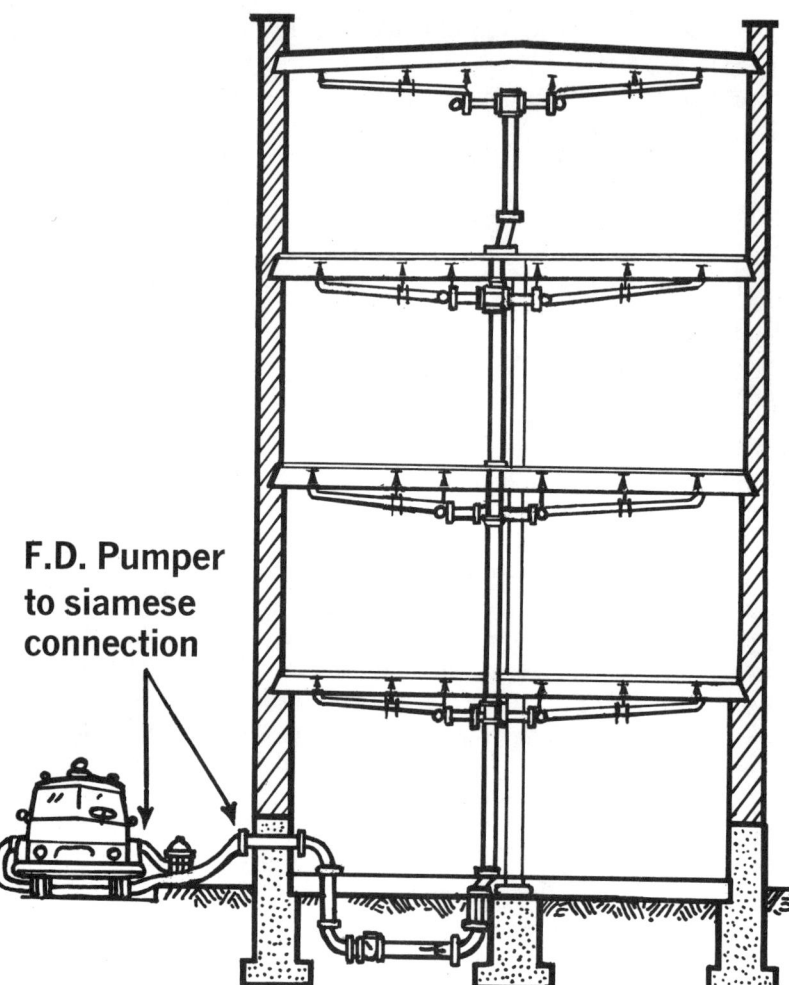

Figure 1. Highly simplified sprinkler system here shows arrangement for use by fire department pumpers. Pipe running up through building would be fed by public main.

ings should not be applied to the heads except by, or on the recommendation of, the sprinkler manufacturer.

In 1952–53, a radical change was made in the pattern of the water discharge from a sprinkler head which improved its effectiveness considerably. This new design was for a time called the "spray sprinkler," but is now designated as the "standard sprinkler." Both new and old-type heads are similar in appearance, but seemingly minor differences in deflector designs brought about major differences in discharge char-

Figure 2. Display case at New York Fire Department training school is indicative of the many and varied types of sprinkler heads available for different occupancies and uses.

acteristics. This change might be referred to as the adoption of the reverse-spray principle, which led to the development of the present upright spray sprinkler. This move has resulted in improved efficiency in the control and extinguishment of fire.

Formerly, research and developments in this area had been concerned with attaining a reasonably uniform distribution of water by a

FIRE SERVICE HYDRAULICS

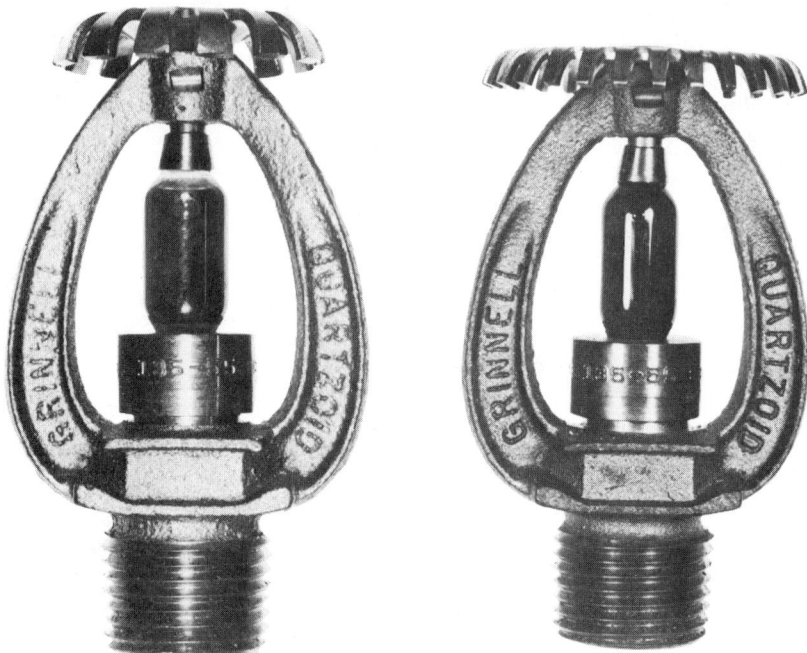

Figure 3. Standard sprinkler (right) replaced old style (left) in 1953. Redesigned deflector has a greater efficiency than old style. Sprinklers can be placed farther apart and flow rates can be appreciably decreased with the same protection as the old style.

single head and also with wetting the ceiling on the assumption that this was essential for efficient fire extinguishment.

Extensive investigation, however, showed that more effective extinguishment and a larger area of coverage could be secured by directing all the water downward and horizontally. It was further shown that this pattern was very effective in controlling fires on the ceiling above the sprinklers, owing to the improved cooling effect of the spray and a better high-level water distribution. There was decreased exposure to the ceiling because of a more effective direct discharge of water on the burning materials.

Because of the new design of the deflector, the solid steam of water issuing from a sprinkler is broken up to form an umbrella-shaped spray, with a pattern roughly that of a half sphere. At a distance of 4 feet

AUTOMATIC SPRINKLER SYSTEMS

below the deflector, the spray covers a circular area having a diameter of approximately 16 feet when the sprinkler is discharging 15 gpm.

These new heads are made for upright or pendant installation. They permit sprinklers to be placed farther apart and allow flow rates to be appreciably decreased with the same protection as the old-type sprinkler. More effective utilization of water flow is reflected in a reduction of water damage, a factor which often outweighs the actual fire damage.

Temperature ratings

Automatic sprinkler heads have various temperature ratings which approximate the temperatures at which they will operate. The temperature rating of all solder-type automatic sprinklers is stamped on the soldered link. For other heat-sensitive units, the temperature rating is stamped on one of the releasing parts. The temperature ratings of various heads are also indicated by different colors.

Where high temperature prevails, such as over boilers, ovens and in drying rooms, a higher degree head must be used than in an ordinary occupancy. If high-degree heads are used where not required, i.e., in an ordinary atmosphere, the value of the sprinkler protection is materially reduced.

Classifications of sprinkler systems

In addition to the wide variety of sprinkler heads manufactured for different kinds of occupancies (light, ordinary, extra hazardous), sprinkler systems themselves are classed as follows:

Automatic wet-pipe system: The piping is always full of water under pressure, so that if for any reason the fusible link in the sprinkler head is released, water will instantly flow from the open head. This type is installed in heated buildings.

Automatic dry-pipe system: This is installed where a wet-pipe system cannot be properly heated and there is danger of freezing. The pipes are filled with air, the release of which when a sprinkler head opens permits the water pressure to open a valve known as a "dry-pipe valve." Water then flows into the piping and out the opened head. In some cases, a separate dry system is installed in an unheated area of a building in which the rest of the system is wet.

Nonautomatic systems: In this type, all pipes are normally dry and water is supplied, when necessary, by pumping into the siamese connection. At times some of these systems are supplied by manual operation of a control valve. Nonautomatic systems include:

1. Perforated pipe systems consisting of single lines of pipe drilled

at intervals for water discharge, and intended to protect basements or other areas which are difficult to reach in fire fighting operations.

2. Open fixed-spray nozzles or distributor nozzle systems for transformer vaults and similarly hazardous areas.

3. Exterior exposure sprinklers (or window sprinklers), using open-type sprinkler heads to form an external water curtain on the walls of a building.

4. Foam supply systems for the protection of special hazardous occupancies into which a foam mixture is pumped, or into which water is supplied to augment that required by foam-mixing apparatus inside the yard or building.

Special systems

There are several types of automatic sprinkler systems, among which may be found any of the following, or a combination of two or more, in an occupancy.

Deluge systems: These may be described as a system of open heads, or a combination of open and closed sprinklers, controlled by a quick-opening mechanical or hydraulic valve known as the deluge valve. This valve may be operated by automatic heat-responsive devices or by manual control upon receipt of a fire alarm. This type of system is installed when it is desirable to immediately wet down an entire area in which a fire may occur. It is done by admitting water to open heads, rather than by using a system in which the automatic sprinklers open independently as the fire gains headway.

Such systems are commonly actuated by the rate of rise of heat within a given space and are used in rooms of very dangerous occupancy, such as explosives manufacturing, film manufacturing, film cutting and packing, lacquer mixing and coating operations. They also are installed in airplane hangars and assembly plants where ceilings are unusually high and ordinary sprinklers would not open quickly enough.

Preaction systems: These systems also operate on a rate-of-rise type detection system and differ from deluge systems in one main respect: All heads have fusible links and the alarm is given before the heat has risen enough to melt the links and permit the flow of water.

Preaction systems are designed to protect properties where there is danger of serious water damage which could result from premature or accidental operation of a sprinkler or a break in the piping system. Here the action of the heat-responsive device, or thermostat, releases the preaction valve, water fills the piping system, and an alarm operates in advance of the fusing of the sprinkler heads.

AUTOMATIC SPRINKLER SYSTEMS

Antifreeze sprinkler systems: These are installed in areas subject to low temperatures, such as cold storage rooms, truck-loading docks, etc. Sprinkler piping is normally filled with an antifreeze solution such as calcium chloride. At times, these systems are small extensions from wet-pipe systems, arranged with check valves and a trap so that the system water does not mix with the antifreeze solution. When a head opens, the solution is lost and must therefore be replaced when the system is again placed in operation.

Combined dry-pipe and preaction systems: These were originally developed for the protection of unheated structures such as the superstructures of piers. Their main advantage is that two dry-pipe valves can be controlled in parallel and supply a system that is larger than would be permitted for a single valve. These valves are located in a normally heated area or in an area where heating equipment can easily be installed. They incorporate the rate-of-rise tripping features of a preaction system, including the sounding of an alarm before sprinklers operate. The two dry-pipe valves are connected so that the tripping of one will cause the other to trip immediately, thereby flooding the entire system from a city water main.

Junior sprinkler systems: This inexpensive system is used in basements of dwellings and similar locations where not more than three or four sprinklers are likely to operate. The system is directly connected to the service water supply pipe where it enters the building. The shutoff valve is commonly found at this connection.

A smaller orifice sprinkler is used, usually $3/8$-inch, discharging about one-half as much water as the standard sprinkler, but designed to cover about the same amount of floor area. This difference must be borne in mind when replacing heads after a fire.

Water supplies for sprinkler systems

The methods used to supply water to sprinkler systems are the same as those for standpipe systems. Sprinklers may be supplied from one or a combination of sources, such as public mains, gravity tanks, pressure tanks, fire pumps, reservoirs, rivers, lakes, wells, etc.

Theoretically, a single water supply would appear to be all that is necessary for satisfactory sprinkler protection, provided the volume and pressure are sufficient. However, a single supply may be temporarily out of service; it may be disabled at the time of the fire or before the fire is extinguished; or the pressure or capacity may fall below normal during an emergency. Adequate secondary or additional water supplies may therefore be advisable or necessary, depending on the strength and reliability of the primary supply, the value and

FIRE SERVICE HYDRAULICS

importance of the property, the height, area and construction of the building, its occupancy and the exposures, or legal requirements.

But despite the added insurance of reliability when two independent water supplies are provided, it is recognized that a single supply of adequate volume and pressure will give satisfactory protection in many instances.

City water connections from large mains, fed two ways, or connections from two mains on a gridiron system with adequate pressure may provide an excellent and completely satisfactory supply for sprinklers.

Valves are automatic

Automatic sprinkler systems operate with two special valves: Wet-pipe and dry-pipe. The wet-pipe valve is essentially a clapper valve. It is so arranged that when a sprinkler head opens, and the water flows, the clapper opens and in so doing activates an alarm.

The dry-pipe valve is similar but more complex since the lines are maintained full of air until a head opens. When this occurs, the dry-pipe valve is "tripped" (unseated), charging the system with water which pushes the escaping air ahead of it. This action also sounds a local alarm, or in some cases transmits the alarm to a supervisory service which notifies the fire department. In certain occupancies where life hazard is very high (schools, hospitals, etc.), the alarm is directly transmitted to the fire department.

When fire conditions indicate the system should be shut down, the supply may be closed off at the individual O.S. & Y. valve for the floor or area, or at the main valve. After the main valve is closed, the drain valve should be opened so that the remaining water in the system beyond the dry-pipe valve will not have to be discharged and possibly cause additional damage in the fire area.

Before the main supply valve is again opened, the drain valve must be closed, the dry-pipe valve reset, air pressure again built up in the system by the compressor, and the alarm system drained and once more placed in its normal position ready to operate.

The dry-pipe valve is designed so that, in the normal position, the air pressure on the top side of the valve will hold a pivoting counter-weighted clapper down over the water supply inlet. A latch holds the clapper in the open position, once the valve trips.

The design is based on the area differential between the air seat and the water seat of the clapper, that enables the air pressure to hold back water pressure which may be six times as great. Some ratios are greater to allow lower air pressures to be used. The latch release is used when resetting the valve.

Excessive air pressure in the system is undesirable as it delays action at the dry-pipe valve. A pressure of 15–20 psi over the normal tripping pressure of the valve has been found to be the maximum necessary in most cases.

According to fire statistics, more sprinkler heads operate on the average with dry-pipe than with wet-pipe systems. This tends to indicate that fire control is not as prompt with the former due to the loss of time between the opening of a sprinkler head and the issuance of water—an interval made necessary to allow the escape of the compressed air in the sprinkler pipes. The problem can be overcome by the installation of quick-opening devices on the valve or the system.

Such quick-opening devices are of two types: accelerators and exhausters. In the first type, when a sprinkler opens and air pressure drops 1 or 2 psi, a diaphragm is unbalanced and by its movement opens an auxiliary valve to admit sprinkler system air pressure to an intermediate chamber beneath the air clapper of the dry-pipe valve. This balances the closing force and allows the water valve to be opened by the water pressure. In the exhauster type, movement of the diaphragm causes an auxiliary valve to open, which discharges system air pressure to the atmosphere, permitting water under pressure to unseat the dry-pipe valve and quickly fill the piping.

Control valves

When a sprinkler system is supplied from a public water main, the entire system may be closed down by operating a control valve between the building and the water main. This shutoff valve is frequently located in a box which is recessed in the sidewalk, with its location designated by a sign on a building or post nearby reading, "Shutoff for Sprinkler System Located 6 Feet from this Sign," or similar instructions. A special key may be required to operate this valve.

The control valve for the building may also be attached to an upright post, known as a post indicator valve (P. I. V.). The building or section of the building controlled by the valve is generally indicated on the post. The condition of this valve (open or closed) is shown through a tell-tale opening in the post. On some posts, a padlock must first be opened or forced to release the operating wrench. On others, an iron strap must first be released by cutting a riveted leather section.

The water main supply for sprinklers may also be controlled by valves of the O.S. & Y. type (outside stem and yoke) which are found just inside the building wall on the main riser, or outside in protected pits. Firemen can tell at a glance if the valve is open or shut because

FIRE SERVICE HYDRAULICS

Figure 4. Location of shut-off valves for sprinklers are often indicated as above. Alert fire companies, however, should know location of all sprinkler shutoffs in their response districts through pre-inspection.

the stem is all the way out when the valve is open and all the way in when it is closed.

Additional valves of this type may be used to control the supply for individual floors, and separate valves may be installed to shut off certain sections of a floor. In many cases, parts of a system subject to freezing are isolated in cold weather by closing the O.S. & Y. valve and draining the pipes.

AUTOMATIC SPRINKLER SYSTEMS

Figure 5. Post indicator control valves have tell-tale openings that show position of valve—open or closed. Padlock on this type must be forced to release wrench. Others have iron strap with riveted leather section that must be cut for release.

Figure 6. Control valves for sprinkler (and standpipe) systems include the O.S. & Y. type—outside stem and yoke. These are found just inside the building wall on the main riser or outside in protected pits. They are also used to control individual floors or sections.

Figure 7. O.S. & Y. valves that are inaccessible because of height are often chain-driven as indicated.

Caution in shutting down

While sprinkler systems are an excellent means of controlling fires, they can add to the loss if they are not shut down at the proper time. However, no control valve of the system should be closed except on orders of the fire officer in charge, who must decide when the fire is under control.

In many instances, considerable time elapses before the proper control valve can be located. This problem can be anticipated when a company visits a premises on an inspection. A small sketch of the sprinkler system including the location of the control valves, should be made and kept in a file on the apparatus. This information is invaluable at a fire because many times it will be found that the con-

AUTOMATIC SPRINKLER SYSTEMS

Figure 8. Poles and tongs may be used for quick shutoff when only a few heads have fused.

trol valve is actually located at some remote point in the building or in another wing or adjoining building.

Generally speaking, the location of the control valve for a particular area of a building may be found by tracing the pipes on which the heads are mounted. It will be found that the diameter of the pipe increases as it approaches the source of supply.

If the fused heads are few in number, they may be closed by using a pole shutoff, sprinkler tongs, or a wooden wedge inserted in the ruptured head. The pole shutoff has the advantage of being operated from a distance, enabling firemen to stay out of the main discharge of water from the sprinkler head while stopping the flow. Poles of this type of sprinkler stopper vary from 6 to 20 feet in length. The tong-type sprinkler stopper is first inserted in the ruptured head in the closed position and held between the strut and the seat of the sprinkler head with one hand. The tongs are then opened by releasing the control lever with the other hand.

Use fire department siamese

Although normally a sprinkler system is connected to an automatic source of water supply, engine companies responding to a sprinklered occupancy should stretch one of their first lines to the siamese connection of the building.

It is the policy of most departments to supply the standpipe system first, if one is present, and then stretch one or more lines to the sprinkler siamese. Care should be taken to see that the proper siamese is selected for the needs of the operation, as those for standpipe systems, sprinkler systems, transformer vaults and other installations are alike in appearance. The exact purpose of each should be indicated nearby or on the siamese itself. Some building codes require that siamese supplying different systems be painted a different color, such as red for standpipes and green for sprinklers.

FIRE SERVICE HYDRAULICS

On wet-pipe systems having a single riser, the siamese connection to the system is made on the system side of the controlling gate valve. On dry-pipe systems having a single riser, the connection to the system is made between the gate valve and the dry-pipe valve. This makes it possible to pump water into the system even if the gate valve is closed.

If there are two or more sprinkler risers connected to a public main, each system should have its own fire department connection attached to the supply side of the gate valve, so that with any one riser shut off, the supply from the pumper will feed the other risers.

Supplying the system

The exact amount of water being discharged from a system depends on the number of heads open, the type of head, and the pressures maintained at the orifice. Seven to 8 psi of working or flowing pressure is generally considered a minimum for the proper action of sprinklers having a nominal ½-inch opening. At this pressure, a sprinkler will discharge about 15 gpm and cover an area of approximately 100 square feet, or more if the newer heads are used.

Discharges, of course, increase as pressures at the orifices increase. For most systems, approximate discharges may be calculated by the following method:

Discharge in gpm equals ½ psi at the sprinkler head plus 15.

Where the fire involves a large area and sprinkler needs are high, more than one pumper should be connected to the system to augment the supply. At a discharge of 20 gpm per head, a 750-gpm pumper can supply only about 35 heads and a 1000-gpm pumper about 50. Generally, a 750-gpm pumper will need more than two lines if located more than 200 feet from the connection and a 1000-gpm pumper will need more than three lines of hose. Fire departments will therefore have to use an additional siamese connection on the building's sprinkler connection to provide additional lines.

Pump pressure required

Exact computations of engine pressures are impractical on the fireground, due to lack of required information on such factors as the number and type of heads opened, and the number, length and size of the supply lines to the building siamese. Where the pumper is stationed near the fire building, most authorities recommend that a pump pressure of 100 psi or more be maintained. On the older systems, consideration must be given to the possibility of the system bursting should the pumper build up excessive pressures. This recommendation is made in spite of the fact that piping in all systems is originally tested at pressures far in excess of those normally required at fire operations.

Naturally, higher pressures may be safely maintained on newer installations. Such pressures may be necessary if the flow is considerable, the supply lines to the siamese long, or the fire located many stories above the street.

If only a few heads open, and the fire is at or near ground level, the losses will be minimal. In such cases, engine pressures approximate nozzle pressures.

If the volume and pressure of the water on the fire is adequate, it makes little difference from a practical standpoint whether the source of standpipe or sprinkler supply is a fire department pumper or another source, such as a gravity or pressure tank, fire pump, or public water main. If the riser is already filled with water from a gravity or pressure tank on the roof, the engine pressure must be sufficient to overcome the head pressure of this column for the pumper to be the actual source of supply. In other words, it will have to unseat the check valve in the basement and close the check valve under the gravity or pressure tank.

This point of information is more than of academic interest, as the pumper may be damaged if it pumps continuously against a closed basement check valve. If no water is discharged from the pumper, the action of the pump will tend to heat up the water considerably, even to the point of creating steam. This hot water may adversely affect the packing or other parts of the pump and is unsuitable for auxiliary cooling purposes. Whether it is directly or indirectly used, it can raise the temperature of the engine coolant. This may then prove inadequate to absorb heat from the engine cylinders and may allow the oil to heat up, thin out or vaporize, and reach a point at which the cylinders may crack or the motor seize up.

These possibilities can be minimized by "cracking" a bleeder valve on a discharge gate, thereby insuring a continuous discharge to the ground and preventing an undesirable increase in the water temperature in the pump.

Whether or not the check valve beyond the siamese connection is actually open in a standpipe or sprinkler system may readily be determined as follows:

Working with caution, partially or completely close the discharge gate(s) of the pumper supplying the siamese. If the check valve is open, i.e., the pumper is actually supplying the water, the engine pressure will rise until the relief valve opens or the pressure governor operates and slows down the engine.

FIRE SERVICE HYDRAULICS

In addition, if the water output was considerable, the pumper intake gage will show an increase in pressure during the closing of the discharge gate(s).

If the pump is not putting water into the system, i.e., the check valve is closed, there will be no change in pump pressure or motor speed upon closing the gate(s).

This same condition and solution would apply if pressure from a building fire pump or city main held the check valve on its seat or if two pumpers were supplying the same system—one pumping into a siamese against the face of a closed check valve, the other keeping the valve on its seat by pressure at its rear, supplied through another siamese or an inside hose outlet.

Sprinkler systems and the fire department

Studies of large-loss fires in sprinkler-protected buildings show that fire departments should follow the suggestions of the American Insurance Association to use sprinkler systems most effectively:

1. Each fire department should have a list of all sprinkler-protected buildings in its area. Each officer should be familiar with at least those buildings in his first-alarm district.

2. Sketches should be prepared showing the locations of alarm valves, control valves and fire department connections. Officers should be familiar with this data on buildings in their district.

3. Fire departments should have information on the available water supply, such as: How many pumpers can be used in addition to those supplying the fire department sprinkler connections; is there a water supply separate from that supplying the sprinkler system that can be used; is there a nearby static water supply, such as a pond, from which pumpers can draft?

4. Fire department connections for sprinkler and standpipe systems should be inspected at regular intervals by the fire department to assure that caps can be readily removed, that threads are in good condition and match those of the fire department, and that the connection is otherwise ready for use.

5. One of the first-alarm engine companies responding to a fire in a sprinkler-protected building should lay two 2½-inch lines (or a 3-inch line) to the fire department connection.

6. Fire departments should impress responsible personnel of sprinkler-protected buildings that once sprinklers have operated for a fire, the system should not be shut off before the fire department arrives.

7. Extreme caution should be exercised by officers when ordering sprinkler systems shut off. The system should be shut off only after

a thorough check reveals the fire will not rekindle. The firemen ordered to close the control valve should remain at the valve so that it can be opened without delay if needed.

When a shutdown is made by the fire department, someone should be left at the valve until the sprinklers can be replaced and the water supply restored. If the system cannot be put back into service, watch service should be maintained on each floor, in addition to the man at the control valve, until all danger is past.

PART FOUR: **foam**

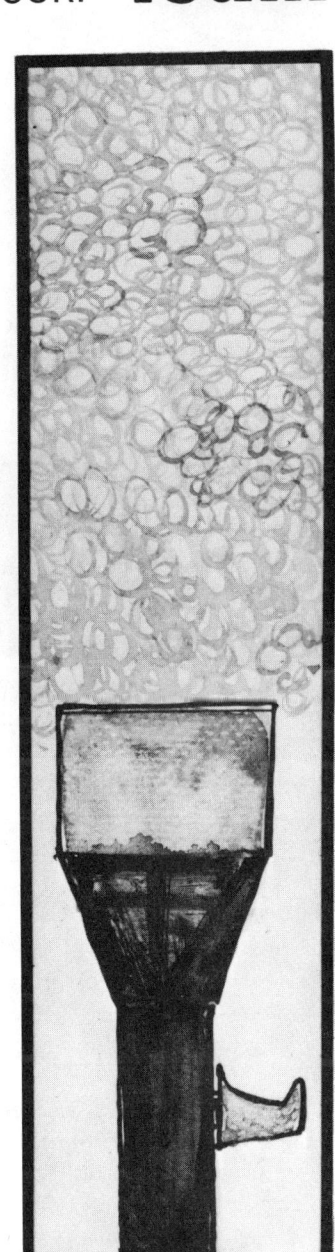

CHAPTER TEN

Fire fighting foams and foam systems

The employment of mixtures of air or gases with water in the form of foam for fire fighting purposes became critically necessary early in the 20th century when liquid petroleum hydrocarbon fuels approached widespread use. The age-old fire extinguishing material—water—could not effectively cope with these new fire hazards because its density was greater than that of these flammable liquids.

Flame propagation upward from the surface of such liquids was hardly affected by the quick transit of water droplets or streams through the burning zone and to the bottom level of the burning liquid where it lay useless. Water was found to even increase the fire problem by displacing the burning liquid fuels and causing them to overflow a tank or container and to spread to where no retaining walls or dikes held them in check.

The obvious solution was simply to change the density of water so that it would float on these relatively light liquids. Foam became the ingenious answer—water in the form of a cooling, surface spreading fluid, lighter than the lightest liquid fuel.

This chapter describes in some detail the composition and characteristics, the methods of generation, the system equipment design and the modes of efficient application for the fire fighting foams generally available in the United States today (1970).

Fire fighting foams have changed a great deal in character and use over the past 50 years. Today we have many different types, some

FIRE FIGHTING FOAMS AND FOAM SYSTEMS

with special properties for specific purposes of fire protection. Others are useful for many fire situations. Still other foams are particularly useful for cavity filling, wet gas cells, rather than as surface spreading blankets of a variant form of water.

Classification of foams

There are several methods of classifying the many modern types of fire fighting foams. Broadly, foams consist of only two categories:

Chemical Foam: Older and almost superseded type. Dependent on a chemical gas-producing reaction for the generation of a foam.

Air Foam or Mechanical Foam: Modern type (since 1939). Dependent on a turbulent mechanical action to mix air into a foaming water solution for the generation of a foam.

Another system of foam categorization would be to consider their action on fire. This divides them into similarly unbalanced classes as follows:

High Expansion Foam: Large-volume masses of air-filled foam are produced to fill cavities and blanket fires.

Surface Spreading Foam: Useful for fire extinguishing by cooling and halting evolution of flammable vapor on burning liquids.

This classification becomes difficult to employ because a clear-cut separation of the actions of the two foams is lacking. It is worth noting that this classification is parallel with one that would use the expansion ratio as a method of typing foams.

These considerations lead to a third and most descriptive method of classification that subdivides fire fighting foams into four principal but separate classes on a basis of the composition and character of the material used for generating the foam:

Class 1. Protein-type agents
Class 2. Synthetic foam-forming agents
Class 3. Surface-film foam-forming agents
Class 4. Chemical foaming agents.

The multiplicity of available foam-forming agents which must be considered in these broad categories can be seen in Table 1.

CHARACTERISTICS OF FOAM

Protein foaming agents

The foam concentrates or foam liquids made from proteinaceous materials are the oldest class of foaming agents now used in the produc-

Table 1 Foam Forming Agent Types-Classified According to Composition

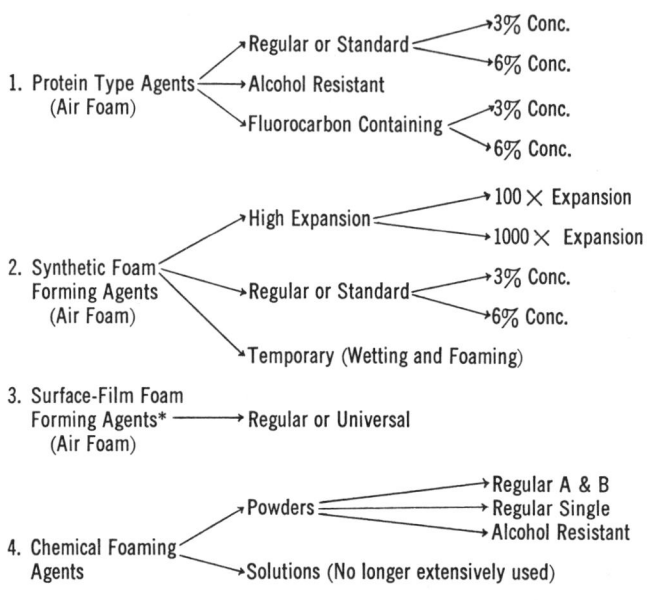

* The surface-film forming agent category is represented by the fluorocarbon synthetic agents known as "Aqueous Film Forming Foam," (AFFF) or "Light Water." At this time the generic term "Light Water" is a registered trade name of one manufacturer of this class of materials.

NOTE: The terms "3% Conc. and 6% Conc." denote the recommended liquid agent concentration by volume in water ready for use: 100× and 1000× are the final expanded volumes of foam generated by Type No. 2 when compared to the original volume of water solution of the agent (100× = 100:1).

tion of air foam. Because they followed the older chemical foam materials, they were designed to produce fire fighting foams in air foam generators which were as close to chemical foam in appearance and characteristics as it was possible to make them.

The development of modern air-foam foaming agents containing protein compounds dates from 1937 in Germany where Weissenborn discovered that waste protein such as the unusable materials from slaughter houses, etc., could be cooked with alkaline water to produce a hydrolysis product which foamed copiously. When this liquid was suitably filtered, it could be mixed with iron salts such as ferrous sulfate, and when foam was generated with this liquid, it formed thick, viscous, semi-rigid bubble masses which were very resistant to flames, heat, or mechanical breakdown. Ratzer, in England, followed with a similar material made from hoof and horn meal which had been reacted with lime to produce the active foaming ingredient, hydrolyzed protein.

Foams from these liquids looked and performed like the older chemical foam.

The important materials in protein foaming agents are the iron salts and the protein hydrolyzate. The latter compounds are high molecular weight polypeptides of many types (depending somewhat on their source of natural protein). They demonstrate an interesting reaction, in which the iron binds up chemically with the protein to give more mechanical strength to the bubble wall of air foams made with solutions of these mixtures. Heat from a fire accelerates this binding action and makes the foam tough and difficult to break down, even by water sprays or wave action on the liquid surface below the foam.

Protein foaming agents are difficult to keep in their original state under storage for long lengths of time. Depending on their method of manufacture, some liquid concentrates become increasingly dark brown in color with time and a black precipitate will settle to the bottom of the storage container in the form of black mud. This is capable of clogging equipment very quickly. Tanks or suction tubes for storing or taking up this liquid must be separated from the bottom of the standing container to prevent drawing up the black viscous material. In some installations it has been the practice to invert storage containers (such as 5-gallon shipping cans) periodically. This re-suspends the black precipitate but does not remove the problem. A good formulation may remain free of these problems for three to four years, whereas other materials may show undesirable storage breakdown within six to eight months. High storage temperatures accelerate these effects. However, the fire fighting capabilities do not seem to be affected by changes in appearance of the liquid.

When protein foaming agents are used on alcohol fires they are especially compounded with materials such as aluminum or zinc stearates (soaps) which confer some lesser water solubility to the foam and give some resistance to solubilizing in the face of alcohol attack. These are proprietary formulations that require careful testing with the solvent hazard it is designed to protect.

The most recent development in protein agents involves the mixture of protein-base materials with certain fluoro-compounds. These mixtures yield a liquid with superior performance, especially when the fuel coats the foam. When air foams are forcefully driven into flammable liquids the burning liquid will coat the upper surface of the foam, continue to burn and thus destroy the foam. When fluorocarbon compounds are carefully combined with protein materials, the foam generated sheds flammable fuels rapidly. They drain down, away from the surface. For the same reasons, these fluoroprotein foaming agents perform in a superior manner when used in the subsurface injection system of fuel tank fire extinguishment.

Fire fighting foams made from protein agents are very efficient and long lasting. They develop a strong vapor seal, and when properly generated they are almost as resistant to a flame front as are the old chemical foams. When extinguished with protein foam, fuel surfaces such as gasoline seldom reflash. Reflash, however, can occur with a severe mechanical disruption of the foam seal. In this case any open areas will be unprotected. Protein types are comparatively slow extinguishing agents because of their poor rate of spreading on burning liquids. Obviously, they are affected by their degree of mixing or emulsification with air; a small bubbled homogeneous foam is a more efficient and satisfactory extinguishing agent than a large-bubbled, poorly mixed, heterogeneous foam. Their efficiency is often related to the application method and expansion value.

Synthetic foam-forming agents

This large class of fire fighting foam concentrates resulted from the discovery in the early 1930s of the chemical synthesis of alkylated fatty alcohols as detergents and foaming materials. "Soapless soaps" of the sodium lauryl sulfate type were formulated with various thickeners and bubble wall strengtheners to produce fire fighting foams and wetting agents. Synthetic sulfonated compounds have also been used for this purpose. Present-day proprietary synthetic mixtures of this type are varied in composition. They are used to make high or low expansion foams depending on the generation device.

In general, these mixtures are stable in storage but they have a tendency to promote rusting because of their detergent action. They foam copiously but are easily washed off equipment.

These foams are low in viscosity and spread quickly over burning liquid fuel surfaces. Their principal deficiency is that water drains off rapidly, leaving them vulnerable to flame attack and rendering them less efficient than other air foams. This drawback is vividly demonstrated when reignition occurs, when most synthetic foams wilt rapidly under flame attack. Flammable vapor suppression of surfaces is only temporary under these conditions.

The synthetic agent foams are useful as white, heat-reflective coatings for vulnerable surfaces exposed to fire. They need to be continuously generated and spread but show an economy of water. Since they are generated from very low surface-tension water solutions, the water which drains from these foams penetrates quickly into materials into which water ordinarily wouldn't penetrate. This confers some

Class A fire fighting capability to these foams not possessed by the protein agents.

Surface film foam-forming agents

This class of foam concentrates for generating air foams is comparatively new. It was discovered by the author and his associates in 1962 while searching for new, more efficient foaming agents. Water solutions of fluorinated compounds of the following type chemical structure possess unique surface and interfacial characteristics when they are properly spread over flammable fuels:

$$[C_8F_{17}SO_2NH(CH_2)_3N(CH_3)_3]I$$

Even though these water solutions of fluorochemicals are heavier in density than flammable fuels, they will float in very thin layers over the exposed surface, restricting the evolution of a flammable concentration of vapors from materials such as gasoline. This occurs because of the capability of these compounds to "line up" or "orient" in water at the interface with a fuel where the "surface skin forces" of the fuel are in action. (A similar action occurs in the familiar floating of a thin steel razor blade on water. The "skin" of the water at the interface with the metal supports the area of the heavier razor blade.)

When air foam is generated with these chemical compounds in a water solution, the foam first acts in the same water-spreading and vapor-excluding fashion as other foams. But any water that drains from the foam bubbles or that results from foam breakdown, slides out from the foam mass, and quickly spreads over the flammable fuel. It aids in the extinguishing process, and is not wasted by sinking below the burning or freshly extinguished fuel surface. Fire fighting efficiencies of two to four times that of other foams are achieved with this "floating water" (hence the name "Light Water") action on fuels. Of course, as long as a reservoir of foam exists on the fuel surface, water drains out and continues to vaporproof the flammable fuel.

The formulation of surface film-forming foam concentrates is somewhat complex but in general these concentrates are water solutions of:

1. Active fluorochemicals which exhibit the necessary surface-tension lowering and orienting characteristics in conjunction with good foaming properties.

2. Foam bubble stabilizers which consist of water "thickeners," or agents which give longer life to the foam bubble wall.

In operating on Class B fires, the surface film-forming foams are our fastest and most efficient foam fire fighting agents, judged on a basis of the gallons of solution necessary to extinguish a number of square feet of burning fuel surface. These foams are comparatively short-lived however, in contrast to the much thicker and more viscous protein foams or the very tenacious and stiff chemical foams. The film of vapor-suppressing water solution which they spread on the surface of extinguished fuel is very thin but will "heal" itself when disturbed mechanically. It is the only foaming agent which exhibits this property.

The fuel fire extinguishing efficiencies of surface film-forming foams do not seem to be critically dependent on the expansion of the foam as is the case with other foaming agents. This operates in favor of poorly designed foam-making devices or substandard performance of equipment in a fire emergency.

The storage of the concentrates of this class of foaming agents is similar to that of the synthetic foam concentrates. Because all constituents in the formulation of surface film-forming concentrates are finished synthetic compounds which do not decompose or change with time, the shelf life of the liquid is practically indefinite.

An important factor to keep in mind is the fact that they are basically detergent in action and are powerful wetting agents. This means that they are capable of removing protective coatings of grease or oil from iron surfaces, promoting the rusting type of corrosion.

Chemical foaming agents

The older chemical foam powders are used today only at fuel tank installations of older design. Given proper maintenance and adequate periodic inspection these systems are reliable and can be expected to perform in an emergency when operated by trained personnel.

The foams from chemical foam-generating devices are very thick and viscous because of their aluminum hydrate or hydroxide content. This makes the foam very resistant to heat and flame but also makes it slow-spreading and capable of crust formation. When this foam is subjected to "baking," it may form "islands of foam" on the burning surface with subsequent open gaps of exposed fuel. Obviously, any fuel splashing on the surface of such stiff, unyielding foam masses will readily burn and prolong the extinguishment process.

One disadvantage of chemically produced foams is that the powders must not be allowed in contact with water in any form prior to actual usage. Metering powdered solids into a flowing water stream becomes difficult under the confusion and stress of fire fighting operations.

For these reasons and others, chemical foam materials and equipment have been largely replaced by the air foam materials, all of

FIRE FIGHTING FOAMS AND FOAM SYSTEMS

which are easily measured and proportioned and demonstrate superior fire fighting performance.

THE USES OF FOAMS

It is important from a fire fighting standpoint that foams be regarded as forms of water which are light in density and useful where the cooling and cohesive vapor-sealing properties of water are needed. All foams are generated with water and the physical properties of water are inseparably associated with the efficiency of the foam as a fire fighting material. In the paragraphs that follow, the alert student will note in each case that foam merely converts water into a useful form by varying its physical limitations so that it efficiently meets the fire situation.

Protection of flammable or combustible liquids (Class B fuels)

To prevent or combat fires in most of the ordinary flammable liquids or fuels, foams with specific gravities or densities of less than 0.7 are needed. (Water has a specific gravity or density of 1.0.) With the exception of liquid carbon disulfide (CS_2) all common flammable liquids have specific gravities between 0.7 to 0.95. The foams produced by any of the agents in the classes listed in Table 1 will have specific gravities from 0.1 (or less) to no more than 0.25. Obviously the light foam mass will float on the surface of the heavier flammable liquid. Since all flammable liquids produce flammable vapors at their burning surface and propagate flame from this source, the floating foam combats burning by first cooling the heated liquid. Foam then prevents the continued escape of vapor by leveling out over the entire exposed surface to exclude air or oxygen.

The most important use of foams is the protection of petroleum hydrocarbon fuels. Application by various equipment at or above the required rate allows the fire fighter to progressively and thus permanently extinguish the fire. If a large surface of fuel is exposed to nearby sources of ignition without becoming ignited, it may be protected from the fire by flowing foam over it, thus halting flammable vapor escape.

Protection of ordinary combustibles (Class A)

Foams may also be used in fire situations involving ordinary combustibles similar to wood, paper, or cloth. In fuels of this type, the water in the foam is required for cooling the surface of the burning solid so that combustion no longer continues and flames will cease.

Foam can often be applied to a burning vertical solid surface to facilitate water absorption before run-off can occur. A solid stream of water in such cases might be wasteful.

Many foams are made with water solutions of very low surface tension. Such solutions will be absorbed very rapidly and deeply into the pores of paper or cloth material and in this way fire is more rapidly extinguished with an economy of water. The synthetic and surface-film forming agents (Classes 2 and 3 of Table 1) demonstrate this property to the greatest extent.

There are certain combustible solid fuels such as rubber, many plastics and foamed polymers such as urethane mattresses, which melt and flow during the combustion process, thus forming a mixed Class A and Class B fire. Foams of all types are very useful in such situations. Both their cooling and surface floating characteristics come into play to efficiently extinguish the flames.

Other uses for fire fighting foams

The water-holding capability of foam is sometimes employed for reducing the spark hazard when an aircraft must land on hard-surfaced runways without normal landing wheel gear. The foam is usually put down many minutes before the aircraft contacts the runway, and acts as a stabilizing matrix for holding a fairly thick layer of water drained from the foam on the runway surface. When the aircraft touches the concrete runway through the foam, the water layer dampens and retards spark production from the metal friction contact with the concrete surface.

The white, heat-reflecting surface of foam protects exposed surfaces from radiant heat during fires. Where adjacent structures are burning fiercely, foam may be applied to exposed vertical surfaces and some measure of heat protection may be achieved as long as the white, water-containing layer of foam remains on the surface.

An important feature of high expansion foam of the synthetic foam-forming class of agents (Class 2, Table 1) is its ability to fill cavities with large volumes of foam that exclude air from the fire. When fires occur in large cavities or rooms, a mass of foam may be generated which slowly diminishes the volume of air in the room. This halts air circulation and diminishes the amount of oxygen needed to sustain the fire. Any foam broken down by the heat lends its water to cool the combustibles. Any steam produced further reduces the amount of air and oxygen available to the fire.

A very important role for foam is in fire prevention rather than in actual fire fighting operations during flame propagation from fuels.

Foams may be used to form semi-permanent vapor blankets or combustible vapor-restricting layers when they are spread on the surface of flammable liquids. Where flammable fuels are accidentally spilled or a large liquid surface becomes vulnerable to ignition, the emission of combustible vapors from the exposed, unignited surface may be halted by spreading a layer of foam over the surface of the fuel. As long as the foam layer is not allowed to open up or dry out, the fuel is safe from any inadvertent ignition of its vapors.

Problems in the use of foams

Even though foams are lighter in density than certain liquids which quickly become flammable gases at ordinary ambient temperatures, they cannot be relied upon to extinguish fires in all such materials or protect them from ignition. Gas and flammable vapor evolution from liquids like pentane and butane cannot be restrained by covering them with foam.

Unless they are especially compounded for the purpose, foams are highly vulnerable to rapid degradation by water-soluble flammable liquids such as the alcohols and acetone, and by polar liquids such as methyl ethyl ketone and isopropyl ether. Foam cannot be relied upon to protect these materials from fire.

Foams are rather delicately balanced physical systems. They may be easily broken down or collapsed by contact with very small amounts of many materials due to rupture by interfacial tension forces at the "skins" of the bubble. Some of the substances which will quickly collapse foams are:

castor oil	acid solutions
turkey red oil	acidic gases
cutting oils	dry chemical powders (incompatible type)
non-petroleum paint solvents	gases of combustion from most solid fuels
furnace soot	certain wetting agents
settling basin waters	certain chemical salt solutions

Obviously, efforts must be directed at all times toward maintaining the high purity of water supplies for the generation of foam.

Solid hose streams of foams are capable of conducting electricity in much the same manner as water streams do. Fires involving charged electrical equipment should be fought with dispersed or wide-pattern foam discharge streams.

Since foams contain varying proportions of water, they should be cautiously used on fires in tanks of heat-wave-producing oils or liquids. When flammable liquids, such as crude oil or partly refined black fuel oil, have been burning for sufficient time to generate a hot oil layer

FIRE SERVICE HYDRAULICS

at the top of the bulk of fuel, the addition of foam will cause a frothing action on the surface and a mixture of steam and burning frothing oil may be violently ejected from the tank. This may be avoided by slow and careful application of foam until "frothing-over" has stopped, indicating successful cooling of the hot oil layer.

The efficiency of each class of foams for extinguishing fires in flammable liquids is governed by the method by which the foam is applied to the burning surface. In succeeding portions of this chapter the careful student will keep this uppermost in his considerations of fire fighting foam.

FOAM GENERATION METHODS

Air aspiration

A large percentage of the devices in general use for making foam are of the portable nozzle aspirator type. These foam generators utilize the kinetic energy of a jet (or jets) of foam-making solution in a crudely constructed venturi. Air is drawn into a mixing tube or chamber that produces a high degree of turbulence. Close contact of the foaming solution with the air results in the formation of small bubbles and foam (Figure 1). The amount of air entrained depends on the hydraulics of the nozzle or aspirator design and the physicochemical nature of the foam-making solution. In general, the synthetic and surface-film foam-forming agents of Table 1 produce more copious (higher expansion value) foams in such devices than do the protein agents. Chemical foaming agents are not employed in nozzle aspirators because they require a mixing action of chemicals and water. Incorporation of air is not necessary.

Usually, nozzle-aspirating foam makers are provided with streamdispersing devices at the outlet end of the foam-mixing section. These may take many different forms but essentially they "spoil" or divert the solid stream of foam into a wider dispersed pattern which may be essential for gentle application of foam to a liquid surface without splashing or submerging the foam. This will be more extensively treated in a later section.

Air aspiration foam makers are also available for generating foam within a system which requires that the foam possess a residual pressure after it has been formed so that it may be forced against a pressure head to the upper surface of the fuel. These are called "pressure foam generators," "high back pressure foam makers" or "inline, forcing

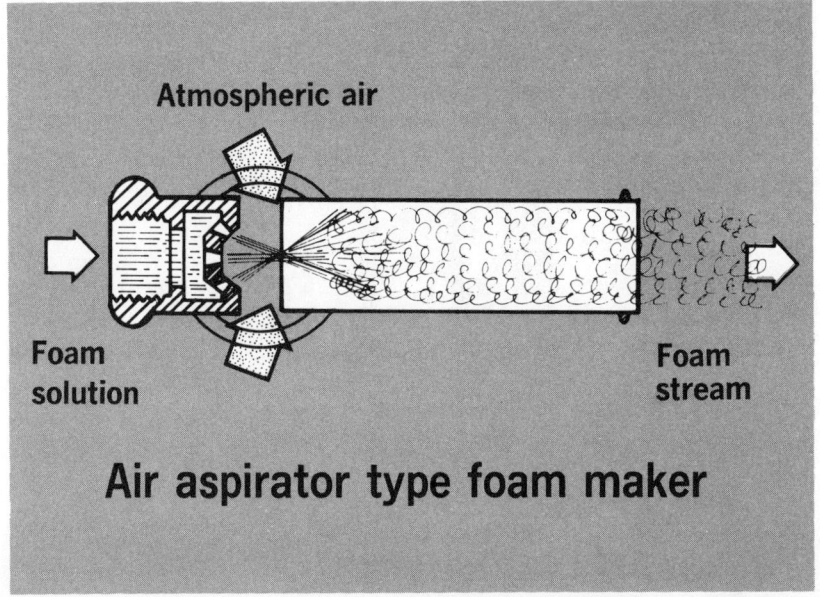

Figure 1

type" generators. They are used mainly for subsurface foam injection systems which will be dealt with in a later section.

The optimum performance characteristics of ordinary nozzle aspirating foam makers require that solutions under pressure be supplied to the nozzle in the range of 80 to 120 psi. The foams generated by them will be in the expansion range of 4 to 11 times the solution volume (4× to 11×) supplied to the nozzle, depending on its design.

The pressure foam aspirator utilizes higher solution pressures from 100 up to 300 psi, and recovers approximately 25 percent of its inlet pressure in residual foam pressure. The expansion of foam in this type of generator will vary from 2.0 to 4.5 depending on the pressure head which it must overcome or which is imposed on it by the distribution system downstream of the inline foam generator.

Variations of air aspirating foam makers include one which is used as a sprinkler head (foam-water) capable of forming a fluid, low expansion foam (about 3 or 4) when foaming solution is supplied to it. Or it may act as a water sprinkler head if the supply of foam concentrate should be exhausted and water only were available (Figure 2). These devices use pressures of about 30 psi at the nozzle and depend on the turbulent action of the solution in the foam-water head for foaming

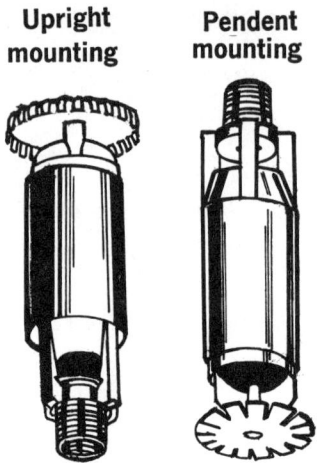

Foam maker sprinklers

Figure 2

action. They are installed in the manner of sprinkler systems for protection of large open areas such as aircraft hangars, etc.

The production of high expansion foam from synthetic foam-forming agents may also be accomplished by an aspirating action in the nozzle of Figure 3. In this nozzle the solution is forced through the interstices of a fabric so that each opening in the weave of the fabric acts as a single bubble producer (much like a wire ring dipped in bubble solution, and then blown upon). A copious stream of bubbles then issues from the nozzle outlet at low velocity and with expansions of $80\times$ to $250\times$.

Foam pump

Research and technological development advanced in the 1940s and early 1950s in the application of air foams to flammable liquid fires. It was found that certain hazards required air foams of a more exacting and constant composition than could be produced by the nozzle air aspirators described in the previous section. To control and extinguish some fuel fires with a minimum of foaming material and manpower, the variability of foams produced by aspirating-type foam makers could not be tolerated. A more exact metering of air and solution yielding a more homogeneous foam was needed.

FIRE FIGHTING FOAMS AND FOAM SYSTEMS

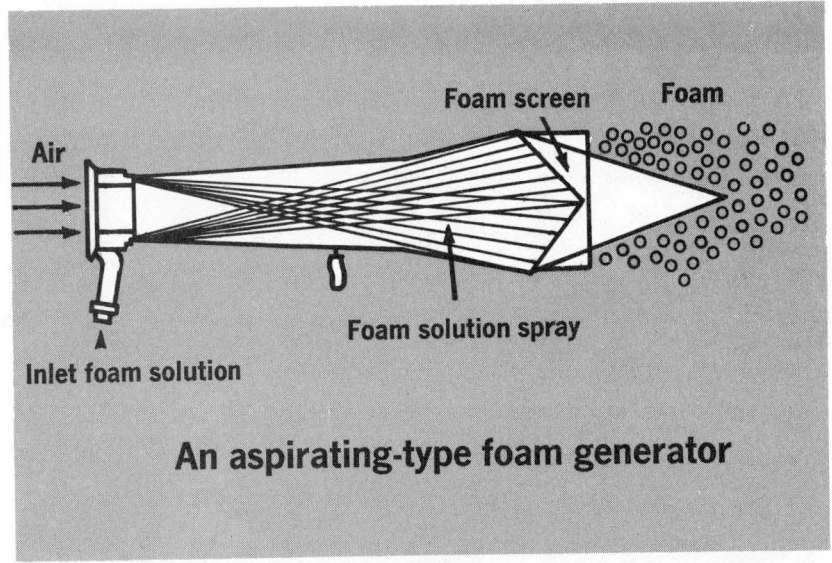

An aspirating-type foam generator

Figure 3

Several types of "constant-volume" foam-making devices were developed to fit this need. The most useful of these was the "foam pump," which is still widely used.

The foam pump (Figure 4) is a positive-displacement rotary pump with sliding plastic vanes sweeping out a constant volume that is generated by off-center operation of a slotted bronze rotor which houses the vanes. In operation, the foam solution is metered into the inlet of the pump at a constant rate governed by an orifice in the supply piping. This inlet is open to the atmosphere and is sufficiently large to serve also as an air intake to the pump. As each blade or vane sweeps across the opening, a definite volume of air is inducted. The air, with the previously determined volume of foam solution, is transported to the discharge port or outlet opening of the pump where it is subjected to a pressure dependent on the system design. As the foam is progressively "squeezed" out of the pump by the decreasing pump cavity volume at the discharge port, a bubble refining action takes place under the pressure. This action is further enhanced by the turbulence in the screen section through which the foam bubbles must pass.

The foam pump is a positive-displacement type, i.e , each revolution "displaces" or transports a fixed volume of air and water—or both—from the entry port to the discharge port. Because of this, the ratio

FIRE SERVICE HYDRAULICS

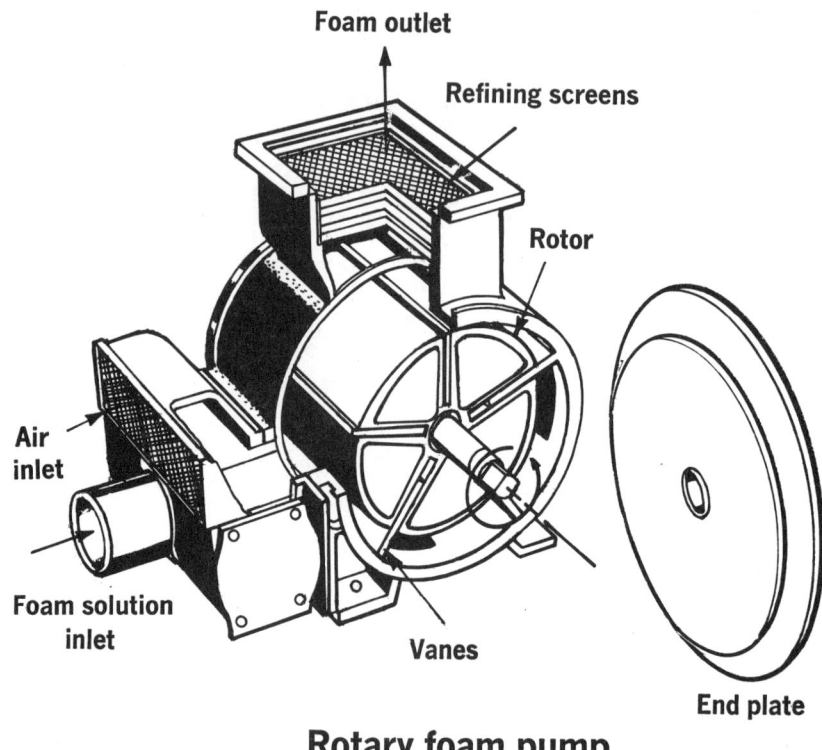

Rotary foam pump

Figure 4

of air to foaming solution may be governed by holding constant the volume of solution fed to it while maintaining a constant rpm of the pump. Thus the air volume inducted at the pump inlet remains constant.

In practice, these pumps supply foam to truck-mounted turret nozzles for air crash fire fighting needs. At 1200 rpm, with a 250-gpm foaming solution input rate, they will generate 3000 gpm of foam with an expansion of 12× at distribution nozzle pressures of about 14 to 18 psi.

Air blower (fan)

The generation of high expansion foams in the region of 1000× expansion requires that a very large volume of air be made available

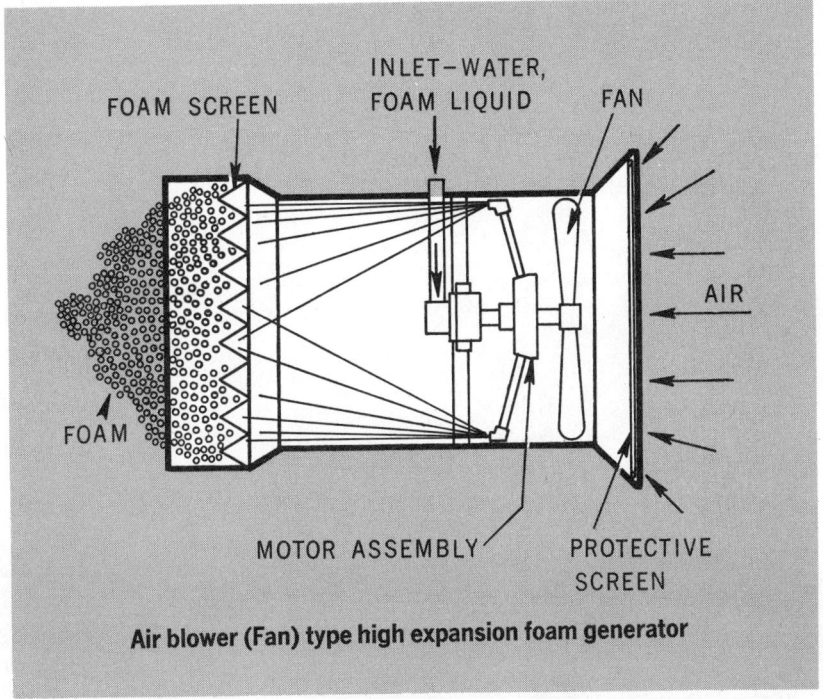

Air blower (Fan) type high expansion foam generator

Figure 5

for mixing with a very small volume of foaming solution. High-velocity fan blowers are used to produce these air volumes. The foam is made by continuously wetting or spraying the small interstices of a loosely woven metal screen or nylon fabric net. This is done in such a manner that each opening forms a film of synthetic foaming solution. As air blows upon the film it forms a bubble which blows through the net and the total action yields very copious volumes of small-bubbled foam (Figure 5).

The high expansion (1000×) foam from this generator may be delivered through large-diameter ducting constructed of fabric or some flexible material. The fan-type generator may also be installed in close proximity to the hazard, provided that vapors or smoke from the fire do not interfere with the foam making. This foam is exceedingly vulnerable to surface contact, or chemical action (a surface problem of bubble incompatibility), and mechanical or wind damage. Most high expansion foam generators develop pressures sufficient for flooding several elevated stories of a building, provided that no large openings exist which offer paths of leakage.

Chemical foam generation

Ordinary chemical foaming agents consist of solids which must be dissolved in water prior to the chemical reaction which produces foam. These agents require an entirely different type of foam generator.

They have almost completely replaced the older two-solution chemical foam system (wet storage system) that used large tanks of dissolved chemical.

There are two principal types of packaging for dry powdered chemical foam materials. One is an "A" and "B" two-container (dual powder) type. Each container or pail holds the requisite amount of the two chemical powders necessary for the production of foam. The "A" composition consists of aluminum sulfate; the "B" composition, sodium bicarbonate plus a foaming agent or stabilizer.

The other is the "A" and "B" single-container (single powder) type, wherein both dry reactants are supplied already mixed in the same pail so that adding water results in an immediate foaming reaction. Needless to say, the latter type must be kept absolutely dry and tightly sealed until used.

The chemical foam generation reaction which produces carbon dioxide-filled foam bubbles coated with a slurry of aluminum hydroxide is as follows:

$$\text{"A"} \quad \text{"B"}$$
$$Al_2(SO_4)_3 + 6NaHCO_3 \rightarrow 3Na_2SO_4 + 2Al(OH)_3 + 6CO_2\uparrow$$

(13 percent aluminum sulfate + 8 percent sodium bicarbonate (containing 3 percent foaming agent + 76 percent water)

To produce this foam the dry powdered solid "A" and "B" chemicals are emptied from their containers into a hopper or hoppers shaped like an inverted cone. They then flow slowly through the small opening at the bottom of the cone into a water stream from a hose supplied by a fire pump (Figures 6 and 7).

Foams so generated show expansions of $10\times$ to $18\times$ at ordinary (50° to 70°F) temperatures. (Since chemical foam generation is a chemical reaction, its reaction rate is temperature dependent.) One pound of powder (A + B) produces 7 to 12 gallons of foam at water "thruput" rates of 60 to 75 gpm. Both types of generators require 75 to 125 psi water pressure. Hose, nozzles or piping connected to

Figure 6

the outlet side of the generator must not impose foam outlet pressures in excess of 40 percent of the inlet water pressure.

There are intrinsic difficulties connected with the process of proportioning and completely solubilizing solids into flowing water streams in the chemical foam generator. Problems attendant to maintaining a continuous flow of powder in the hopper, if lumps of material are encountered or water backs up in the hopper, make the process questionable in operation at a fire emergency. As a consequence, this foam is being replaced by modern air foam.

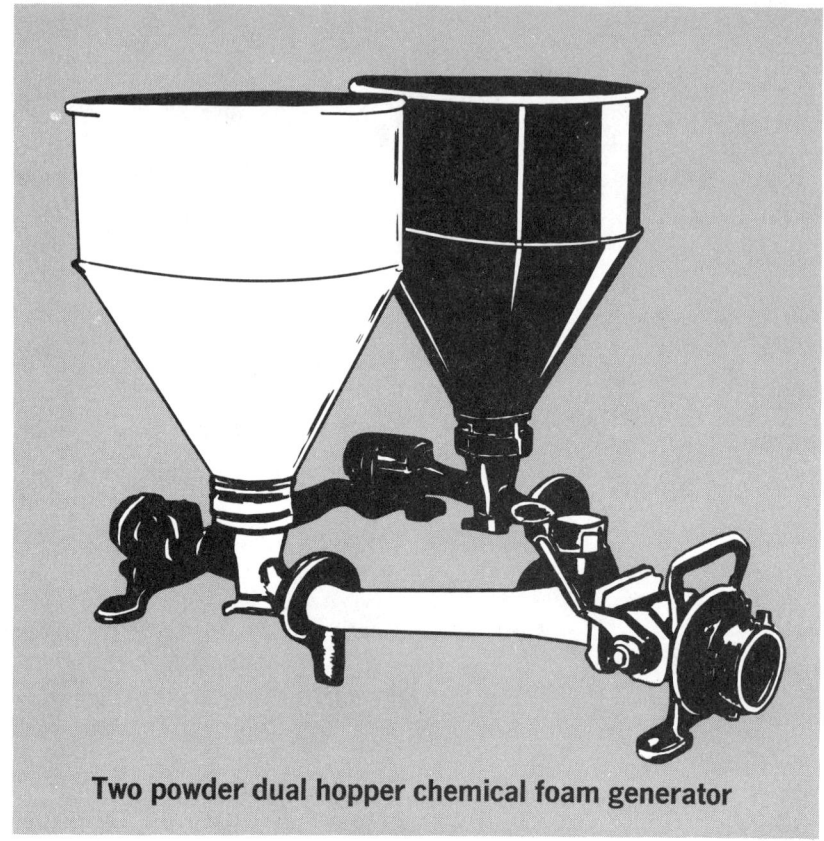

Figure 7

PROPORTIONING DEVICES

All air foam generating systems require admixture of a concentrated liquid foam-forming composition with water at some point prior to the generation of foam. Early systems employed premixed solutions in storage tanks so that the foam-making water solution was ready for supplying the foam-generating device or nozzle. Storage difficulties were encountered and except for small portable "batch-type" pressurized systems, the premix storage of air foam solutions is no longer used. Present-day systems utilize liquid volume concentrate proportioning devices such as those described in the following section.

Nozzle eductor*

The portable air-aspirating nozzles shown in Figure 1 are the most widely used type of equipment for foam-producing purposes. They are quickly and easily put into action and require only that water under pressure be supplied to the nozzle. In order to make foam they must

* Considerable confusion has occurred over the years concerning terminology applicable to those devices which utilize the familiar water jet ejector venturi action for their operation. The mechanics of such designs involves the drawing out or *educing* of a substance from one place and putting it in another place or duct. The substance is introduced or *inducted* into the desired place or duct. From this it may be seen that both an *eduction* action and an *induction* action occur simultaneously. In this text the use of *eductor* and its *eduction* action has been arbitrarily selected as the terminology most fitting the circumstances involved.

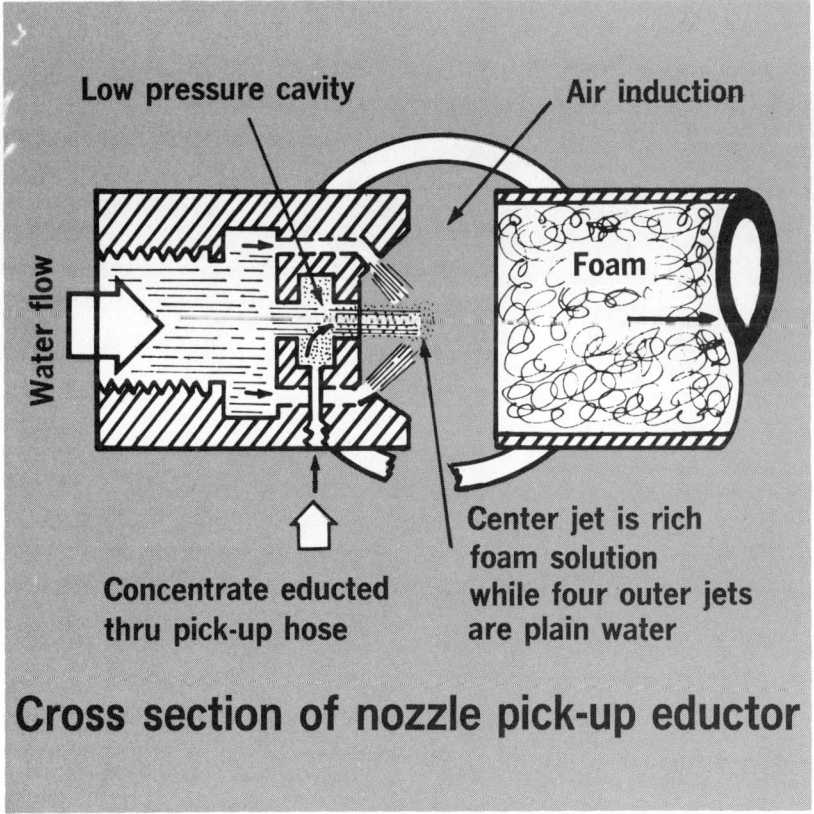

Cross section of nozzle pick-up eductor

Figure 8

FIRE SERVICE HYDRAULICS

also be supplied with foam concentrate and a means for mixing it with the water stream. This is done with a built-in venturi section within the nozzle. Figure 8 illustrates this type of construction. As water is forced through the nozzle, the venturi or jet action of the flowing stream creates a negative pressure. This negative pressure is conferred to the tube passing down to the source of concentrate and the liquid is educted into the nozzle. Here the concentrate mixes with the water in the correct proportions governed by the orifices in the venturi section of the nozzle.

This type of proportioner is relatively foolproof and easily kept clean. If the screen at the bottom of the concentrate pick-up tube is not clogged by particles in the concentrate container, it will function continuously as long as concentrate is supplied to it. This latter operation is often a critical one due to difficulties in estimating the degree of exhaustion of liquid concentrate in a partially open 5-gallon

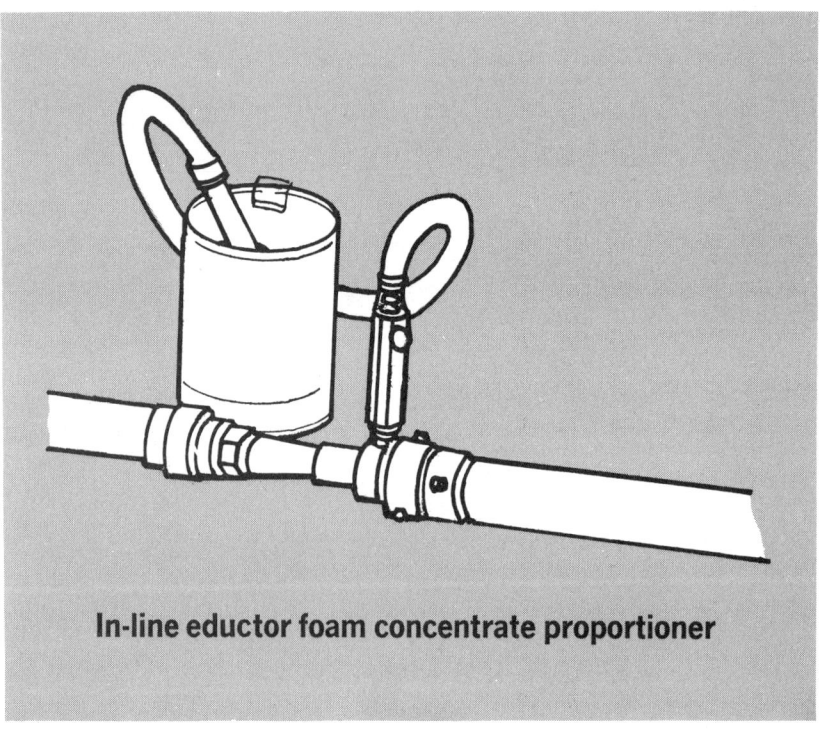

In-line eductor foam concentrate proportioner

Figure 9

pail. Care must be exercised to keep the pick-up tube continuously immersed.

"In-line" eductor

Under certain circumstances it may be desirable to introduce foam concentrate directly into the water line which supplies the foam maker. This involves placing an eductor or venturi section "in-line" at a point upstream from the foam-making device. Figure 9 illustrates this type of eduction apparatus. It may be connected to a tank containing foam concentrate. Or it may be provided with a transferable flexible pick-up tube for use with 5-gallon containers.

The in-line eductor imposes a pressure drop of about 30 percent on the system water supply. It is usually designed with specified lengths of hose or pipe between it and the foam maker. It cannot be used with foam makers having variable or open-and-closed valve operation at the nozzle or foam applicator.

The "around-the-pump" proportioner

The most satisfactory and flexible-in-operation foam concentrate proportioning devices are those which operate under positive flow conditions that utilize pump pressure for injection of the correct volume of concentrate. The around-the-pump scheme of injection of liquid provided an early method for accomplishing this. Figure 10 illustrates the flow method by which this is done. A small portion of the pump discharge flows through a bypass line to the suction side of the pump. An eductor in this line produces a negative pressure on the suction line leading from the foam concentrate container. Foam concentrate is thus drafted into the suction side of the pump. A multiported metering valve calibrated for the desired flow of solution to single or multiple foam makers is usually placed in the foam concentrate suction line.

The positive flow proportioner requires that the volume delivery of the pump be adequate to allow a bypass flow of from 10 to 40 gpm of water in the concentrate educting section. Variation in bypass volume is dependent on variation in concentrate educting volume requirements and pump discharge pressure.

To operate correctly, the pressure at the suction side of the pump must always be on the vacuum or negative pressure side. A positive gage pressure at this point could result in reversed flow of water into the concentrate container. As is always the case with venturi eductor devices, the maximum lift required for drafting of concentrate from a container must not exceed 6 feet at any time.

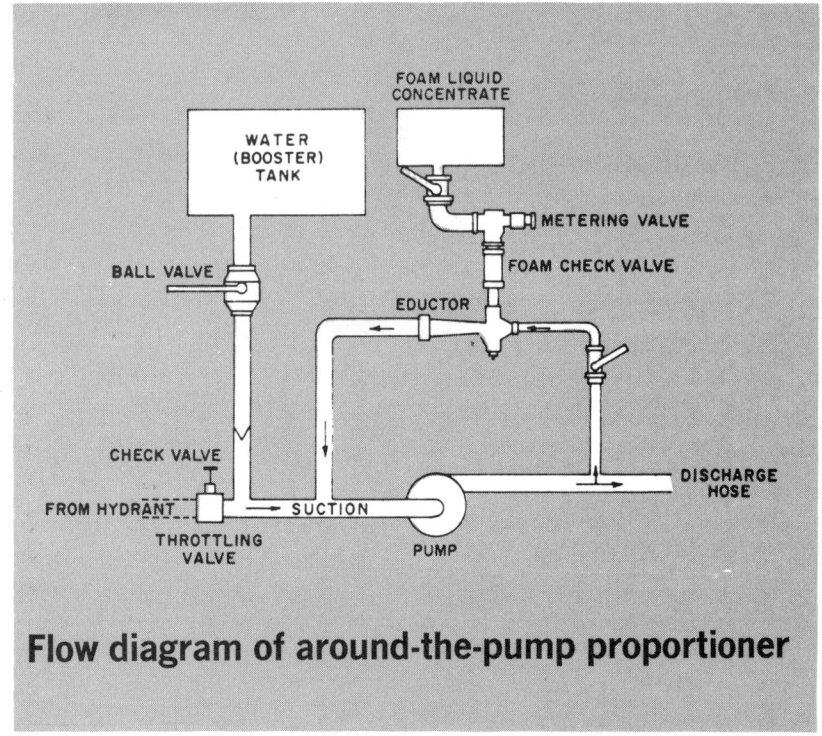

Figure 10

Foam concentrate pump proportioners

In fixed-installation foam-making systems the variation in flow demand for a foam-forming solution must be considered in the design of the system. A separate pump (used exclusively to supply foam concentrate under pressure to the water mains supplying the system) is often chosen for proportioning purposes. Design of such a proportioning system may take several different forms:

Type 1. A fixed-pressure, fixed-metering orifice, constant-ratio-type pump pressurized proportioning system.

Type 2. A balanced-pressure, variable rate-of-flow eductor, demand-type pressurized proportioning system with automatic pressure-control valve or duplex gage.

Type 3. A balanced-pressure, variable-metering orifice, variable-flow, demand-type pump pressurized proportioning system.

FIRE FIGHTING FOAMS AND FOAM SYSTEMS

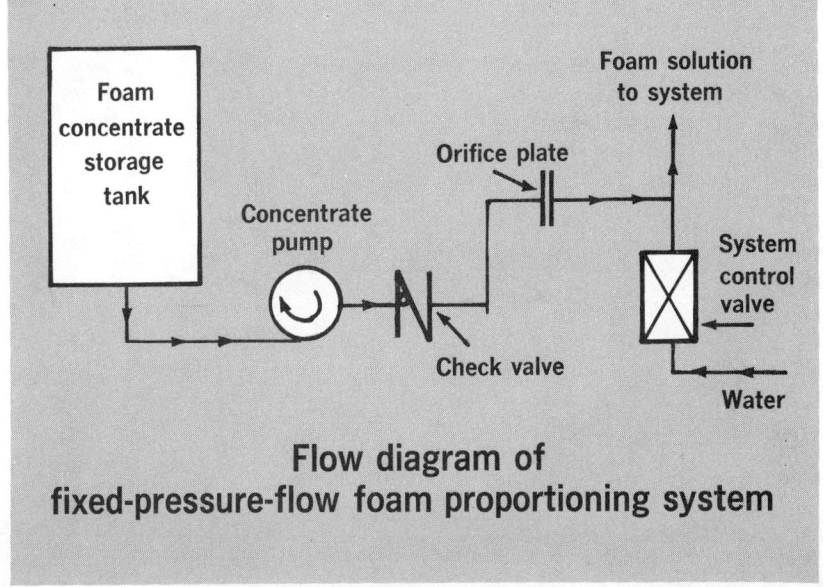

Flow diagram of fixed-pressure-flow foam proportioning system

Figure 11

Figure 11 is a flow diagram of Type 1—the fixed-pressure system—which is designed for fixed-flow rate proportioning used in a fixed foam-making system such as a foam-water sprinkler grid of a specific size and capacity.

Operation of such a proportioner requires only that a constant pressure of foam concentrate be supplied by the pump. The orifice plate—downstream of the valve that opens the system—then governs the constant flow volume of concentrate into the water line leading to the system foam makers.

Figure 12 is a flow diagram of the Type 2 pump proportioning system that uses automatic pressure control. A variable rate of flow is possible with this type of control.

In this system a relatively constant pressure drop exists between the upstream and downstream sides of the proportioner. As the flow increases, the pilot line to the valve diaphragm that is hydraulically connected to the pressure control valve closes the foam concentrate pump bypass valve. Therefore less foam concentrate is bypassed back to the storage tank and more liquid is forced through the metering orifice into the foam solution input line and into the foam-making system.

FIRE SERVICE HYDRAULICS

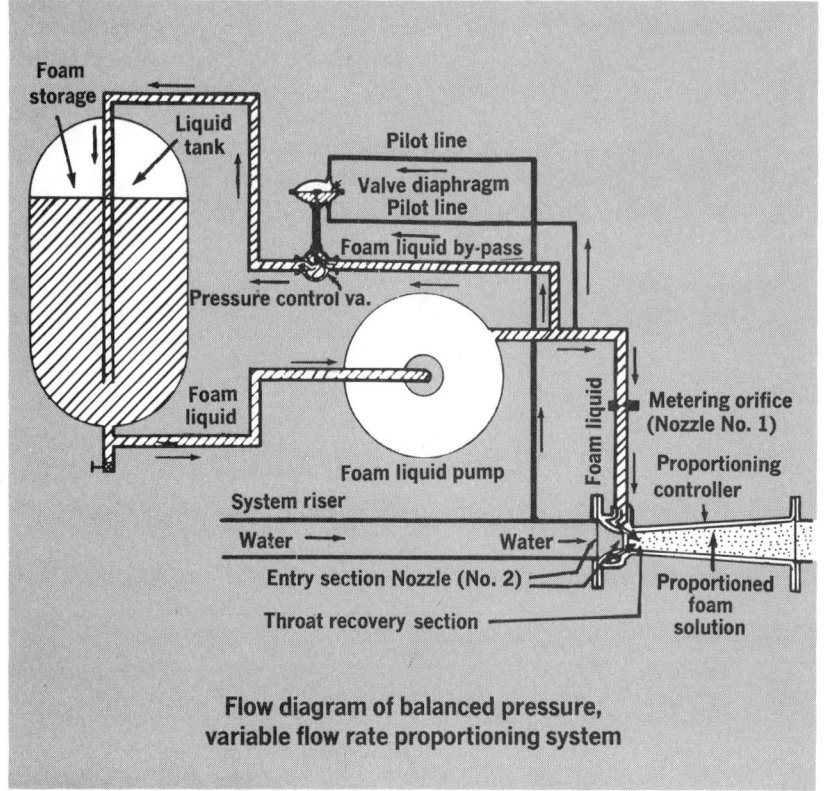

Figure 12

This type of "demand" proportioning system is capable of operating with accuracy over wide ranges in flow, such as 20 to 1 between maximum and minimum. In other words, with a single unit, accurate proportioning could be obtained with total solution flow rates of from 60 to 1200 gpm. Another type of demand proportioning system is illustrated in Figure 13. Here manual control is necessary to maintain concordance of pressure readings on a duplex gage that has two indicating needles. To maintain foam concentrate percentage, the operator matches the pressure readings on the upstream side of the venturi inductor with the foam concentrate discharge pressure by opening or closing the pump bypass line.

The range and other characteristics of this system are similar to the automatically controlled Type 2 system.

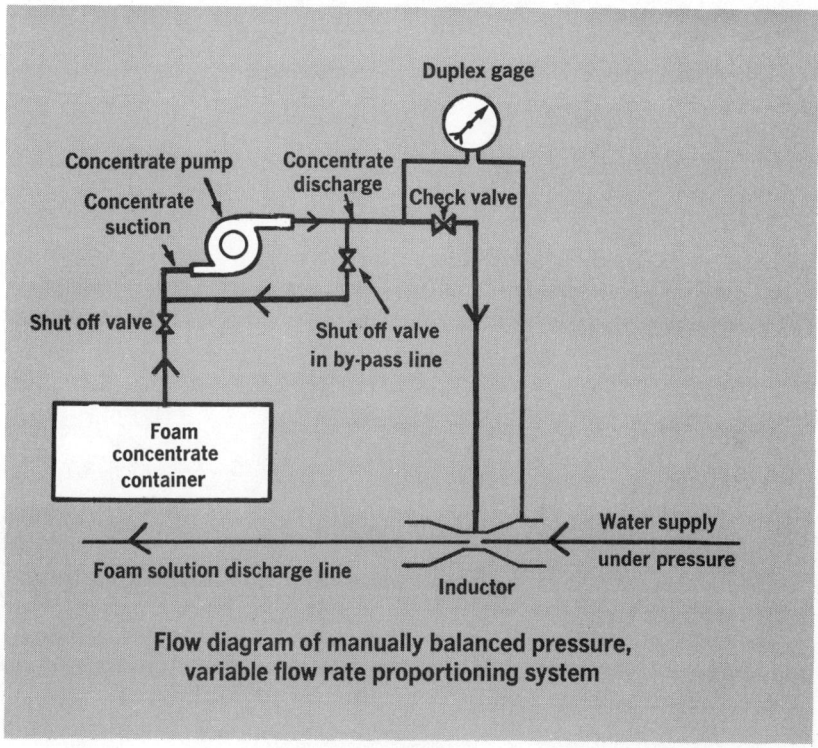

Figure 13

Figure 14 is a schematic flow diagram of the variable-orifice, balanced-pressure type of variable-flow-demand proportioning system.

Referring to the diagram of the variable-orifice device, it will be seen that the flow of water displaces the piston to the right against the pressure of a spring. As flow increases, the piston moves further to the right so that a constant pressure drop continues between the inlet and outlet of the waterway. The movement of the piston also moves the concentrate inlet opening to a larger or smaller metering port.

Obviously, accurate proportioning depends on identical input pressure of the water and the foam concentrate. This may be accomplished by using automatic pressure-regulating valves and pumps or by providing the storage tank of each fluid with identical hydraulic head pressure.

Satisfactory proportioning of this system—±16 percent variation—is from 5.2 to 6.8 percent concentrate, obtained with ranges of from

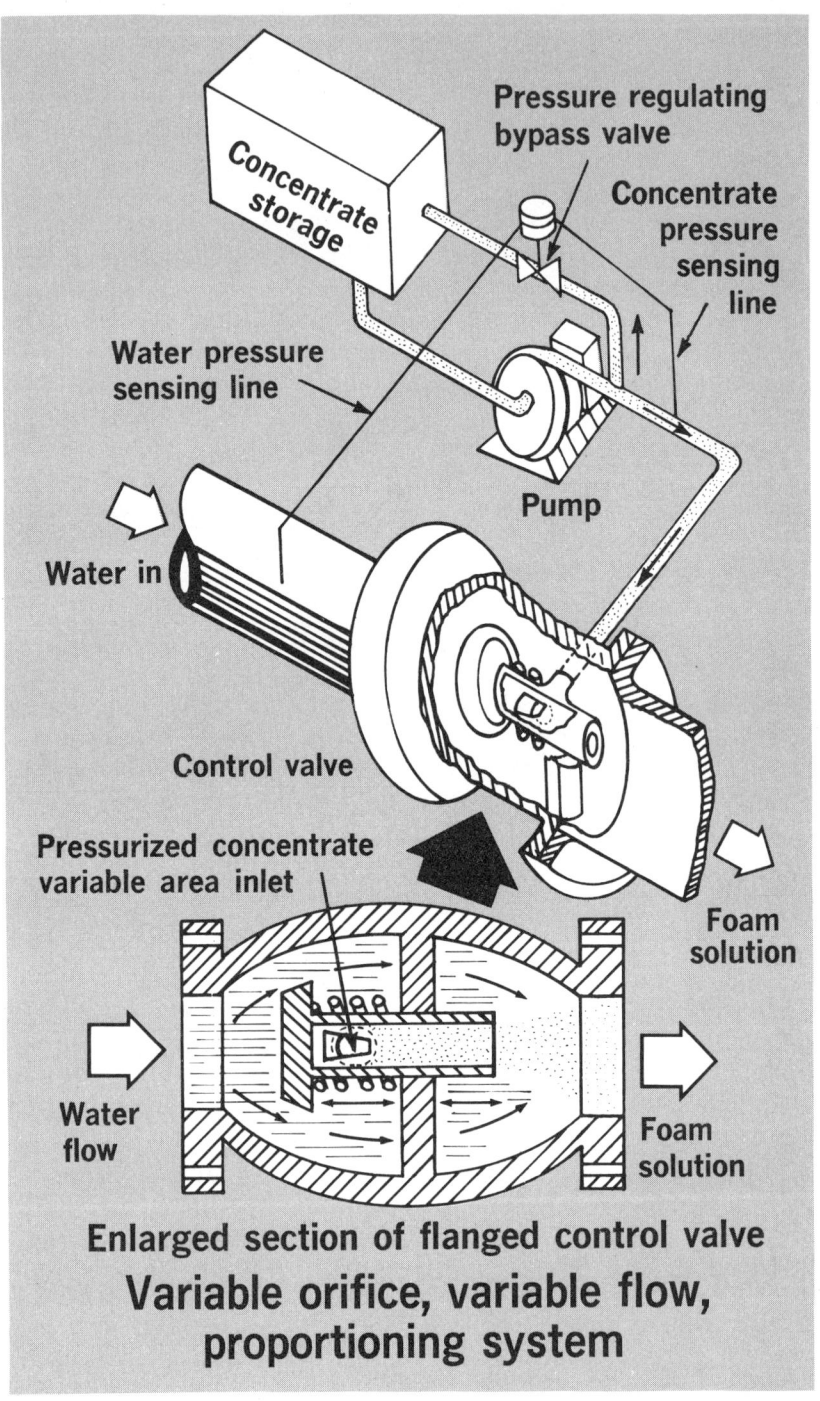

Figure 14

FIRE FIGHTING FOAMS AND FOAM SYSTEMS

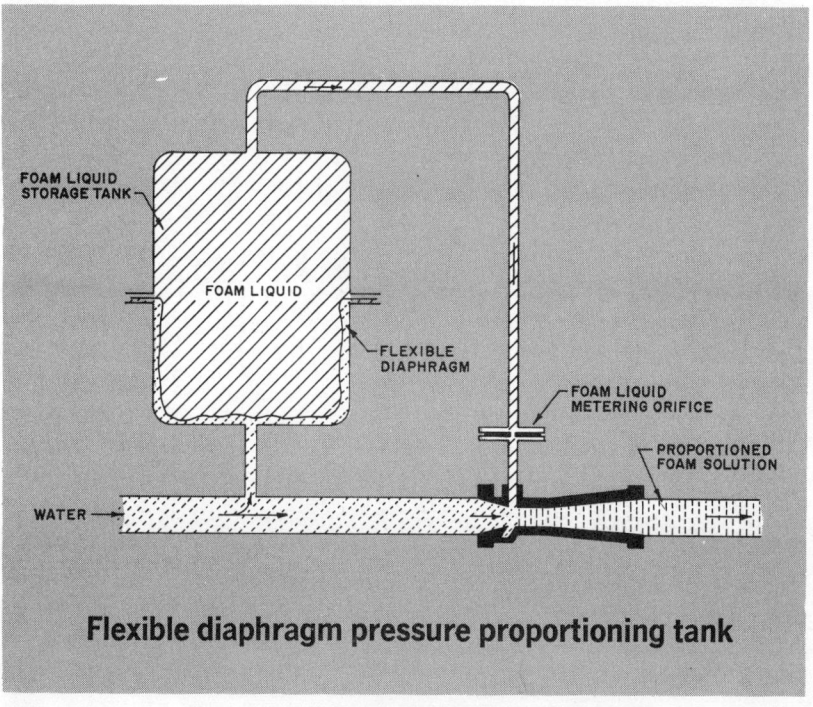

Figure 15

three-fourths to one and three-fourths times the rated flow of the system.

The pressure-proportioning tank

An interesting type of foam-concentrate proportioning is represented by the flow diagram of Figure 15. This system depends on the displacement of foam concentrate in a closed tank by water taken from the supply line of the foam-making system. As water enters the tank, the flexible diaphragm transmits pressure to the foam concentrate contained in the tank (without admixture with water). This causes the contents of the tank to pass through the metering orifice and into the venturi eductor section in the system water supply line.

This proportioner has a limited range of operation and suffers from problems of resupply. It is a "batch" type operation. The periodic refilling of the tank requires drainage of water, refilling with foam concentrate and repositioning the diaphragm before re-use of the system.

FIRE SERVICE HYDRAULICS

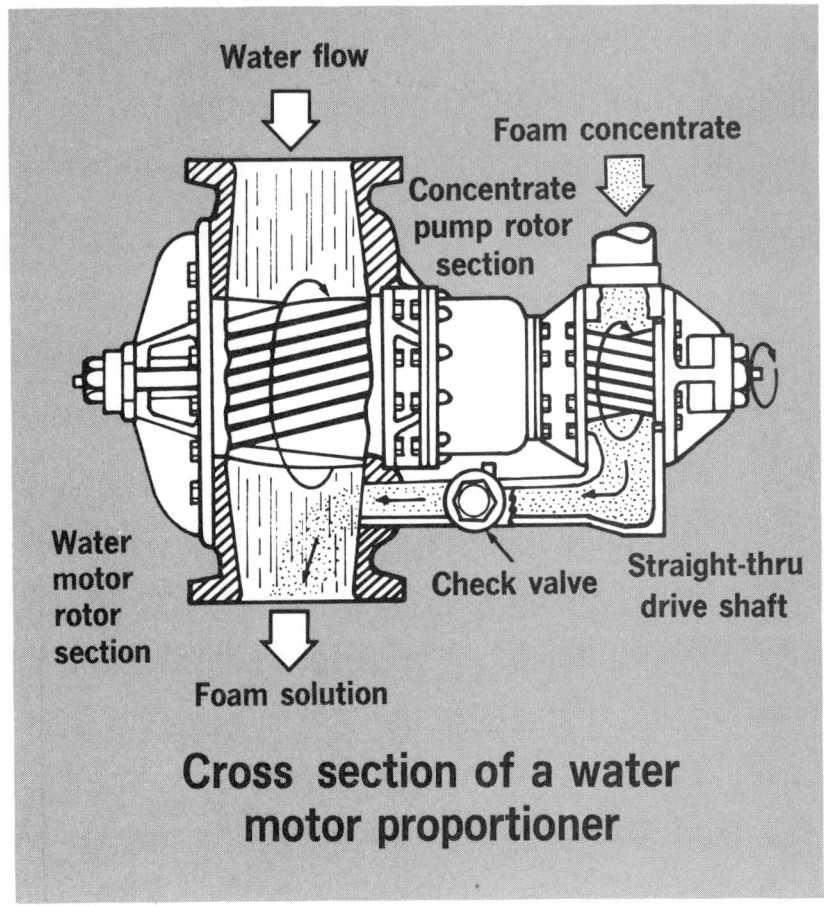

Figure 16

The water-motor proportioner

In positive-volume proportioning devices, a positive-displacement water-motor proportioner is geared to a positive-displacement foam-concentrate pump. The arrangement is unique in that it is not dependent on external pressure or power for successful metering by volume. The cutaway drawing of Figure 16 illustrates the construction and fluid flow of this system.

Water delivered under pressure to the inlet of the large positive-displacement, synchronous-lobed pump causes it to act as a motor with rpm in proportion to water volume "thruput." Connected to the shaft of this motor is a smaller positive-displacement, synchronous, multi-lobed pump. Its inlet is connected to a source of foam concentrate and its outlet to the outlet of the water motor, which is the system. Variation in speed of the small pump (and thus variation in volume pumped) is governed by variation in speed of the larger-volume water motor. With ordinary commercial units of this device the relationship is for proper proportioning of 6 percent foam concentrates (6 parts by volume concentrate in 94 perts water). Hence, the volume displacement of the small pump to the large motor is arranged to be 1 to 15.6.

This type of proportioning system requires no external power input for operation. But the process of pumping the foam concentrate from 0 gage (or below) to the fluid system pressure requirements of the downstream side of the water motor imposes a pressure drop of 25 percent. This is at the maximum flow of each of the two sizes of proportioners now being manufactured (ranges of 60 to 180 gpm and 200 to 1200 gpm). The accuracy of these devices is very close to the required 6 percent volume concentration at all operational ranges.

FOAM SYSTEMS

Portable nozzle aspirators

Perhaps the most widely used foam device is the portable, hand-operated nozzle-aspirating foam maker. It is called a "foam tube" or "foam branchpipe" in Europe and other countries. When attached to a 1½-inch hose line or even a 2½-inch hose, it can be quickly put into action. The foam issuing from the nozzle in the form of a solid stream or in a wide, dispersed pattern can be easily and efficiently directed at the burning surface and then moved forward to consolidate gains made in extinguishing the fire.

When foam is applied to a burning fuel surface with a nozzle of this type, the solid-stream pattern can be used to provide "reach." The solid-stream pattern can also be directed to a vertical surface near the fire. This permits the foam to flow down gently, contact the burning fuel and spread over it.

The splashing or submergence of foam into the burning fuel will cause the liquid to coat the top surface of the foam and destroy it. For this latter reason, the widely dispersed foam pattern is more desirable for rapid fire control, but, of course, reach and projection are compromised. In most cases the gentle, wide-area, dispersed-pattern coverage foam produces maximum extinguishing efficiency and rapidity of control of flame and heat.

The use of portable foam applicators requires that the operator carefully watch the progress of his surface flame control. He must move or even oscillate his stream or pattern to utilize foam to its greatest advantage. And when flames cease to exist he must realize there is nothing to be gained, in most cases, in building up thick layers of foam on the extinguished fuel.

Portable nozzle-aspirating foam makers are very useful for combatting all types of flammable fuel fires such as ground spills or large burning surfaces not elevated above the level of the operator of the nozzle. Visual target control of the foam is a requirement of the method. Burning fuel dropping from an elevation above the operator is difficult to extinguish with streams or patterns from ordinary portable foam nozzles.

Nozzles are available in 45, 60, 120 and 250-gpm sizes with operating pressures of from 60 to 100 psi. They may be provided with pick-up tubes for aspirating the correct proportion of foam concentrate into the water stream passing through the nozzle. Proportioning may also be accomplished by other means upstream from the nozzle itself.

Fixed foam makers and distributors

One of the most important categories of fire problems that employ foam as a first line of defense is the large-volume storage of flammable fuel in outside tanks. These are called tank "farms" when more than two tanks are located in a single area. Petroleum refineries and bulk transfer depots use such installations to hold and store large liquid quantities prior to shipment or use.

Storage tanks for ordinary flammable liquids are of two principal types: cone-roof and floating-roof tanks. Each requires certain fire protection systems. For the cone-roof tank, foam fire protection systems are usually mounted integrally with the tank construction. The floating-roof tank depends on portable fire fighting equipment, principally because of the its lesser incidence of ignition and fire.

There are many different types of equipment for introducing foam to burning surfaces of flammable liquid confined in a storage tank. Each of them has particular advantages.

Tank-side Fixed Foam Makers: The most widely chosen foam fire protection for ordinary cone-roof steel fuel storage tanks over the past 20 years has been the aspirating foam maker which is installed at the top rim of the tank shell. There are several proprietary designs for such fixed foam makers. Figure 17 illustrates a typical installation.

Regardless of their varying designs, each foam maker consists of a nozzle and modified venturi section for aspiration of air that is con-

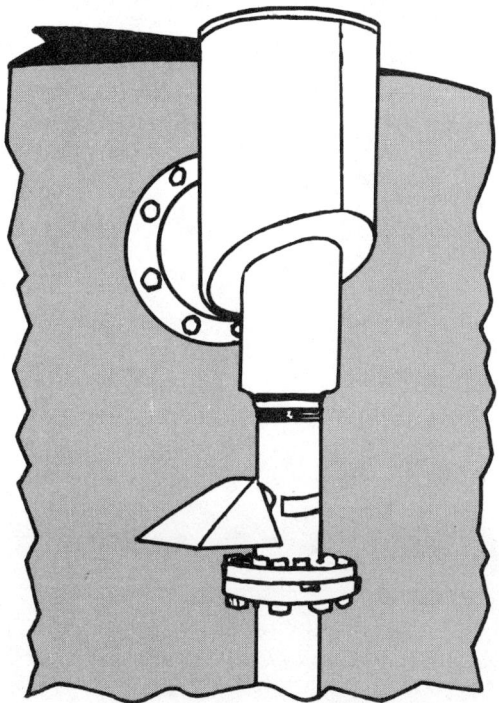

**Fixed foam maker installed
at top rim of fuel storage tank**

Figure 17

nected to a foam-mixing section and a large pipe connection for entry of the foam to the inside of the tank. The correctly proportioned (3 percent or 6 percent) foam solution is supplied to the nozzle by piping from a suitable pump or pressure supply, remote from the tank and its protecting dike (Figures 18 and 19). Swing joints are provided in the supply line that is installed on the tank so that expansion or warpage of the tank shell during fire exposure will not rupture piping. Protective screens are provided over air inlets to prevent entry of foreign matter and to discourage bird nesting.

These devices are maintained in a dry condition with suitable frangible diaphragm closures at the tank for restriction of product vapor escape. At the time of a fire, water supplies, proportioners and pumps are activated to feed foam solution to the foam makers

FIRE SERVICE HYDRAULICS

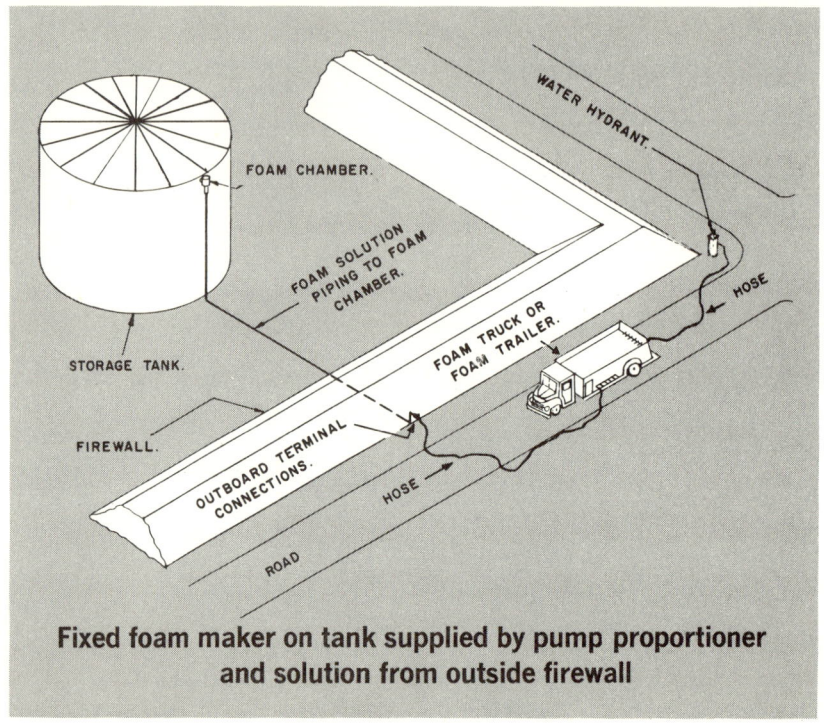

Figure 18

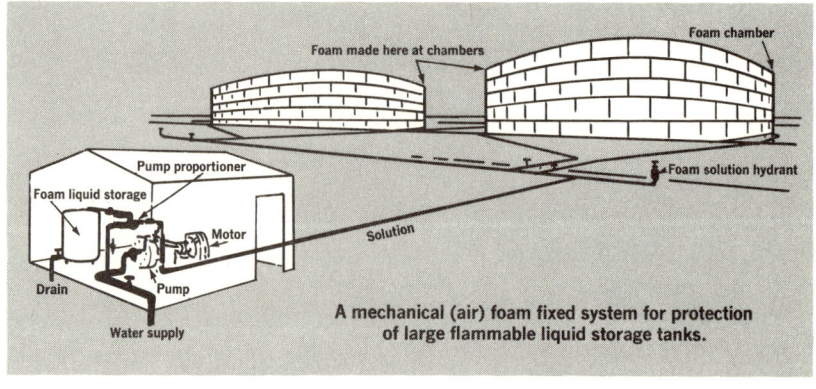

Figure 19

FIRE FIGHTING FOAMS AND FOAM SYSTEMS

at the necessary pressures and flow. The foam pressure bursts the vapor diaphragm (if it has not fractured during ignition or explosion) and foam delivery to the burning surface proceeds without further attention. The design of foam delivery devices internal to the tank will be covered at a later point.

Testing and fire experience over the years have resulted in certain design requirements for tanks using these devices. The following table lists some of these recommendations:

Table 2 Design Requirements for Tank-Side Installed Fixed Foam Makers

Tank Diameter (Feet)	Minimum Number of Foam Discharge Inlets to Tank
Up to 80	1
Over 80 to 120	2
Over 120 to 140	3
Over 140 to 160	4
Over 160 to 180	5
Over 180 to 200	6

Taken from NFPA Pamphlet No. 11—"Foam Extinguishing Systems"—1969.

These requirements are for cone-roofed tanks containing flammable liquid hydrocarbon fuels (below 140°F flash point), which require a minimum foam solution delivery rate of 0.1 gpm per square foot of liquid surface. Where flammable liquids (such as methyl or ethyl alcohol, butyl alcohol, methyl ethyl ketone, etc.), which break down ordinary foams, are stored, a special "alcohol-type" foam must be used.

Fixed foam equipment when it is properly installed, maintained and activated, can be very effective for protecting flammable liquid storage. It comes in the required sizes for single or multi-point foam application at optimum solution rates for all tank sizes and diameters. The system has one vulnerable feature: it may be damaged by explosion or fire even before it is used or during the extinguishment period.

Subsurface Injection Systems ("Protected System"): This type of "protected" foam system injects foam at the base of a tank. This method utilizes the fuel product line as a protected conduit and allows the foam to move upward through the liquid contents of the tank to the burning surface. Figure 20 shows the typical design characteristics of such a system. Again, the foam is generated in a protected position outside the dike by a forcing-type foam maker, then injected into the product line of the tank. The foam at the bottom of the tank may be distributed by "spider" manifold array piping, or

FIRE SERVICE HYDRAULICS

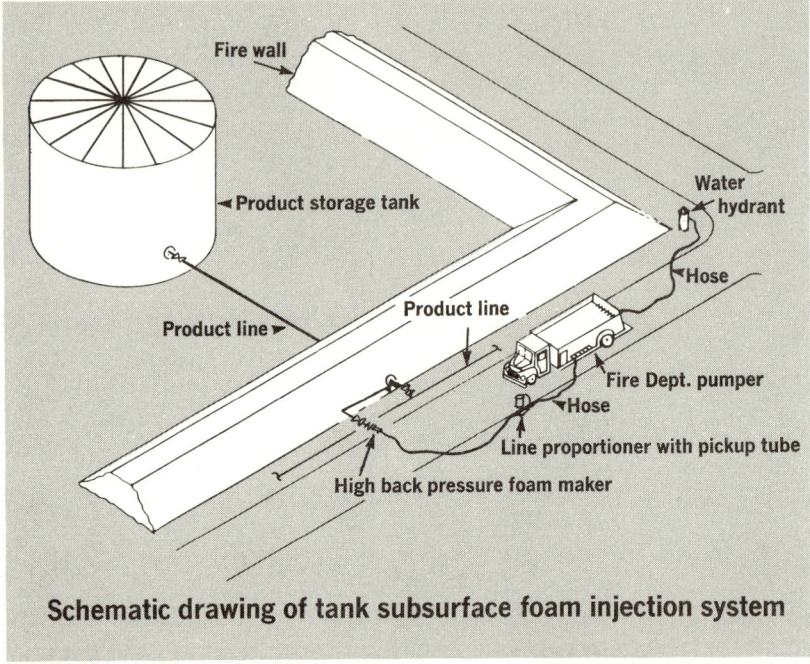

Schematic drawing of tank subsurface foam injection system

Figure 20

Courtesy National Foam System, Inc.

it may be piped to several inlet points on the bottom periphery of the tank shell.

In general, the design requirements for subsurface injection foam systems are similar to those of other fixed systems. They require a minimum calculated solution delivery rate of 0.1 gpm per square foot of liquid surface area of the tank. Foam inlets at the base of the tank are similar in number to those required for topside fixed foam maker installation. However, there is an added requirement that the inlets must be so sized that the foam velocity into the liquid contents of the tank does not exceed 10 feet per second for flammable liquids with flash points below 73°F, and no greater than 20 feet per second for all other type liquids. Alcohols or other flammable liquids which attack foams cannot be protected by the subsurface injection method.

Chemical Foam Systems: The oldest type of foam system used for fire protection of flammable liquid storage tanks employs chemical foam in the form of dual-solution piping arrangements. One leg of the piping carries a water solution of "A" ingredients and the other a

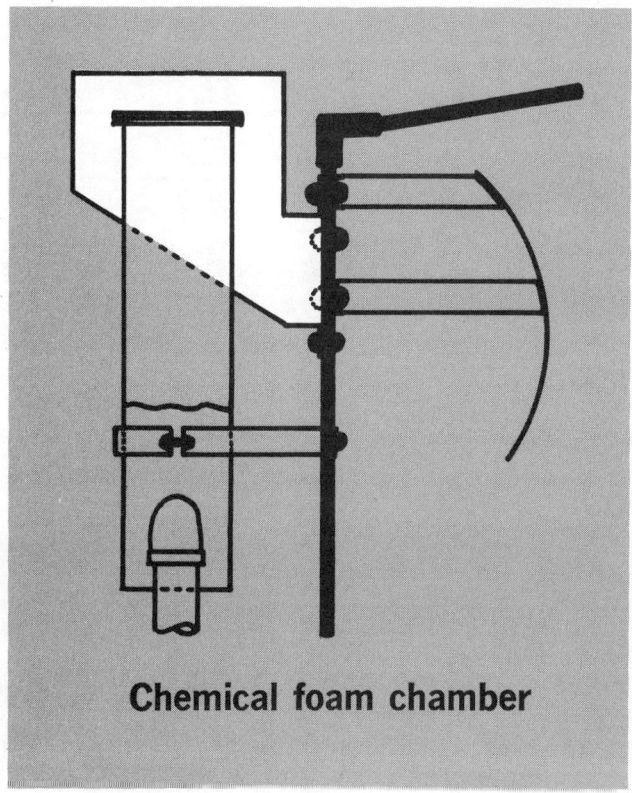

Figure 21

solution of "B" ingredients. The dual piping terminates in a common mixing chamber at the top periphery of the tank (Figure 21).

The dry chemical foam powder ingredients for each solution are put into flowing water streams by hopper devices (Figures 6 and 7). The earliest systems utilized separate storage tanks of "A" and "B" materials. Very few of these "wet storage" systems still exist.

Internal Fixed Foam Distributors or Pourer Types: Applying foam in a gentle manner that avoids any submergence in the burning liquid surface of a fuel in a tank may be accomplished in several ways.

Figures 22, 23 and 24 are foam distributing devices used in present tank fire protection installations. The ingenious semi-subsurface foam outlet hose in Figure 24 is a comparative newcomer in the field and is used in many European installations, especially in Sweden.

FIRE SERVICE HYDRAULICS

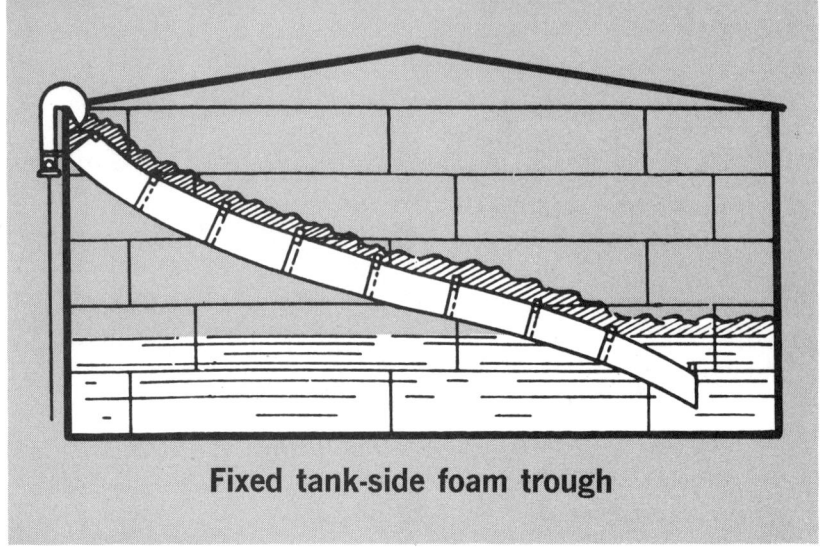

Fixed tank-side foam trough

Figure 22

It is imperative that these devices operate automatically, even if distortion of the tank has occurred.

The foregoing devices are denoted as "Type I discharge outlets" in the National Fire Codes. They all possess the designed capability of applying foam gently without submergence. There is an additional classification known as "Type II" outlets which are not particularly designed for gentle foam addition. The Type II installations require that more foam be provided for longer expected fire extinguishment operations.

Portable Foam Towers: Fuel tank fire fighting experience over the past 50 years has pointed out the large number of variable conditions under which fires occur and how they must be combatted. Explosions or fire exposure may completely or partially disable fixed foam equipment installed on tanks. Foam piping in the areas between the flaming tank and the surrounding containment dikes may become inoperative or ruptured. Central pumping or foam proportioning stations in the tank farm complex may be damaged so that they cannot be employed. Many situations arise which require a portable, quickly activated foam device, which is elevated above grade at some distance from the working level (Figure 25).

The portable foam tower is a relatively old device which may

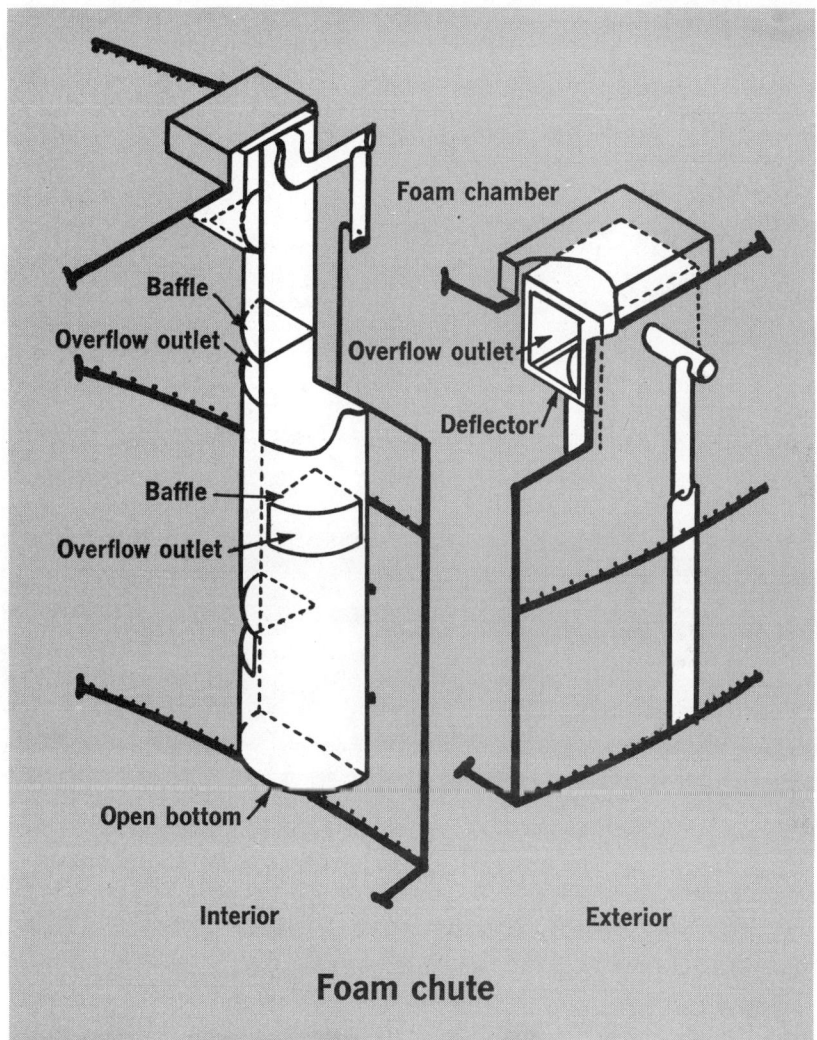

Figure 23

be used effectively in the space immediately adjacent to the burning tank if no danger to personnel is involved. Usually these towers are connected by flexible hose to a source of foam solution. The foam maker is at the top of the elevated section. Foam monitor nozzles, such as the one illustrated in Figure 26, are usually designed with high solution flows that produce a larger-volume solid stream of foam and

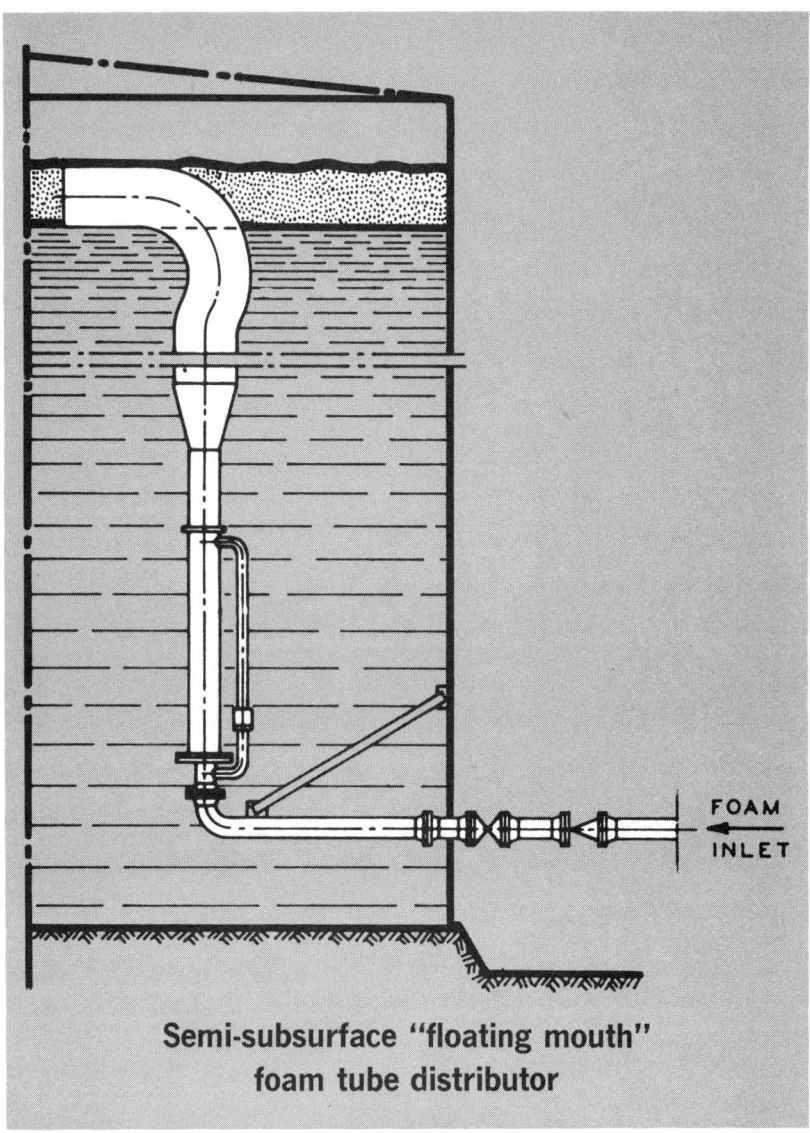

Semi-subsurface "floating mouth" foam tube distributor

Figure 24

Figure 25

FIRE SERVICE HYDRAULICS

Courtesy National Foam System, Inc.

Figure 26

project it for distances up to 160 feet horizontally and up to 60 feet vertically. The hydraulic-powered foam nozzle boom (articulated water tower) is a modern device for pinpoint application of foam on difficult to reach fires.

High expansion foam systems

An important use of foam is to halt air circulation and convection in enclosures. The use of high expansion foam for this purpose and its methods of generation have been covered earlier in this text (Figures 3 and 5). These high expansion foam systems employ suitable hoses or tubes from the generator to enclosure openings or to the fire. They may also be held by hand. Three types of application are utilized:

1. Total flooding systems (fixed equipment discharging into an enclosed space)

2. Local application systems (fixed equipment arranged to discharge foam directly on to the fire)

FIRE FIGHTING FOAMS AND FOAM SYSTEMS

3. Portable foam generating devices (portable high expansion foam generators connected to the necessary supply of water, foam concentrate proportioners and foam concentrate supply)

The design characteristics of these systems are difficult to determine because of the number of variables involved. However, in the fixed, total flooding system (Type I above), various empirical factors have been developed so that a numerical relationship can be set up for the one critical value needed to determine the protection capability of a system. This value is the minimum rate of discharge or generator capacity for fire extinguishment of a hazard in an enclosure of a given volume.

The following equation is used to calculate R, the minimum required rate of foam discharge of a high expansion foam generator in cubic feet per minute:

$$R = \left(\frac{V}{T} + R_s\right) C_n C_L$$

where: V = submergence volume (cubic feet)
(A minimum depth of foam of not less than 1.1 times the highest hazard, multiplied by the floor area of the space to be protected.)

T = submergence time (minutes)
(This factor is empirical and varies from 3 for flammable liquid hazards to 8 for rubber tires, etc. See Reference 4.)

R_s = rate of foam breakdown by sprinklers (if installed) (cubic feet per minute)
(This has to be derived by test, or may be calculated by multiplying the total discharge (gpm) of the number of sprinklers expected to discharge by 10.)

C_n = compensation for normal foam shrinkage. (It is 1.15.)

C_L = compensation for foam leakage.
(This is estimated by inspection. It can vary from 1.0 to 1.2.)

Performance requirements for other types of high expansion foam systems depend on time of operation and the general foam submergence capability of the hazard to be protected.

Fixed foam-water sprinklers and foam spray systems

Overhead water sprinkler systems have been used for protecting ordinary combustibles and combustible construction since before the turn of the century. In almost every case of correct operation, these systems (mostly automatic in design), have proved their worth, controlling and extinguishing Class A fires. However, with the advent of large-volume handling and storage of equipment using flammable and combustible liquids as fuels, a Class B fire hazard was introduced into areas protected by water sprinklers. The sprinklers, of course, could not handle such materials. Aircraft storage and repair hangars and large storage areas for automotive vehicles needed an automatic system similar to the old water sprinkler grid systems but to which foam could be added to combat the Class B fire encountered.

Depending on their design they produce foam patterns over the area to be protected. When water is discharged from such a device, its pattern is different from the foam pattern from the same device and is somewhat different from ordinary water sprinkler head patterns.

Such systems require trustworthy proportioning devices for adding the foam concentrate to the water. Also, specific minimum pattern densities of discharge are required of no less than 0.16 gallons of foam concentrate per minute per square foot for a period of 10 minutes. A suitable time reduction can be made under certain circumstances.

Flush-mounted foam sprinkler systems

Under certain circumstances fuel fires have occurred on horizontal plane surfaces due to overfill or rupture of aircraft fuel tanks. The resulting fuel fire requires foam application by portable devices to control and extinguish it. Recently (1968) it was proven by large fire tests that a type of fixed foam sprinkler system, flush-mounted in a horizontal surface could automatically extinguish surface flammable fuel fires. When aqueous film-forming (Light Water) foam solutions are employed in such a system, foam is generated by the upward direction of the sprinkler discharge, falling back on the burning surface and extinguishing it. When these solutions are used in subsurface piping grids supplying these nozzles, a minimum density of discharge of 0.06 gallons per minute per square foot will control and extinguish surface fuel fires.

Crash-rescue foam vehicles

The advent of air transport operations involving large quantities of flammable fuels in close proximity to a large number of passengers made necessary the design of mobile foam systems for fire emergencies at airports.

The modern airport crash-rescue fire fighting vehicle is the result of many developments over the past 35 years. It is actually a completely mobile and trustworthy foam-generating system capable of

FIRE FIGHTING FOAMS AND FOAM SYSTEMS

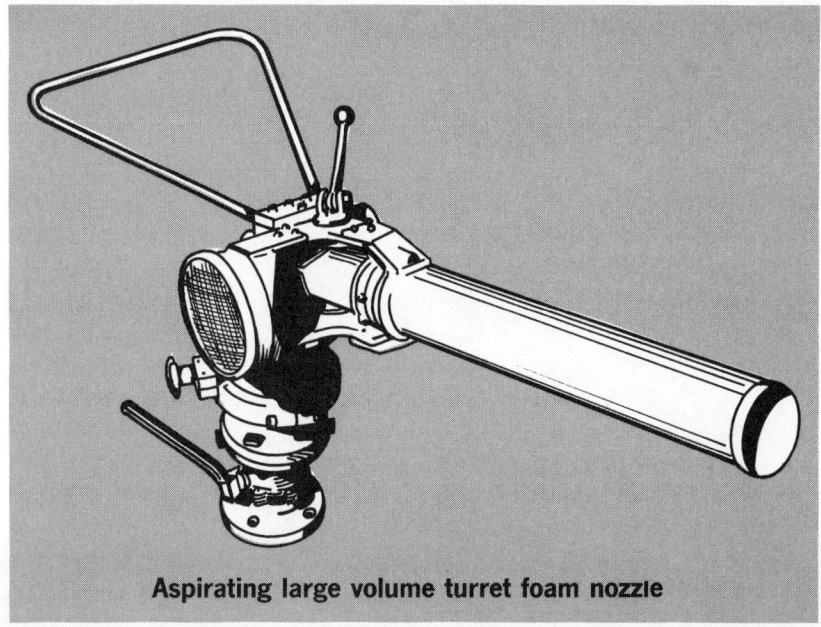

Aspirating large volume turret foam nozzle

Figure 27
Courtesy Bliss-Portland, Inc.

deluging flammable fuel fires with large volumes of foam almost instantaneously and efficiently.

Because this use of foam involves the safeguarding of human lives, it is very important that the system be completely reliable. Airport fire emergencies have occurred under many conditions and the system must be flexible in foam application to meet these highly variable situations.

In general, such a foam system must be completely self-sufficient, with its own source of power, pumping capacity, foam concentrate proportioning and storage tanks for water and foam concentrate. The vehicle must incorporate high-volume foam deluge or turret nozzles capable of long-distance projection of foam in a solid stream. The nozzle must also be instantly variable to a wide-pattern foam spray. There must be hand-operated hose line foam nozzles on the unit that are also capable of variable-pattern foam discharge. And the system must be capable of continuous foam generation if a source of replenishment of foam concentrate and water can be supplied.

For these systems there are two principal types of foam-generating methods in use in the United States: the air aspirating nozzle foam maker, exemplified by Figure 1 and shown in Figure 27, and the foam

FIRE SERVICE HYDRAULICS

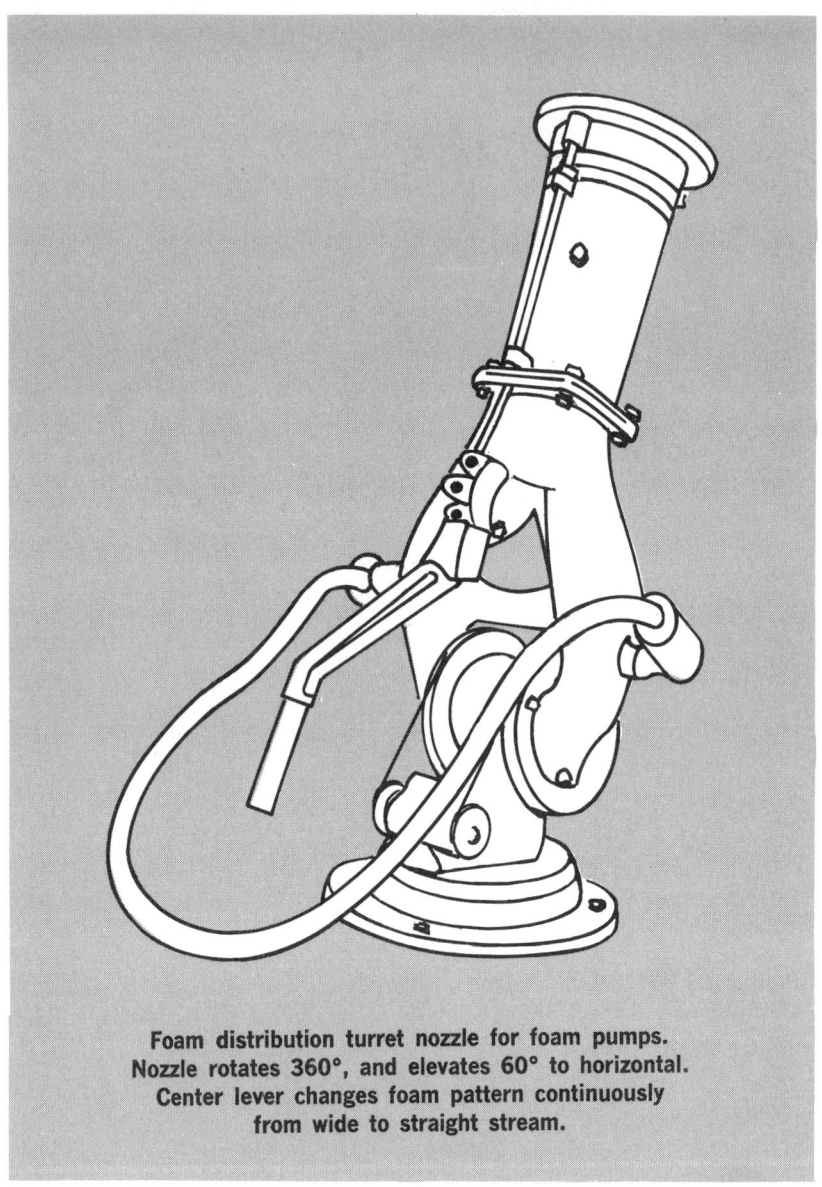

Foam distribution turret nozzle for foam pumps. Nozzle rotates 360°, and elevates 60° to horizontal. Center lever changes foam pattern continuously from wide to straight stream.

Figure 28

FIRE FIGHTING FOAMS AND FOAM SYSTEMS

pump system using the foam pump of Figure 4 supplying a foam distribution turret nozzle of the type shown in Figure 28. Hand line nozzles employ air aspirator foam makers in both types of systems. Many varieties of proportioning devices are used in these systems to accurately proportion the foam concentrate during all variable conditions of operation. Depending on the design of the foam system, the in-line proportioner, the positive-pressure pump proportioner, the flexible diaphragm type or the balanced pressure proportioning system may be employed.

The piping arrangement and control valving are complex. Of prime importance are the instantaneous activation of the system and its complete reliability. Controls are usually powered by air or hydraulic activation systems so that a single operator in the vehicle may exercise complete control.

The sizes and capacities of this foam system cover a wide range because of the variability in size and load-carrying capacity of the vehicle on which it is installed. Water tanks may vary from 400 to 2500-gallon capacities with corresponding sizes of foam concentrate tanks. Turret foam nozzles may be designed in sizes from 250 to 1000 gpm. Foam solution supply and hand line foam nozzles are manufactured with 55 to 100 gpm at 60 to 100 psi capacities.

The determination of recommended quantities and rates of application of foam for crash-rescue operations at airports depends on many factors. The size and passenger capacity of the largest aircraft using an airport can act as a guide to indicate the largest fire (and the most fuel) which must be combatted. When this is established, the number of required gallons of water can be specified for foam-making operations to control and extinguish this size fire. This has been done in NFPA No. 403. Many contributing characteristics of crash-rescue foam extinguishment devices are discussed at length in other references on this subject.

Combined agent or "twinned" equipment

Using foam for the control and extinguishment of flammable and combustible liquid fires is not a perfect solution to all the problems in this area. Although foam is a reasonably permanent and complete agent for these fires, it is slow in action and the mode of delivery to a burning liquid surface is critical. Because of its light weight it cannot be projected to distant target areas and it requires simultaneous operation of specially designed equipment to generate it.

Quite early in the evolution of airport crash-rescue vehicles using foam, many fire fighting situations arose for which a gaseous-type

FIRE SERVICE HYDRAULICS

Figure 29

Official U.S. Navy Photo

extinguishing agent was superior in action. Early designs employed carbon dioxide gas as a "secondary" or "clean-up" agent, with foam as the primary fire-attacking agent. The gaseous agent was capable of quick flame quenching and deep penetration within conglomerate masses. This action was of considerable aid to the blanketing and cooling action of foam. As developments in fire science took place, sodium bicarbonate dry chemical, an agent twice as effective as carbon dioxide, replaced it as the "secondary" agent.

The discovery of aqueous, film-forming foam (Light Water) in 1962 made it possible to employ potassium bicarbonate simultaneously with Light Water foam without degrading the action of the foam. Potassium bicarbonate is twice as effective an agent as the sodium powder but it cannot be made in a "foam-compatible" form so it cannot be efficiently used in conjunction with foams other than the aqueous film-forming type.

The result of this union was the so-called "twinned agent" equipment, with dual containers for the Light Water foam solution and the Purple-K dry chemical. Hose lines or turret nozzles are supplied by separate hoses which are tied together by the rubber-molding process during manufacture so that they may be reeled and run out as a single hose. Each hose supplied a distribution nozzle which has a pistol grip, trigger valve, on-off control. When using it, the fire fighter has a separate agent in each hand for fighting the fire—dry chemical powder for instantaneously quenching flame, and foam for sealing the surface of the liquid fuel and preventing its reflash. The pistol grip nozzles are combined in a single holder to further strengthen the twinned concept. Figure 29 shows details of this equipment.

To activate this device quickly without resorting to pumps or engines, the two agents are stored in separate tanks capable of quick pressurization by high-pressure gas bottles.

These twinned devices are manufactured in various sizes for varying discharge times. Flow rates of the nozzles are customarily 60 gpm for Light Water and 5 pounds per second for Purple-K. In general, a twinned agent unit containing 100 gallons of Light Water solution and 450 pounds of Purple-K will extinguish a motor gasoline fuel fire of 3,000 square feet in 90 seconds.

REFERENCES

1. *History and Development of Foam as a Fire Extinguishing Medium*, A. F. Ratzer, Industrial and Engineering Chemistry, Volume 48, November 1956, page 2013.
2. *Foam Extinguishing Systems*, 1969, National Fire Protection Association, No. 11.
3. *Studies with High Expansion Foams for Controlling Experimental Coal-Mine Fires*, I. Hartman, J. Nagy, R. W. Barnes, E. M. Murphy, Report of Investigations No. 5415, U. S. Bureau of Mines.
4. *High Expansion Foam Systems (Expansion Ratios from 100:1 to 1000:1)*, 1970, National Fire Protection Association, No. 11A.
5. *Fire Protection Handbook*, Thirteenth Edition, Section 15, Chapter IV; Section 17, Chapter III, National Fire Protection Association, Boston.
6. *Characterization of Foams for Fire Extinguishment*, R. L. Tuve et al, Industrial and Engineering Chemistry, Volume 48, November 1956, page 2024.
7. *Foam-Water Sprinkler and Spray Systems*, 1968, National Fire Protection Association, No. 16.
8. *Research Studies in Foam-Generating Equipment*, H. B. Peterson et al, Industrial and Engineering Chemistry, Volume 48, November 1956, page 2031.
9. *The Fire Chief's Handbook*, Third Edition, Chapter 1, The Reuben H. Donnelley Corporation, New York.
10. *Storage and Handling of Petroleum Liquids*: Practice and Law, John R. Hughes, Charles Griffin and Co., Ltd., London, England.
11. *S. O. P. Aircraft Rescue and Fire Fighting*, 1969, National Fire Protection Association, No. 402.
12. *Aircraft Rescue and Fire Fighting Services at Airports*, 1969, National Fire Protection Association, No. 403.
13. *Evaluating Foam Fire Equipment (Aircraft Rescue and Fire Fighting Vehicles)*, 1969, National Fire Protection Association, No. 412.
14. *Aircraft Rescue and Fire Fighting Vehicles*, 1969, National Fire Protection Association, No. 414.

appendix

FIRE SERVICE HYDRAULICS

DISCHARGE TABLE FOR SMOOTH NOZZLES.

NOZZLE PRESSURE MEASURED BY PITOT GAGE.

Nozzle Pressure in lbs. per sq. inch.	Nozzle Diam. in Inches				Nozzle Pressure in lbs. per sq. inch.	Nozzle Diam. in Inches			
	1/4	5/16	3/8	7/16		1/4	5/16	3/8	7/16
	Gallons per Minute.					Gallons per Minute.			
5	4	6	9	13	60	14	22	31	43
6	4	6	10	14	62	14	22	32	44
7	4	7	11	15	64	14	22	32	45
8	5	7	11	16	66	14	23	33	46
9	5	8	12	17	68	14	23	33	46
10	6	9	13	18	70	15	24	34	47
12	6	10	15	19	72	15	24	34	48
14	7	11	15	21	74	15	24	35	48
16	7	12	16	22	76	15	24	35	49
18	7	12	17	24	78	15	24	36	50
20	8	13	18	25	80	16	25	36	50
22	8	13	19	26	82	16	25	37	51
24	8	13	20	27	84	16	25	37	51
26	9	14	21	29	86	16	26	37	52
28	9	14	21	30	88	16	26	38	52
30	10	15	22	31	90	17	27	39	53
32	10	15	23	32	92	17	27	39	54
34	11	16	23	33	94	17	27	39	54
36	11	16	24	34	96	17	27	40	55
38	11	17	25	35	98	17	27	40	55
40	11	18	26	35	100	18	28	41	56
42	11	18	26	36	105	18	29	42	57
44	12	18	27	37	110	19	29	43	59
46	12	19	28	38	115	19	30	43	60
48	12	19	28	39	120	19	31	44	61
50	13	20	29	40	125	20	31	45	62
52	13	20	29	40	130	20	32	46	64
54	13	20	30	41	135	21	33	47	65
56	13	21	30	42	140	21	33	48	66
58	13	21	31	43	145	21	34	49	68
60	14	22	31	43	150	22	34	50	69

Assumed coefficient of discharge = .98¼ .98¼ .98½ .98½

APPENDIX

DISCHARGE TABLE FOR SMOOTH NOZZLES.

NOZZLE PRESSURE MEASURED BY PITOT GAGE.

Nozzle Pressure in lbs. per sq. inch.	Nozzle Diam. in Inches				Nozzle Pressure in lbs. per sq. inch.	Nozzle Diam. in Inches			
	1/2	5/8	3/4	7/8		1/2	5/8	3/4	7/8
	Gallons per Minute.					Gallons per Minute.			
5	16	26	37	50	60	57	89	130	174
6	18	28	41	55	62	58	90	132	177
7	19	30	44	59	64	59	92	134	180
8	21	32	47	64	66	60	93	136	182
9	22	34	50	67	68	60	95	138	185
10	23	36	53	71	70	61	96	140	188
12	25	40	58	78	72	62	97	142	191
14	27	43	63	84	74	63	99	144	193
16	29	46	67	90	76	64	100	146	196
18	31	49	71	95	78	65	101	148	198
20	33	51	75	101	80	66	103	150	201
22	34	54	79	105	82	66	104	152	204
24	36	56	82	110	84	67	105	154	206
26	37	59	85	115	86	68	107	155	208
28	39	61	89	119	88	69	108	157	211
30	40	63	92	123	90	70	109	159	213
32	41	65	95	127	92	70	110	161	215
34	43	67	98	131	94	71	111	162	218
36	44	69	100	135	96	72	113	164	220
38	45	71	103	138	98	73	114	166	223
40	46	73	106	142	100	73	115	168	225
42	47	74	109	146	105	75	118	172	230
44	49	76	111	149	110	77	121	176	236
46	50	78	114	152	115	79	123	180	241
48	51	80	116	156	120	80	126	183	246
50	52	81	118	159	125	82	129	187	251
52	53	83	121	162	130	84	131	191	256
54	54	84	123	165	135	85	134	195	262
56	55	86	125	168	140	87	136	198	266
58	56	87	128	171	145	88	139	202	271
60	57	89	130	174	150	90	141	205	275

Assumed coefficient of discharge = .98½ .98⅝ .98¾ .98⅞

FIRE SERVICE HYDRAULICS

DISCHARGE TABLE FOR SMOOTH NOZZLES.

NOZZLE PRESSURE MEASURED BY PITOT GAGE.

Nozzle Pressure in lbs. per sq. inch.	Nozzle Diam. in Inches					Nozzle Pressure in lbs. per sq. inch.	Nozzle Diam. in Inches				
	1	1⅛	1¼	1⅜	1½		1	1⅛	1¼	1⅜	1½
	Gallons per minute.						Gallons per Minute.				
5	66	84	103	125	149	60	229	290	357	434	517
6	72	92	113	137	163	62	233	295	363	441	525
7	78	99	122	148	176	64	237	299	369	448	533
8	84	106	131	158	188	66	240	304	375	455	542
9	89	112	139	168	200	68	244	308	381	462	550
10	93	118	146	177	211	70	247	313	386	469	558
12	102	130	160	194	231	72	251	318	391	475	566
14	110	140	173	210	249	74	254	322	397	482	574
16	118	150	185	224	267	76	258	326	402	488	582
18	125	159	196	237	283	78	261	330	407	494	589
20	132	167	206	250	298	80	264	335	413	500	596
22	139	175	216	263	313	82	268	339	418	507	604
24	145	183	226	275	327	84	271	343	423	513	611
26	151	191	235	286	340	86	274	347	428	519	618
28	157	198	244	297	353	88	277	351	433	525	626
30	162	205	253	307	365	90	280	355	438	531	633
32	167	212	261	317	377	92	283	359	443	537	640
34	172	218	269	327	389	94	286	363	447	543	647
36	177	224	277	336	400	96	289	367	452	549	654
38	182	231	285	345	411	98	292	370	456	554	660
40	187	237	292	354	422	100	295	374	461	560	667
42	192	243	299	363	432	105	303	383	473	574	683
44	196	248	306	372	442	110	310	392	484	588	699
46	200	254	313	380	452	115	317	401	495	600	715
48	205	259	320	388	462	120	324	410	505	613	730
50	209	265	326	396	472	125	331	418	516	626	745
52	213	270	333	404	481	130	337	427	526	638	760
54	217	275	339	412	490	135	343	435	536	650	775
56	221	280	345	419	499	140	350	443	546	662	789
58	225	285	351	426	508	145	356	450	556	674	803
60	229	290	357	434	517	150	362	458	565	686	817

Assumed coefficient of discharge = .99   .99   .99   .99¼   .99½

Note.—Coefficients of discharge are based on experiments by Mr. John R. Freeman, Transactions Am. Soc. C. E. Vols. XXI and XXIV.

APPENDIX

DISCHARGE TABLE FOR SMOOTH NOZZLES.

NOZZLE PRESSURE MEASURED BY PITOT GAGE.

Nozzle Pressure in lbs. per sq. inch.	Nozzle Diam. in Inches					Nozzle Pressure in lbs. per sq. inch.	Nozzle Diam. in Inches				
	1⅝	1¾	1⅞	2	2¼		1⅝	1¾	1⅞	2	2¼
	Gallons per Minute.						Gallons per Minute.				
5	175	208	234	266	337	60	607	704	810	920	1167
6	192	228	256	292	369	62	617	716	823	936	1187
7	207	241	277	315	399	64	627	727	836	951	1206
8	222	257	296	336	427	66	636	738	850	965	1224
9	235	273	314	357	452	68	646	750	862	980	1242
10	248	288	330	376	477	70	655	761	875	994	1260
12	271	315	362	412	522	72	665	771	887	1008	1278
14	293	340	391	445	564	74	674	782	900	1023	1296
16	313	364	418	475	603	76	683	792	911	1036	1313
18	332	386	444	504	640	78	692	803	924	1050	1330
20	350	407	468	532	674	80	700	813	935	1063	1347
22	367	427	490	557	707	82	709	823	946	1076	1364
24	384	446	512	582	739	84	718	833	959	1089	1380
26	400	464	533	606	769	86	726	843	970	1102	1396
28	415	481	554	629	799	88	735	853	981	1115	1412
30	429	498	572	651	826	90	743	862	992	1128	1429
32	443	514	591	673	854	92	751	872	1002	1140	1445
34	457	530	610	693	880	94	759	881	1012	1152	1460
36	470	546	627	713	905	96	767	890	1022	1164	1476
38	483	561	645	733	930	98	775	900	1032	1176	1491
40	496	575	661	752	954	100	783	909	1043	1189	1506
42	508	589	678	770	978	105	803	932	1070	1218	1542
44	520	603	694	788	1000	110	822	954	1095	1247	1579
46	531	617	710	806	1021	115	840	975	1120	1275	1615
48	543	630	725	824	1043	120	858	996	1144	1303	1649
50	554	643	740	841	1065	125	876	1016	1168	1329	1683
52	565	656	754	857	1087	130	893	1036	1191	1356	1717
54	576	668	769	873	1108	135	910	1056	1213	1382	1750
56	586	680	782	889	1129	140	927	1076	1235	1407	1780
58	596	692	796	905	1149	145	944	1095	1257	1432	1812
60	607	704	810	920	1168	150	960	1114	1279	1456	1842

Assumed coefficient of discharge = .995 .995 .996 .997 .997

1-INCH SMOOTH NOZZLE.—

Nozzle Pressure Indicated by Pitot Gage.	Discharge, Gallons per Minute.	Pressures Required at Hydrant or maintain Nozzle Pressures given Lengths of Best Quality Single 2½-inch Lines.							
		100 Feet.	200 Feet.	300 Feet.	400 Feet.	500 Feet.	600 Feet.	700 Feet.	800 Feet.
20	132	25	30	35	39	44	49	53	58
25	148	31	37	43	49	55	60	66	72
30	162	38	44	51	58	65	72	78	85
35	175	44	52	59	67	75	83	91	98
40	187	50	59	68	77	86	94	103	112
45	198	56	66	76	86	96	106	115	125
50	209	62	73	84	95	106	117	128	139
55	219	68	80	92	104	116	128	140	152
60	229	75	88	101	114	127	140	153	166
65	238	81	95	109	123	137	151	165	179
70	247	87	102	117	132	147	162	177	192
75	256	93	109	125	141	157	173	189	205
80	264	99	116	133	150	167	183	200	217
85	272	105	123	141	159	177	195	212	230
90	280	111	130	149	167	186	205	224	243
95	287	117	137	157	177	196	216	236	256
100	295	123	144	165	185	206	227	247	268

2 1/2- AND 3-INCH HOSE.

Fire Engine, while stream is flowing, to
in First Column, through various
2½- and 3-inch Rubber Lined Hose.

2½-inch Lines.		Single 3-inch Lines.				Two 2½-inch Lines Siamesed.			Nozzle Pressure Indicated by Pitot Gage.
1,000 Feet.	1,200 Feet.	800 Feet.	1,000 Feet.	1,200 Feet.	1,500 Feet.	1,000 Feet.	1,500 Feet.	2,000 Feet.	
68	77	35	39	42	48	33	40	46	20
84	95	43	48	52	59	41	49	57	25
99	112	52	57	62	70	49	59	68	30
114	130	60	66	72	81	57	68	79	35
130	148	68	75	82	92	65	78	90	40
145	165	77	84	92	103	72	86	99	45
160	182	85	93	102	114	80	95	110	50
175	199	93	102	112	125	88	105	121	55
192	218	102	112	122	137	96	114	132	60
207	235	110	121	131	148	103	122	141	65
222	252	118	130	141	159	111	132	152	70
237	269	127	139	151	170	120	142	164	75
251	285	135	148	161	181	128	151	175	80
266	302	143	156	170	191	135	159	184	85
280	....	151	165	180	202	143	169	195	90
295	...	158	173	189	211	150	177	204	95
310	...	167	183	199	223	157	186	215	100

FIRE SERVICE HYDRAULICS

1 1/8-INCH SMOOTH NOZZLE.—

PRESSURES REQUIRED AT HYDRANT OR FIRE NOZZLE PRESSURES GIVEN IN FIRST QUALITY 2½- AND

Single 2½-inch Lines.

Nozzle Pressure Indicated by Pitot Gage.	Discharge, Gallons Per Minute.	100 Feet.	200 Feet.	300 Feet.	400 Feet.	500 Feet.	600 Feet.	700 Feet.	800 Feet.	1,000 Feet.	1,200 Feet.
20	167	28	35	42	49	56	64	71	78	92	107
25	187	35	44	53	62	71	79	88	97	115	133
30	205	42	52	63	73	84	95	105	116	137	158
35	221	49	61	73	85	97	110	122	134	158	183
40	237	55	69	83	96	110	124	138	151	179	206
45	251	62	77	93	108	123	139	154	169	200	230
50	265	69	86	103	120	137	154	171	188	222	256
55	277	76	94	112	131	149	168	186	204	241	278
60	290	83	103	123	143	163	183	203	223	263	304
65	301	89	111	132	154	175	197	218	240	283	326
70	313	96	119	142	165	188	211	234	257	303	....
75	324	103	128	152	177	202	227	252	276	325	....
80	335	110	136	162	188	215	241	267	294	....	....
85	345	116	144	171	199	226	254	282	309	....	....
90	355	123	152	181	210	240	269	298	327	....	....
95	365	130	160	191	222	252	283	314	...	....	....
100	374	136	168	201	233	265	297	329	..	....	...

2 1/2- AND 3-INCH HOSE.

Engine, while stream is flowing, to maintain Column, through various Lengths of Best 3-inch Rubber Lined Hose.

Single 3-inch Lines.							Two 2½-inch Lines Siamesed.					Nozzle Pressure Indicated by Pitot Gage.
400 Feet.	600 Feet.	800 Feet.	1,000 Feet.	1,200 Feet.	1,500 Feet.	1,800 Feet.	800 Feet.	1,000 Feet.	1,200 Feet.	1,500 Feet.	1,800 Feet.	
32	37	43	48	54	62	71	38	42	46	53	60	20
40	46	53	60	67	77	87	45	50	55	63	70	25
47	55	63	71	79	91	103	53	59	65	74	82	30
55	65	74	83	93	107	121	62	69	76	86	96	35
63	73	84	95	105	121	137	70	78	86	97	108	40
70	82	94	106	118	135	153	79	87	95	108	121	45
78	91	104	117	130	150	169	88	98	107	121	135	50
86	100	114	128	142	164	185	96	107	117	132	147	55
93	109	124	139	155	178	201	105	116	127	143	160	60
101	117	134	151	167	192	217	114	126	138	156	174	65
108	126	144	162	180	206	233	122	135	148	167	186	70
116	135	154	173	192	221	249	130	144	157	178	198	75
124	144	165	185	206	236	267	138	153	167	189	210	80
131	153	174	195	217	249	281	147	163	178	201	224	85
139	161	184	207	229	263	297	156	172	188	212	237	90
146	170	194	218	242	277	313	164	181	198	224	249	95
154	178	203	228	253	290	....	172	190	208	235	261	100

1 1/4-INCH SMOOTH NOZZLE.—

Pressures Required at Hydrant or Fire Pressures given in First Column, 2½- AND 3-INCH

Single 2½-inch Lines.

Nozzle Pressure Indicated by Pitot Gage.	Discharge, Gallons per Minute.	100 Feet.	200 Feet.	300 Feet.	400 Feet.	500 Feet.	600 Feet.	700 Feet.	800 Feet.	1,000 Feet.	1,200 Feet.
20	206	32	42	53	64	75	85	96	107	128	149
25	230	40	53	66	79	92	105	118	131	158	184
30	253	48	63	79	95	110	126	142	157	189	220
35	273	55	73	91	109	127	145	163	181	217	253
40	292	63	83	104	124	144	165	185	206	246	287
45	309	70	93	116	138	161	183	206	229	274	319
50	326	78	103	128	153	178	203	228	253	303	....
55	342	86	113	140	167	194	222	249	276	330	....
60	357	93	123	152	182	211	241	270	300	....	....
65	372	101	133	164	196	228	260	292	323	....	....
70	386	108	142	176	210	244	278	312	....	....	....
75	399	116	152	188	224	261	297	333	....	....	....
80	413	124	163	201	240	279	318	....	....	....	....
85	425	131	172	213	254	295	....	....	....	....	....
90	438	139	182	225	269	312	....	....	....	....	....
95	449	146	191	236	282	327	....	....	....	....	....
100	461	153	201	248	295	....	....	....	....	....	....

2 1/2- AND 3-INCH-HOSE.

Engine, while stream is flowing, to maintain Nozzle through various Lengths of Best Quality Rubber Lined Hose.

Single 3-inch Lines.							Two 2½-inch Lines Siamesed.						Nozzle Pressure Indicated by Pitot Gage.
400 Feet.	600 Feet.	800 Feet.	1,000 Feet.	1,200 Feet.	1,500 Feet.	1,800 Feet.	600 Feet.	800 Feet.	1,000 Feet.	1,200 Feet.	1,500 Feet.	1,800 Feet.	
37	46	54	62	70	83	95	39	45	51	57	67	76	20
47	57	67	77	87	102	117	48	55	62	70	80	91	25
56	68	81	93	105	123	142	57	66	74	83	96	109	30
65	79	92	106	120	141	161	66	76	86	95	110	125	35
74	89	105	120	136	159	183	75	87	99	110	127	144	40
83	100	117	135	152	178	204	84	96	109	121	140	158	45
91	111	130	149	168	197	226	93	107	121	135	155	176	50
100	121	142	163	184	216	247	102	117	132	147	169	192	55
109	132	155	178	201	235	270	111	128	144	160	185	210	60
118	143	167	192	217	254	291	120	137	155	173	199	225	65
127	154	180	206	233	272	....	129	147	166	185	213	241	70
136	164	192	220	248	290	....	137	157	177	197	227	257	75
145	175	205	235	265	....	....	147	169	190	212	244	276	80
153	184	216	247	279	....	....	156	179	201	224	258	292	85
162	195	228	261	295	....	....	165	189	213	237	273	309	90
170	205	240	275	....	....	....	173	198	223	248	286	323	95
179	215	252	288	....	....	....	182	208	235	261	300	..	100

1 3/8-INCH SMOOTH NOZZLE.—

Nozzle Pressure Indicated by Pitot Gage.	Discharge, Gallons per Minute.	Pressures Required at Hydrant or Fire Nozzle Pressures given in First Quality 2½- and									
		Single 2½-inch Lines.									
		100 Feet.	200 Feet.	300 Feet.	400 Feet.	500 Feet.	600 Feet.	700 Feet.	800 Feet.	200 Feet.	400 Feet.
20	250	37	52	68	83	98	113	128	144	34	45
25	280	46	64	83	102	121	139	158	177	41	56
30	307	55	77	99	121	144	166	188	210	50	67
35	331	64	89	115	140	166	191	217	242	58	78
40	354	73	102	131	160	189	218	247	276	67	89
45	376	81	114	146	178	211	243	275	307	74	99
50	396	90	125	161	196	222	257	293	328	82	109
55	415	99	137	176	215	254	292	331	....	90	121
60	434	107	149	191	233	276	318	....	....	98	131
65	451	116	161	206	251	297	....	....	....	106	141
70	469	125	173	222	270	319	....	....	....	114	152
75	485	134	185	237	289	....	....	....	....	122	162
80	500	142	196	251	305	....	....	....	....	130	172
85	516	151	209	267	325	....	....	....	....	138	183
90	531	159	220	281	....	....	....	....	....	146	194
95	546	168	232	297	....	....	....	....	....	153	203
100	560	177	244	312	....	....	....	....	....	162	215

APPENDIX

2 1/2- AND 3-INCH HOSE.

Engine, while stream is flowing, to maintain Column, through various Lengths of Best 3-inch Rubber Lined Hose.

Single 3-inch Lines.					Two 2½-inch Lines Siamesed.							Nozzle Pressure Indicated by Pitot Gage.
600 Feet.	800 Feet.	1,000 Feet.	1,200 Feet.	1,500 Feet.	400 Feet.	600 Feet.	800 Feet.	1,000 Feet.	1,200 Feet.	1,500 Feet.	1,800 Feet.	
57	68	80	92	109	37	46	54	63	71	84	96	20
70	85	99	113	135	46	57	67	78	88	104	119	25
84	101	118	135	161	56	68	81	93	106	124	143	30
97	117	137	157	187	65	80	94	108	122	143	165	35
112	134	157	180	214	74	90	106	122	138	162	186	40
125	150	175	200	238	83	101	119	137	155	182	209	45
137	164	192	220	267	92	111	131	151	171	201	230	50
151	182	212	242	288	100	122	144	165	187	219	252	55
163	196	229	262	....	109	133	156	180	203	238	273	60
177	212	247	282	....	118	143	168	194	219	257	294	65
189	227	265	303	....	128	155	182	209	236	277	317	70
203	243	283	....	....	137	165	194	223	252	295	....	75
215	257	300	....	....	145	175	206	236	266	312	....	80
229	274	....	....	....	153	186	218	250	282	331	....	85
241	289	....	....	....	162	196	230	264	298	....	....	90
254	304	....	....	....	170	206	241	277	313	..	....	95
267	....	....	....	....	179	217	254	291	329	....	....	100

1 1/2-INCH SMOOTH NOZZLE.—

Pressures Required at Hydrant or Fire Nozzle Pressures given in First Quality 2½- and

Nozzle Pressure Indicated by Pitot Gage.	Discharge, Gallons per Minute.	Single 2½-inch Lines.								Single		
		100 Feet.	200 Feet.	300 Feet.	400 Feet.	500 Feet.	600 Feet.	700 Feet.	800 Feet.	200 Feet.	400 Feet.	600 Feet.
20	298	44	65	86	107	128	149	170	191	39	55	71
25	333	54	80	106	132	158	184	210	236	48	68	88
30	365	65	95	126	157	188	219	250	280	58	81	105
35	394	75	110	145	181	216	251	287	322	67	94	122
40	422	85	126	166	206	246	286	327	....	76	107	139
45	447	96	141	185	230	275	320	....	....	85	120	155
50	472	106	155	205	254	304	..	....	...	95	133	171
55	494	116	170	224	278	332	....	....	....	104	145	187
60	517	126	184	242	301	....	....	...,	....	113	158	203
65	537	136	198	261	324	....	....	....	....	122	170	218
70	558	146	213	281	....	....	....	....	....	131	183	235
75	578	156	228	299	....	....	....	....	....	140	196	251
80	596	166	242	318	....	....	....	....	....	149	208	267
85	614	176	257	337	....	....	....	....	....	158	220	282
90	633	187	272	....	....	....	....	....	....	167	233	298
95	650	197	286	....	....	....	....	....	....	176	245	314
100	667	207	300	....	....	....	....	....	....	185	257	....

2 1/2- AND 3-INCH HOSE.

Engine, while stream is flowing, to maintain Column, through various Lengths of Best 3-inch Rubber Lined Hose.

3-inch Lines.				Two 2½-inch Lines Siamesed.								Nozzle Pressure Indicated by Pitot Gage.
800 Feet.	1,000 Feet.	1,200 Feet.	1,500 Feet.	200 Feet.	400 Feet.	600 Feet.	800 Feet.	1,000 Feet.	1,200 Feet.	1,500 Feet.	1,800 Feet.	
87	104	120	144	33	45	56	68	79	91	108	126	20
108	128	148	178	41	56	70	84	99	113	135	156	25
129	153	177	212	49	66	83	100	117	134	160	185	30
149	177	204	245	57	77	96	116	135	155	184	214	35
170	201	232	279	65	88	110	132	155	177	211	244	40
189	224	258	....	73	97	122	146	171	196	233	269	45
209	247	286	....	81	108	136	163	190	218	259	300	50
228	270	....	....	88	118	148	178	208	237	282	327	55
248	293	....	....	96	128	161	193	225	257	305	....	60
267	....	....	....	104	139	174	208	243	278	....	....	65
287	....	....	....	112	149	186	223	261	298	....	....	70
307	....	....	....	120	160	199	239	279	319	....	....	75
....	....	....	....	127	170	212	254	296	....	....	....	80
....	....	....	....	135	179	224	268	313	....	....	....	85
....	....	....	....	143	190	237	284	....	....	....	....	90
....	....	....	....	152	201	251	301	....	....	....	....	95
....	....	....	....	160	212	264	316	....	....	....	....	100

1 5/8-INCH SMOOTH NOZZLE.—

PRESSURES REQUIRED AT HYDRANT OR FIRE NOZZLE PRESSURES GIVEN IN FIRST QUALITY 2½- AND

Nozzle Pressure Indicated by Pitot Gage.	Discharge, Gallons per Minute.	Single 2½-inch Lines.						Single 3-inch			
		100 Feet.	200 Feet.	300 Feet.	400 Feet.	500 Feet.	600 Feet.	200 Feet.	400 Feet.	600 Feet.	800 Feet.
20	350	52	80	108	136	165	193	46	68	90	112
25	392	65	100	135	170	205	240	57	84	111	138
30	429	77	118	160	201	242	284	68	100	132	164
35	463	89	136	184	231	279	326	78	115	152	189
40	496	101	155	208	262	316	....	89	131	173	215
45	525	113	173	233	293	....	....	100	146	193	239
50	554	125	192	258	324	....	....	111	162	214	265
55	581	137	210	282	....	....	....	121	178	234	290
60	607	149	228	306	....	....	....	132	193	254	....
65	631	162	246	330	....	....	....	143	209	275	....
70	655	173	263	....	....	....	....	153	223	294	....
75	678	184	281	....	....	....	....	163	237	312	....
80	700	197	299	....	....	....	....	174	253	....	....
85	722	209	317	....	....	....	....	184	269	....	....
90	743	220	....	....	....	....	....	195	284	....	....
95	763	232	....	....	....	....	....	205	299	....	....
100	783	244	....	....	....	....	....	216	314	....	....

2 1/2- AND 3-INCH HOSE.

Engine, while stream is flowing, to maintain Column, through various Lengths of Best 3-inch Rubber Lined Hose.

Lines.		Two 2½-inch Lines Siamesed.								Nozzle Pressure Indicated by Pitot Gage.
1,000 Feet.	1,200 Feet.	200 Feet.	400 Feet.	600 Feet.	800 Feet.	1,000 Feet.	1,200 Feet.	1,500 Feet.	1,800 Feet.	
134	156	37	53	68	84	100	115	139	162	20
165	192	47	66	85	104	123	143	171	200	25
196	228	56	79	102	125	148	171	205	240	30
226	263	65	91	117	144	170	197	236	276	35
257	299	74	104	134	164	194	224	269	314	40
286	....	82	116	149	182	215	248	298	....	45
....	....	91	128	165	202	239	275	331	....	50
....	....	100	140	181	221	261	301	....	....	55
....	....	109	153	196	240	283	327	....	....	60
....	....	118	164	211	258	305	....	....	....	65
....	....	126	176	226	276	326	....	....	....	70
....	....	135	189	242	295	....	....	....	....	75
....	....	144	201	258	314	....	....	....	....	80
....	....	153	213	273	....	....	....	....	....	85
....	....	162	225	289	....	....	....	....	....	90
....	....	170	237	303	....	....	....	....	....	95
....	....	179	249	319	....	....	....	....	....	100

1 3/4-INCH SMOOTH NOZZLE.—

Pressures Required at Hydrant or Fire Nozzle Pressures given in First Quality 2½- and

Nozzle Pressure Indicated by Pitot Gage.	Discharge Gallons per Minute.	Single 2½-in. Lines.				Single 3-inch					
		100 Feet.	200 Feet.	300 Feet.	400 Feet.	100 Feet.	200 Feet.	300 Feet.	400 Feet.	500 Feet.	600 Feet.
20	407	63	100	138	175	40	55	71	86	101	116
25	455	77	123	169	215	49	67	84	102	120	138
30	498	91	145	199	253	58	79	100	121	142	163
35	538	106	169	231	294	68	92	117	141	166	190
40	575	120	191	262	333	77	104	132	159	187	215
45	609	135	215	294	....	87	118	149	180	211	241
50	643	150	237	325	....	96	130	164	199	233	267
55	674	164	259	....	....	105	142	179	216	254	291
60	704	177	280	....	....	114	154	194	234	274	314
65	732	191	302	....	....	123	166	209	252	296	....
70	761	206	325	....	....	133	180	227	273	....	....
75	787	220	....	....	....	143	192	242	291	....	....
80	813	234	....	....	....	152	204	257	309	....	....
85	838	247	....	....	....	160	215	270	....	....	....
90	862	261	....	....	....	169	228	286	....	....	....
95	885	274	....	....	....	178	240	301	....	....	....
100	909	....	....	....	....	188	253	317	....	....	....

2 1/2- AND 3-INCH HOSE.

ENGINE, WHILE STREAM IS FLOWING, TO MAINTAIN COLUMN, THROUGH VARIOUS LENGTHS OF BEST 3-INCH RUBBER LINED HOSE.											Nozzle Pressure Indicated by Pitot Gage.
Lines.		Two 2½-inch Lines Siamesed.									
800 Feet.	1,000 Feet.	100 Feet.	200 Feet.	300 Feet.	400 Feet.	500 Feet.	600 Feet.	800 Feet.	1,000 Feet.	1,200 Feet.	
147	177	33	43	53	64	74	84	105	125	146	20
173	209	40	53	65	78	91	103	128	154	179	25
205	247	49	64	79	94	110	125	155	185	215	30
239	288	56	74	91	109	126	143	178	213	248	35
270	325	64	84	103	123	143	162	201	241	280	40
303	....	73	95	117	139	161	183	227	271	315	45
....	....	80	104	128	152	177	201	249	297	....	50
....	....	88	114	140	167	193	219	272	324	....	55
....	....	96	125	153	182	210	239	296	....	....	60
....	....	104	134	165	195	226	257	318	....	....	65
....	....	111	144	177	210	243	275	....	....	....	70
....	....	118	153	188	223	258	293	....	....	....	75
....	....	127	164	201	239	276	313	....	....	....	80
....	....	135	174	214	253	293	....	....	....	....	85
....	....	142	183	225	266	308	....	....	....	....	90
....	....	150	194	237	281	....	....	....	....	....	95
....	....	158	204	250	296	....	....	....	....	....	100

2-INCH SMOOTH NOZZLE.—

Nozzle Pressure Indicated by Pitot Gage.	Discharge, Gallons per Minute.	Pressures Required at Hydrant or Fire Nozzle Pressures given in First Quality 2½- and							
		Single 2½-inch Lines.			Single 3-inch Lines.				
		100 Feet.	200 Feet.	300 Feet.	100 Feet.	200 Feet.	300 Feet.	400 Feet.	500 Feet.
20	532	90	152	214	52	76	100	124	148
25	594	111	187	263	65	94	123	152	182
30	651	132	222	312	77	112	147	181	216
35	703	152	255	.....	89	129	169	209	249
40	752	173	290	.....	102	147	193	238	283
45	797	193	323	.....	113	163	213	263	314
50	841	214	.....	.....	126	182	237	293	.....
55	881	.....	.....	.....	138	199	260	321	.....
60	920	.....	.....	.....	150	216	282	.....	.....
65	958	.....	.....	.....	162	233	304	.....	.....
70	994	.....	.....	.....	175	251	327	.....	.....
75	1,029	.....	.....	.....	187	268	.....	.....	.....
80	1,063	.....	.....	.....	199	285	.....	.....	.....
85	1,095	.....	.....	.....	211	302	.....	.....	.....
90	1,128	.....	.....	.....	223	319	.....	.....	.....
95	1,158	.....	.....	.....	235	335	.....	.....	.....
100	1,189	.....	.....	.....	247	.....	.....	.....	.....

2 1/2- AND 3-INCH HOSE.

Engine, while stream is flowing, to maintain Column, through various Lengths of Best 3-inch Rubber Lined Hose.

		Two 2½-inch Lines Siamesed.								Nozzle Pressure Indicated by Pitot Gage.
600 Feet.	800 Feet.	100 Feet.	200 Feet.	300 Feet.	400 Feet.	500 Feet.	600 Feet.	800 Feet.	1,000 Feet.	
172	220	41	58	75	92	110	127	161	195	20
211	270	51	72	93	114	135	156	198	240	25
251	321	61	86	110	135	160	185	234	284	30
289	.....	71	100	128	157	186	214	271	329	35
.....	.....	81	113	146	178	211	243	308	.....	40
.....	.....	90	126	162	198	234	270	.....	.....	45
.....	.....	100	140	180	220	260	300	.....	.....	50
.....	.....	110	153	197	240	284	.....	.....	.....	55
.....	.....	119	166	213	260	308	.....	.....	.....	60
.....	.....	129	180	230	281	.....	.....	.....	.....	65
.....	.....	139	193	248	302	.....	.....	.....	.....	70
.....	.....	148	206	264	322	.....	.....	.....	.....	75
.....	.....	158	219	280	.....	.....	.....	.....	.....	80
.....	.....	167	232	297	.....	.....	.....	.....	.....	85
.....	.....	177	245	314	.....	.....	.....	.....	.....	90
.....	.....	186	258	.....	.....	.....	.....	.....	.....	95
.....	.....	196	272	.....	.....	.....	.....	.....	.....	100

FRICTION LOSS IN FIRE HOSE.
BASED ON TESTS OF BEST QUALITY RUBBER LINED FIRE HOSE.*

Flow, Gallons per Minute.	Pressure Loss in Each 100 Feet of Hose, Pounds per Sq. Inch.				Flow, Gallons per Minute.	Pressure Loss in Each 100 Feet of Hose, Pounds per Sq. Inch.		
	2½" Hose.	3" Hose.	3½" Hose.	2 Lines of 2½" Siamesed.		3" Hose.	3½" Hose.	2 Lines of 2½" Siamesed.
140	5.2	2.0	0.9	1.4	525	23.2	10.5	16.6
160	6.6	2.6	1.2	1.9	550	25.2	11.4	18.1
180	8.3	3.2	1.5	2.3	575	27.5	12.4	19.0
200	10.1	3.9	1.8	2.8	600	29.9	13.4	21.2
220	12.0	4.2	2.1	3.3	625	32.0	14.4	23.0
240	14.1	5.4	2.5	3.9	650	34.5	15.5	24.8
260	16.4	6.3	2.9	4.5	675	37.0	16.6	26.5
280	18.7	7.2	3.3	5.2	700	39.5	17.7	28.3
300	21.2	8.2	3.7	5.9	725	42.3	18.9	30.2
320	23.8	9.3	4.2	6.6	750	45.0	20.1	32.2
340	26.9	10.5	4.7	7.4	775	47.8	21.4	34.2
360	30.0	11.5	5.2	8.3	800	50.5	22.7	36.2
380	33.0	12.8	5.8	9.2	825	53.5	24.0	38.4
400	36.2	14.1	6.3	10.1	850	56.5	25.4	40.7
425	40.8	15.7	7.0	11.3	875	59.7	26.8	43.1
450	45.2	17.5	7.9	12.5	900	63.0	28.2	45.2
475	50.0	19.3	8.7	13.8	1,000	76.5	34.3	55.0
500	55.0	21.2	9.5	15.2	1,100	91.5	41.0	65.5

*Rough rubber lining is liable to increase the losses given in the table as much as 50 per cent.

APPENDIX

CONVERSION TABLE—CUBIC FEET TO GALLONS
1 cubic foot = 7.480519 U. S. gallons;
1 gallon = 231 cubic inches = .13368056 cubic foot

Cubic Feet	Gallons	Cubic Feet	Gallons	Cubic Feet	Gallons
0.1	0.75	50	374.0	8,000	59,844.2
0.2	1.50	60	448.8	9,000	67,324.7
0.3	2.24	70	523.6	10,000	74,805.2
0.4	2.99	80	598.4	20,000	149,610.4
0.5	3.74	90	673.2	30,000	224,415.6
0.6	4.49	100	748.0	40,000	299,220.8
0.7	5.24	200	1,496.1	50,000	374,025.9
0.8	5.98	300	2,244.2	60,000	448,831.1
0.9	6.73	400	2,992.2	70,000	523,636.3
1	7.48	500	3,740.3	80,000	598,441.5
2	14.96	600	4,488.3	90,000	673,246.7
3	22.44	700	5,236.4	100,000	748,051.9
4	29.92	800	5,984.4	200,000	1,496,103.8
5	37.40	900	6,732.5	300,000	2,244,155.7
6	44.88	1,000	7,480.5	400,000	2,992,207.6
7	52.36	2,000	14,961.0	500,000	3,740,259.5
8	59.84	3,000	22,441.6	600,000	4,488,311.4
9	67.32	4,000	29,922.1	700,000	5,236,363.3
10	74.80	5,000	37,402.6	800,000	5,984,415.2
20	149.6	6,000	44,883.1	900,000	6,732,467.1
30	224.4	7,000	52,363.6	1,000,000	7,480,519.0
40	299.2				

CONVERSION TABLE—GALLONS TO CUBIC FEET

Gallons	Cubic Feet	Gallons	Cubic Feet	Gallons	Cubic Feet
1	.134	1,000	133.681	1,000,000	133,680.6
2	.267	2,000	267.361	2,000,000	267,361.1
3	.401	3,000	401.042	3,000,000	401,041.7
4	.535	4,000	534.722	4,000,000	534,722.2
5	.668	5,000	668.403	5,000,000	668,402.8
6	.802	6,000	802.083	6,000,000	802,083.3
7	.936	7,000	935.764	7,000,000	935,763.9
8	1.069	8,000	1,069.444	8,000,000	1,069,444.4
9	1.203	9,000	1,203.125	9,000,000	1,203,125.0
10	1.337	10,000	1,336.806	10,000,000	1,336,805.6

FIRE SERVICE HYDRAULICS

CONTENTS IN CUBIC FEET AND U.S. GALLONS OF PIPES AND CYLINDERS OF VARIOUS DIAMETERS AND ONE FOOT IN LENGTH

1 gallon = 231 cubic inches.   1 cubic foot = 7.4805 gallons

For 1 Foot in Length			For 1 Foot in Length			For 1 Foot in Length		
Diameter in Inches	Cu. Ft. also Area in Sq. Ft.	U. S. Gals., 231 Cu. In.	Diameter in Inches	Cu. Ft. also Area in Sq. Ft.	U. S. Gals., 231 Cu. In.	Diameter in Inches	Cu. Ft. also Area in Sq. Ft.	U. S. Gals., 231 Cu. In.
1/4	0.0003	0.0025	3	.0491	.3672	8	.3491	2.611
5/16	.0005	.004	3 1/4	.0576	.4309	8 1/4	.3712	2.777
3/8	.0008	.0057	3 1/2	.0668	.4998	8 1/2	.3941	2.943
7/16	.001	.0078	3 3/4	.0767	.5738	8 3/4	.4176	3.125
1/2	.0014	.0102	4	.0873	.6528	9	.4418	3.305
9/16	.0017	.0129	4 1/4	.0985	.7369	9 1/4	.4667	3.491
5/8	.0021	.0159	4 1/2	.1104	.8263	9 1/2	.4922	3.682
11/16	.0026	.0193	4 3/4	.1231	.9206	9 3/4	.5185	3.879
3/4	.0031	.0230	5	.1364	1.020	10	.5454	4.08
13/16	.0036	.0269	5 1/4	.1503	1.125	10 1/4	.5730	4.286
7/8	.0042	.0312	5 1/2	.1650	1.234	10 1/2	.6013	4.498
15/16	.0048	.0359	5 3/4	.1803	1.349	10 3/4	.6303	4.715
1	.0055	.0408	6	.1963	1.469	11	.66	4.937
1 1/4	.0085	.0633	6 1/4	.2131	1.594	11 1/4	.6903	5.164
1 1/2	.0123	.0918	6 1/2	.2304	1.724	11 1/2	.7213	5.396
1 3/4	.0167	.1249	6 3/4	0.2485	1.859	11 3/4	.7530	5.633
2	.0218	.1632	7	.2673	1.999	12	.7854	5.875
2 1/4	.0276	.2066	7 1/4	.2867	2.145	12 1/2	.8522	6.375
2 1/2	.0341	.2550	7 1/2	.3068	2.295	13	.9218	6.895
2 3/4	.0412	.3085	7 3/4	.3276	2.45	13 1/2	.994	7.436

CONTENTS IN CUBIC FEET AND U. S. GALLONS OF PIPES AND CYLINDERS OF VARIOUS DIAMETERS AND ONE FOOT IN LENGTH

1 gallon = 231 cubic inches. 1 cubic foot = 7.4805 gallons

Diameter in Inches	For 1 Foot in Length		Diameter in Inches	For 1 Foot in Length		Diameter in Inches	For 1 Foot in Length	
	Cu. Ft. also Area in Sq. Ft.	U.S. Gals., 231 Cu. In.		Cu. Ft. also Area in Sq. Ft.	U.S. Gals., 231 Cu. In.		Cu. Ft. also Area in Sq. Ft.	U.S. Gals., 231 Cu. In.
14	1.069	7.997	21½	2.521	18.86	34	6.305	47.16
14½	1.147	8.578	22	2.640	19.75	35	6.681	49.98
15	1.227	9.180	22½	2.761	20.66	36	7.069	52.88
15½	1.310	9.801	23	2.885	21.58	37	7.467	55.86
16	1.396	10.44	23½	3.012	22.53	38	7.876	58.92
16½	1.485	11.11	24	3.142	23.50	39	8.296	62.06
17	1.576	11.79	25	3.409	25.50	40	8.727	65.28
17½	1.670	12.49	26	3.687	27.58	41	9.168	68.58
18	1.768	13.22	27	3.976	29.74	42	9.621	71.97
18½	1.867	13.96	28	4.276	31.99	43	10.085	75.44
19	1.969	14.73	29	4.587	34.31	44	10.559	78.99
19½	2.074	15.51	30	4.909	36.72	45	11.045	82.62
20	2.182	16.32	31	5.241	39.21	46	11.541	86.33
20½	2.292	17.15	32	5.585	41.78	47	12.048	90.10
21	2.405	17.99	33	5.940	44.43	48	12.566	94.00

FIRE SERVICE HYDRAULICS

CYLINDRICAL VESSELS, TANKS, CISTERNS, ETC.
Diameter in Feet and Inches, Area in Square Feet, and
Capacity in U. S. Gallons for One Foot in Depth

$$1 \text{ gallon} = 231 \text{ cubic inches} = \frac{1 \text{ cubic foot}}{7.4805} = 0.13368 \text{ cubic feet}$$

Diam.		Area	Gals.	Diam.		Area	Gals.	Diam.		Area	Gals.
Ft.	In.	Sq. Ft.	1 Foot Depth	Ft.	In.	Sq. Ft.	1 Foot Depth	Ft.	In.	Sq. Ft.	1 Foot Depth
1		0.7854	5.87	3		7.069	52.88	5	1	19.63	146.88
1	1	0.922	6.89	3	1	7.467	55.86	5	2	20.29	151.82
1	2	1.069	8.00	3	2	7.876	58.92	5	3	20.97	156.83
1	3	1.227	9.18	3	3	8.296	62.06	5	4	21.65	161.93
1	4	1.396	10.44	3	4	8.727	65.28	5	5	22.34	167.12
1	5	1.576	11.79	3	5	9.168	68.58	5	6	23.04	172.38
1	6	1.767	13.22	3	6	9.621	71.97	5	7	23.76	177.72
1	7	1.969	14.73	3	7	10.085	75.44	5	8	24.48	183.15
1	8	2.182	16.32	3	8	10.559	78.99	5	9	25.22	188.66
1	9	2.405	17.99	3	9	11.045	82.62	5	10	25.97	194.25
1	10	2.640	19.75	3	10	11.541	86.33	5	11	26.73	199.92
1	11	2.885	21.58	3	11	12.048	90.13	6		27.49	205.67
2		3.142	23.50	4		12.566	94.00	6	3	28.27	211.51
2	1	3.409	25.50	4	1	13.095	97.96	6	6	30.68	229.50
2	2	3.687	27.58	4	2	13.635	102.00	6	9	33.18	248.23
2	3	3.976	29.74	4	3	14.186	106.12	7		35.78	267.69
2	4	4.276	31.99	4	4	14.748	110.32	7	3	38.48	287.88
2	5	4.587	34.31	4	5	15.321	114.61	7	6	41.28	308.81
2	6	4.909	36.72	4	6	15.90	118.97	7	9	44.18	330.48
2	7	5.241	39.21	4	7	16.50	123.42	8		47.17	352.88
2	8	5.585	41.78	4	8	17.10	127.95	8	3	50.27	376.01
2	9	5.940	44.43	4	9	17.72	132.56	8	6	53.46	399.88
2	10	6.305	47.16	4	10	18.35	137.25	8	9	56.75	424.48
2	11	6.681	49.98	4	11	18.99	142.02	9		60.13	449.82

APPENDIX

9		63.62	475.89	17		226.98	1697.9	25		490.87	3672.0
9	3	67.20	502.70	17	3	233.71	1748.2	25	3	500.74	3745.8
9	6	70.88	530.24	17	6	240.53	1799.3	25	6	510.71	3820.3
9	9	74.66	558.51	17	9	247.45	1851.1	25	9	520.77	3895.6
10		78.54	587.52	18		254.47	1903.6	26		530.93	3971.6
10	3	82.52	617.26	18	3	261.59	1956.8	26	3	541.19	4048.4
10	6	86.59	647.74	18	6	268.80	2010.8	26	6	551.55	4125.9
10	9	90.76	678.95	18	9	276.12	2065.5	26	9	562.00	4204.1
11		95.03	710.90	19		283.53	2120.9	27		572.56	4283.0
11	3	99.40	743.58	19	3	291.04	2177.1	27	3	583.21	4362.7
11	6	103.87	776.99	19	6	298.65	2234.0	27	6	593.96	4443.1
11	9	108.43	811.14	19	9	306.35	2291.7	27	9	604.81	4524.3
12		113.10	846.03	20		314.16	2350.1	28		615.75	4606.2
12	3	117.86	881.65	20	3	322.06	2409.2	28	3	626.80	4688.8
12	6	122.72	918.00	20	6	330.06	2469.1	28	6	637.94	4772.1
12	9	127.68	955.09	20	9	338.16	2529.6	28	9	649.18	4856.2
13		132.73	992.91	21		346.36	2591.0	29		660.52	4941.0
13	3	137.89	1031.5	21	3	354.66	2653.0	29	3	671.96	5026.6
13	6	143.14	1070.8	21	6	363.05	2715.8	29	6	683.49	5112.9
13	9	148.49	1110.8	21	9	371.54	2779.3	29	9	695.13	5199.9
14		153.94	1151.5	22		380.13	2843.6	30		706.86	5287.7
14	3	159.48	1193.0	22	3	388.82	2908.6	30	3	718.69	5376.2
14	6	165.13	1235.3	22	6	397.61	2974.3	30	6	730.62	5465.4
14	9	170.87	1278.2	22	9	406.49	3040.8	30	9	742.64	5555.4
15		176.71	1321.9	23		415.48	3108.0	31		754.77	5646.1
15	3	182.65	1366.4	23	3	424.56	3175.9	31	3	766.99	5737.5
15	6	188.69	1411.5	23	6	433.74	3244.6	31	6	779.31	5829.7
15	9	194.83	1457.4	23	9	443.01	3314.0	31	9	791.73	5922.6
16		201.06	1504.1	24		452.39	3384.1	32		804.25	6016.2
16	3	207.39	1551.4	24	3	461.86	3455.0	32	3	816.86	6110.6
16	6	213.82	1599.5	24	6	471.44	3526.6	32	6	829.58	6205.7
16	9	220.35	1648.4	24	9	481.11	3598.9	32	9	842.39	6301.5

FIRE SERVICE HYDRAULICS

CIRCUMFERENCES AND AREAS OF CIRCLES

Dia.	Circum.	Area	Dia.	Circum.	Area
1/8	.39270	.01227	7.	21.991	38.485
1/4	.78540	.04909	8.	25.133	50.265
3/8	1.1781	.11045	9.	28.274	63.617
1/2	1.5708	.19635	10.	31.416	78.540
5/8	1.9635	.30680	11.	34.558	95.033
3/4	2.3562	.44179	12.	37.699	113.10
7/8	2.7489	.60132	13.	40.841	132.73
1.	3.1416	.7854	14.	43.982	153.94
1/8	3.5343	.9940	15.	47.124	176.71
1/4	3.9270	1.2272	16.	50.265	201.06
3/8	4.3197	1.4849	17.	53.407	226.98
1/2	4.7124	1.7671	18.	56.549	254.47
5/8	5.1051	2.0739	19.	56.690	283.53
3/4	5.4978	2.4053	20.	62.832	314.16
7/8	5.8905	2.7612	21.	65.973	346.36
2.	6.2832	3.1416	22.	69.115	380.13
1/8	6.6759	3.5466	23.	72.257	415.48
1/4	7.0686	3.9761	24.	75.398	452.39
3/8	7.4613	4.4301	25.	78.540	490.87
1/2	7.8540	4.9087	26.	81.681	530.93
5/8	8.2467	5.4119	27.	84.823	572.56
3/4	8.6394	5.9396	28.	87.965	615.75
7/8	9.0321	6.4918	29.	91.106	660.52
3.	9.4248	7.0686	30.	94.248	706.86
1/8	9.8175	7.6699	31.	97.389	754.77
1/4	10.210	8.2958	32.	100.531	804.25
3/8	10.603	8.9462	33.	103.673	855.30
1/2	10.996	9.6211	34.	106.814	907.92
5/8	11.388	10.321	35.	109.956	962.11
3/4	11.781	11.045	36.	113.097	1017.9
4.	12.566	12.566	37.	116.239	1075.2
1/4	13.352	14.186	38.	119.381	1134.1
1/2	14.137	15.904	39.	122.522	1194.6
3/4	14.923	17.728	40.	125.664	1256.6
5.	15.708	19.635	41.	128.805	1320.3
1/4	16.493	21.648	42.	131.947	1385.4
1/2	17.279	23.758	43.	135.088	1452.2
3/4	18.064	25.967	44.	138.230	1520.5
6.	18.850	28.274	45.	141.372	1590.4
1/4	19.635	30.680	46.	144.513	1661.9
1/2	20.420	33.183	47.	147.655	1734.9
3/4	21.206	35.785	48.	150.796	1809.6

CIRCUMFERENCES AND AREAS OF CIRCLES — Continued

Dia.	Circum.	Area	Dia.	Circum.	Area
49.	153.938	1885.7	75.	235.619	4417.9
50.	157.080	1963.5	76.	238.761	4536.5
51.	160.221	2042.8	77.	241 903	4656 6
52.	163.363	2123.7	78.	245.044	4778.4
53.	166.504	2206.2	79.	248.186	4901.7
54.	169.646	2290.2	80.	251.327	5026.5
55.	172.788	2375.8	81.	254.469	5153.0
56.	175.929	2463.0	82.	257.611	5281.0
57.	179.071	2551.8	83.	260.752	5410.6
58.	182.212	2642.1	84.	263.894	5541.8
59.	185.354	2734.0	85.	267.035	5674.5
60.	188.496	2827.4	86.	270.177	5808.8
61.	191.637	2922.5	87.	273.319	5944.7
62.	194.779	3019.1	88.	276.460	6082.1
63.	197.920	3117.2	89.	279.602	6221.1
64.	201.062	3217.0	90.	282.743	6361.7
65.	204.204	3318.3	91.	285.885	6503.9
66.	207.345	3421.2	92.	289.027	6647.6
67.	210.487	3525.7	93.	292.168	6792.9
68.	213.628	3631.7	94.	295.310	6939.8
69.	216.770	3739.3	95.	298.451	7088.2
70.	219.911	3848.5	96.	301.593	7238.2
71.	223.053	3959.2	97.	304.734	7389.8
72.	226.195	4071.5	98.	307.876	7543.0
73.	229.336	4185.4	99.	311.018	7697.7
74.	232.478	4300.8	100.	314.16	7854.

SQUARE ROOTS

No.	Square Root	No.	Square Root	No.	Square Root	No.	Square Root
1	1.0000	43	6.5574	85	9.2195
2	1.4142	44	6.6332	86	9.2736
3	1.7321	45	6.7082	87	9.3274
4	2.0000	46	6.7823	88	9.3808
5	2.2361	47	6.8557	89	9.4340
6	2.4495	48	6.9282	90	9.4868
7	2.6458	49	7.0000	91	9.5394
8	2.8284	50	7.0711	92	9.5917
9	3.0000	51	7.1414	93	9.6437
10	3.1623	52	7.2111	94	9.6954
11	3.3166	53	7.2801	95	9.7468
12	3.4641	54	7.3485	96	9.7980
13	3.6056	55	7.4162	97	9.8489
14	3.7417	56	7.4833	98	9.8995
15	3.8730	57	7.5498	99	9.9499
16	4.0000	58	7.6158	100	10.0000
17	4.1231	59	7.6811	101	10.0499
18	4.2426	60	7.7460	102	10.0995
19	4.3589	61	7.8102	103	10.1489
20	4.4721	62	7.8740	104	10.1980
21	4.5826	63	7.9373	105	10.2470
22	4.6904	64	8.0000	106	10.2956
23	4.7958	65	8.0623	107	10.3441
24	4.8990	66	8.1240	108	10.3923
25	5.0000	67	8.1854	109	10.4403
26	5.0990	68	8.2462	110	10.4881
27	5.1962	69	8.3066	111	10.5357
28	5.2915	70	8.3666	112	10.5830
29	5.3852	71	8.4261	113	10.6301
30	5.4772	72	8.4853	114	10.6771
31	5.5678	73	8.5440	115	10.7238
32	5.6569	74	8.6023	116	10.7703
33	5.7446	75	8.6603	117	10.8167
34	5.8310	76	8.7178	118	10.8628
35	5.9161	77	8.7750	119	10.9087
36	6.0000	78	8.8318	120	10.9545
37	6.0828	79	8.8882	121	11.0000
38	6.1644	80	8.9443	122	11.0454
39	6.2450	81	9.0000	123	11.0905
40	6.3246	82	9.0554	124	11.1355
41	6.4031	83	9.1104	125	11.1803
42	6.4807	84	9.1652		

INDEX

Aerial stream calculations for engine pressure, 280
Altitude, effects of, on lift, 47
Appliances, friction loss in, 250
Atmospheric pressure, 14, 45
  measure of, 16

Back pressure, 247
Barometer, 16, 206
Bernoulli's Equation, 19, 20, 25, 27
Bourdon gage, 17, 206

Capacity
  mains, 91
  pumps, 138
Carrying capacity
  pipes, 96
  relative, of pipes, 99
Cavitation
  description, 147
  pumps, 147
Clappers
  action under different pressures, 257
  in siamese, 257
Coefficients
  discharge for orifices, 50
  C values for small hose friction loss formula, 235
  for velocity pressure, hydrants, 116
  Hazen Williams equation, 97
  of discharge, hydrants, 114
  typical, 52
Combined agent or twinned foam equipment, 389
Comparison of nozzles, table of, 264
Control valves, standpipe systems, 310
Conversion rules of thumb to 2½-inch hose, 233
Conversion table
  other sizes to 2½-inch hose, 234
  siamesed line friction losses, 234
  2½-inch hose friction loss, 232
Couplings, 2½-inch on 3-inch hose, 238

Deck guns, hydraulic problems of, 282
Deluge sets, friction loss in, 250
Distribution system, water, 90
Drafting, 44, 45
  maximum lift, 48

Elevating platform stream calculations for engine pressure, 281
Energy
  conservation of, 25
  kinetic, 21
  potential, 21
  relationship to hydraulics, 20
Engine pressure
  determining by rule of thumb, 250
  formula for, 239
  leeway in calculations, 291
  required for standpipe systems, 320
  spacing relay pumpers for equal engine pressures, 286
Equation
  determining discharge from a nozzle or orifice, 34
  Hazen Williams, 92, 97
  use of Hazen Williams, 100
Equivalent lengths, conversion to 2½-inch hose, 233

Factors
  hose conversion, 256
  to determine equivalent length of 2½-inch hose, 256
Fire streams
  calculation of reach, 299
  defective, 305
  description, 293
  direction of, 300
  directed from street, 301
  effects of air resistance, 294
  effects of wind, 295
  from turret nozzles, 303
  horizontal, 296
  limits of reach, 295
  nozzle size and pressures, 305
  range, 296
  reach, 297
  use of heavy, 304
  vertical, 297
Fixed foam makers and distributors
  chemical foam, 378
  internal fixed, 379
  portable tower, 380
  subsurface injection, 377
  tank-side installed, 374
Fixed foam water sprinklers, 385

FIRE SERVICE HYDRAULICS

Flow
  capabilities of fire hose, 237
  determining, from friction loss, 235
  hydrants, 92
  measuring, from hydrants, 114
  velocity, 49
Fluid pressure principles, 9
Flush-mounted foam sprinkler systems, 386
Foam
  characteristics, 345
  classification
    air or mechanical, 345
    chemical, 345
    chemical foaming agents, 345, 346, 350
    high expansion, 345
    surface-film foam-forming agents, 345, 346, 349
    surface spreading, 345
    synthetic foam-forming agents, 345, 346, 348
  composition, 346
  fluoroprotein, 347
  generation
    air aspiration, 354
    air blower, 358
    chemical foam, 360
    foam pump, 356
  Light Water, 349, 386, 390
  proportioning devices
    around-the-pump proportioner, 365
    in-line eductor, 365
    foam concentrate pump, 366
    pressure-proportioning tank, 371
    water-motor, 372
  systems
    combined agent equipment, 389
    crash-rescue vehicles, 386
    fixed foam makers and distributors, 374
    fixed foam water sprinklers, 385
    high-expansion, 384
    portable nozzle aspirators, 373
  uses, 351
    Class A combustibles, 351
    Class B fuels, 351
    other, 352
    problems, 353
Force, explanation, 6
Formula
  AIA, 50
  basic flow, 49
  discharge through pipes, 92
  engine pressure, 239, 241
  flow from friction loss, 236
  flow, 50
  Freeman's, 50
  friction loss, 64
  friction loss for 2½-inch hose, 231
  friction loss for 2½-inch hose, flows under 100 gpm, 231
  hydrant discharge, 113
  nozzle pressure, 240, 242
  small hose friction loss, 235
  small lines and nozzles, 241, 242
  Underwriters, 241, 242

Formula (cont'd)
  Underwriters engine pressure, 240, 241
  Underwriters nozzle pressure, 241
Forward pressure, 248
Friction, hydraulic laws governing, 227
Friction loss
  advantages of big hose, 236
  calculating by rule of thumb, 249
  causes, 63
  coefficient C values for small hose, 235
  definition, 226
  determination of, 230
  effect of flow pattern, 60
  effect in hose line, 229
  finding flow from, 235
  formula, for 2½-inch hose with flows under 100 gpm, 231
  formula for 2½-inch hose, 231
  in appliances, 250
  in fire hose, 62
  in suction hose, 67
  in unlined linen hose, 235
  rubber lined fire hose, 65
  small diameter rubber or rubber lined fire hose, 66
  small hose, 235
  small hose formula, 235
  standpipe systems, 320
  table, rule-of-thumb, 249
  2½-inch hose conversion table, 232
  various sources of, 228

Gage
  adjustment and repair of, 207
  Bourdon, 111, 206
  compound, 205
  pump, 205
Gage pressure, 14
Gated outlets, NFPA requirements, 169
Governors, pump, 220
Gravity tanks, 315

Head, 43
  elevation, 24, 25
  pressure, 23, 24, 25
  total static, 24
High expansion foam systems, 384
Horsepower, 141
  brake, 142
Hose
  advantages of big, 236
  butt calculations for flow, 246
  conversion factors for equivalent 2½-inch, 256
  conversion table, other sizes to 2½-inch, 234
  flow capabilities, 237
  lines stretched to standpipe, 321
  standpipe, 314
  3-inch with 2½-inch couplings, 238
Hydrants
  branch connection, 84
  breakable, 85

424

INDEX

Hydrants (cont'd)
  calculation of results in testing, 116
  caps, 81
  coefficient for velocity pressure, 116
  computing fire flow test results, 119
  discharge formula, 92
  drains, 78
  dry barrel, 77
  early development, 75
  flush, 85
  high pressure, 89
  loss of head, 82
  outlet nozzles, 79
  pipe materials, 92
  principal types, 76
  private yard, 89
  reporting test results, 124
  supply from system, 79
  testing, 77
  using test results, 124
  wet barrel, 79
Hydraulic losses, effects on lift, 47
Hydraulics
  definition, 3
  history, 3
Hydrokinetics, 19
Hydrostatics, 5

Impeller, double suction, 152

K values
  large hose and nozzles, 241
  small hose and nozzles, 242
  solving for in Underwriters formula, 243

Ladder pipe calculations, engine pressure, 280
Ladder pipes, friction loss in, 250
Leakage, pump seal, 153
Lift, 45
  cause of failure, 48
Light Water, 349, 386, 390
Linen hose
  friction loss, 235
  Underwriters formula for engine and nozzle pressures, 242

Mains
  equipment needed for testing, 111
  fire flow tests, 104
  pipe standards, 96
  temporary service, 95
  test on dead-end, 110
  test procedure, 112
Manifold
  roof, 311
  standpipe systems, 311
Manometer, mercury, 38

NEPA Specification, 19
  pump performance, 139
  requirements for gated outlet, 169
Nomograph, Hazen-Williams equation, 101
Nozzle
  basic formula, 4
  comparisons, table of, 264
  diameter, effect of changing on nozzle pressure and flow, 243
  discharge, 49
  solving for equivalent single size, 243
Nozzle pressure
  average change when engine pressure changes, 292
  equation for, 68
  formula for, 240

Open hose butt calculations, 246
Outlets, standpipe, 314

Parallel lines
  calculating friction loss in, 253
  converting to single line equivalent, 256
  determining engine pressure, 251
  determining nozzle pressure, 252
  equal length, 251, 252
  siamesed, 251, 252
  use in relay pumping, 285
Piezometer, 18
Pipe
  materials, 92
  protective coating, 94
  relative carrying capacity, 96
Pitot, 39, 115
Potential energy
  due to elevation, 23
  due to pressure, 23
Pressure
  atmospheric, 206
  explanation of, 6
  function of height and density, 13
  negative, 44
  residual, 108
Pressure control devices, 214
Pressure tanks for standpipe systems, 316
Pressures, ladder pipe, Table of, 281
Primers, pump, 210, 211
  engine vacuum system, 213
  rotary, 211
Pumps
  ability to utilize positive pressure, 146
  back-pack, 198
  booster, 184, 186
  brake horsepower, 142
  capabilities and limitations, 144
  cavitation, 147
  centrifugal, 140, 144, 146, 150, 158, 162, 163, 164, 168, 186, 190, 192, 193, 194
  classification, 138
  description, 140

FIRE SERVICE HYDRAULICS

Pumps (cont'd)
  displacement, 182
  double-volute design, 155
  duplex multistage centrifugal, 175
  dynamic suction lift, 150
  efficiency rating, 168
  fire service, 138
  flow, 162
  front mounting, 139
  gages, 205
  gear ratios, piston, 179
  governors, 220
  high pressure, 189, 190, 192, 193, 194
  horsepower, 141
  impeller, double suction, 152
  lobe-rotor, 181
  loss of prime, 210
  midship mounting, 139
  mobile, 138
  parallel-series operation, 162
  parallel-series 2-stage, 160
  parallel-series 3-stage, 170
  parallel-series 4-stage, 173
  performance affected by water temperature, 148
  performance in agreement with NFPA 19, 139
  piston, 176, 178, 179, 180
  portable, 195, 196
  pressure control, 214
  pressure required for sprinklers, 338
  priming, 208
  priming systems, 210
  relief valves, 214
  requirements for drafting, 48
  rotary, 181, 182, 184
  seal leakage, 183
  series operation, 164
  single and double-acting, piston, 176
  single-stage centrifugal, 184
  slippage, 180
  standard capacities, 138
  static suction lift, 149
  stationary, 200
  strainers, 209
  tanker, 199
  tests, 139
  theoretical displacement of, 178
  transfer valve, 160, 163
  two-stage centrifugal, 159
  UL test, 139
  vacuum reading, 209
  vapor pressure, 149
  water horsepower, 141

Reaction
  effects of size nozzle on, 56
  nozzle, 54
Reducers
  for standpipes, 314
  pressure, 314

Relay pumping, 282
  advantage of large diameter hose, 283
  determining distance between pumpers, 286
  examples of spacing pumpers, 288
  standard engine pressure method, 283
  standard engine pressure for 200 gpm, 284
  suction relief valves, 283
  use of parallel lines, 285
Risers, standpipe systems, 309
Rule-of-thumb calculations
  engine pressure, 250
  friction loss, 249

Series operation, centrifugal pumps, 164
Siamese connections, 313
  care and maintenance, 323
  clapper valve action under different pressures, 257
  fire department use, 337
  friction loss in, 250
  pressures to service ladder pipes, table of, 281
  standpipe systems, 313, 319
  when out of order, 319
Siamesed lines
  changing to equivalent 2½-inch hose, 255
  complicated layouts, 255
  friction loss conversion table, 234
Slippage
  booster pump, 186
  piston, 180, 186
Sprinkler systems
  antifreeze, 330
  automatic, 324, 332
  caution in shutting down, 336
  checking flow from fire department pumpers, 339
  classifications of, 329
  combined dry pipe and preaction, 330
  control valves, 333
  deluge, 230
  design and construction, 325
  dry pipe, 332
  fire department use, 337
  heads, design, 325, 329
  junior, 330
  occupancies using, 324
  operation of, 325
  outside stem and yoke, 333
  preaction, 330
  post indicator valve, 333
  recommended use by fire departments, 340
  shutoff, 333
  siamese, 337
  supplying by fire department, 338
  temperature rating of, 329
  water supplies for, 331
  valves, 332
  wet-pipe, 332
Standards, pipe, 96

INDEX

Standpipe systems, 308
  automatic fire pumps for, 316
  classification, 309
  components, 309
  engine pressures needed at fires, 320
  fire department operations at fires, 318, 322
  fire department operations in theaters, 322
  fire hose requirements and care, 323
  friction loss, 320
  gravity tanks, 315
  hose lines stretched to, 321
  hose outlet and drip valves, 314
  pressure tanks, 316
  pumps, 316
  siamese, 313, 319
  water supplies, 315
  when out of order, 319
  uses of, 308
Static head, 23
Static pressure, 23
Strainers, suction, 209
Suction, 44
Suction lift
  dynamic, 150
  static, 149
Suction relief valves in relay pumping, 283

Tanker, pumps for, 199
Temperature, water
  effect on lift, 47
  effect on pump performance, 148
Tests
  fire flow, 104, 106
  dead-end mains, 110
  mains, 111, 112
  pumps, 139
  water distribution systems, 109
Threads
  National Standard fire hose coupling, 82
  hydrant nozzle, 80
Tip pressure, change when engine pressure changes, 292
Tuberculation in mains, 93

Underwriters formulas
  engine pressure, 240, 241
  engine pressure, equal parallel lines into deluge, 251
  linen hose, engine and nozzle pressure, 242
  nozzle pressure, 241, 242
  nozzle pressure, equal parallel lines into deluge set, 252
  small lines and nozzles, 241, 242
Vacuum, 46

Valves-pump
  relief, 214
  transfer, 160, 163
Valves, sprinkler and standpipe
  automatic, 332
  check, 311
  control, for sprinkler systems, 333
  dry pipe, 332
  outside stem & yoke, 333
  post indicator valve, 333
  shutoff, 333
  sprinkler, 332
  standpipe, globe, 310
  standpipe, outside stem & yoke, 310
  standpipe systems, 311
  wet pipe, 332
Vapor pressure, effect on lift, 149
Velocity
  coefficient, 51
  effects on friction loss, 61
  from orifices and nozzles, 31
  head, relationship to discharge, 32
Venturi tube, 38
Viscosity, effect on friction loss, 61

Water
  characteristics, 5
  consumption, 4
  density, 5
  weight, 5
Water hammer, 56
Water horsepower, pumps, 141
Water supplies
  public, 74
  sprinkler systems, 331
  to sprinkler system by fire department, 338
  standpipe systems, 315
  water works, 90
Wyed lines
  calculating individual nozzle pressure, 267, 272
  determining average nozzle pressure, 260, 266
  determining engine pressure, 259, 265, 270, 274
  different size nozzles, 270, 272, 274
  equal diameters and lengths, 259, 260, 263
  equal diameters, unequal lengths, 274
  equal hose and nozzle diameters, 265, 266
  equal size and length, 270, 272
  mental calculation for engine pressure, 278
  rapid calculation for engine pressure, 263
  rule-of-thumb methods for solving problems, 277
  solving problems of, 259
  unequal lengths, 265, 266
Wyes, friction loss in, 250

Zero, absolute, 206

427